EARTHWORMS IN WASTE AND ENVIRONMENTAL MANAGEMENT

EARTHWORMS IN WASTE AND ENVIRONMENTAL MANAGEMENT

edited by
Clive A. Edwards and Edward F. Neuhauser

SPB Academic Publishing bv 1988

CIP-DATA KONINKLIJKE BIBLIOTHEEK, DEN HAAG

Earthworms

Earthworms in waste and environmental management / ed. by
Clive A. Edwards and Edward F. Neuhauser. – The Hague :
SPB Academic Publishing. – Ill.
With ref.
ISBN 90-5103-017-7
SISO 570.3 UDC 595.1:504.064 NUGI 821
Subject heading: earthworms ; environmental management.

ISBN 90-5103-017-7

CONTENTS

IV. Earthworms in production of plant growth media

V. Earthworms in land reclamation, soil amelioration and land improvements

VI. Earthworms as indicators of environmental contamination

PREFACE

More than 100 years have passed since Charles Darwin wrote his book (1881) 'The Formation of Vegetable Mould Through the Action of Worms' (1881) which was the sequel to 40 years of careful personal observations of the activities of earthworms. Prior to this, many scientists and laymen considered earthworms to be pests and few appreciated their contribution to soil formation and fertility. He established the importance of these animals in the maintenance of fertility, demonstrated how much soil they turn over, and aroused the interests of many scientists in their habits and role in soil. Nevertheless, our knowledge of these ubiquitous animals progressed only slowly over the following 60 years, and it was not until after the Second World War, when Soil Zoology first came into its own, at the first International Conference which was held in the University of Nottingham, U.K. in 1955, that earthworm research began to increase again. Thereafter, the regular International Soil Zoology conferences encouraged a strong interest in soil-inhabiting invertebrates as a whole, and especially earthworms, so that the publication of papers on earthworms began to increase significantly. In 1971, Edwards and Lofty published the first book on the Biology of Earthworms that summarized our knowledge on earthworms up to that time and this was expanded to a Second Edition in 1977. In 1980, the first Applied Workshop, on the Role of Earthworms in the Stabilization of Organic Residues, was held in Kalamazoo, Michigan and the Proceedings were published in 1981. This set the scene for the use of earthworms in applied problems for the first time and stimulated a great deal of interest in the applied aspects of earthworm ecology and utilization.

In the same year, the centenary of Darwin's book, the first International Conference on Earthworms, attended by 150 scientists, was held in Grange-over-Sands, U.K., and the Proceedings were published under the title 'Earthworm Ecology', edited by J.E. Satchell. This was followed four years later in 1985 by the Second International Conference held in Bologna, Italy and published under the title, the 'On Earthworms', edited by A.M. Pagliai and P. Omadeo in 1987. Contributions to the present book include most of the papers presented at the First International Conference on Earthworms in Waste and Environmental Management, held in Cambridge, U.K. This Conference was held in response to the great expansion of interest and research into the application of earthworms into waste management and land reclamation. This started with investigations by R. Hartenstein and his associates, into the use of earthworms in stabilization of sewage sludge at the State University of New York, Syracuse in 1976, and was followed up by C.A. Edwards and his associates at Rothamsted Experimental Station in England and elsewhere, by research aimed at using earthworms to break down a wide range of organic wastes. During the same period, there was an increasing interest in the use of earthworms in land reclamation and soil improvement and also into the use of earthworms as key bioindicators of soil contamination by chemicals. The present volume gives a comprehensive overview of these studies and demonstrates the great potential for using the earthworm in a multitude of applied ways, some of which may ultimately form the basis for extensive commercial and industrial development. This kind of commercial application has already started in the U.K. and many other parts of the world. We hope that this book will provide a source book and a basis for rapid expansion of practical applications, because we are convinced of the potential of earthworm ecology for solving a range of waste and environmental problems.

Clive A. Edwards
The Ohio State University

and

Edward F. Neuhauser
Mohawk Power Company
2 August 1988

I. Processing of animal and human wastes by earthworms

CONTENTS

VERMICULTURE: BRING ON THE FUTURE

JOHN R. SABINE
University of Adelaide, Australia

Vermiculture comes in two parts – as a science, and as an industry. While these two parts are necessarily interdependent, the industry in particular being dependent upon the science, quite different and even independent factors will determine the future growth of the two parts.

The growth of the science of vermiculture seems clearly assured, that of the industry less obviously so. First, let us look at the science. Any interest in vermiculture as a science, both current and especially future interest, will depend very much upon a satisfactory answer to one very simple question – are earthworms intrinsically interesting as biological organisms? In other words, do earthworms perform any role or task or function that is scientifically interesting and/or important? Patently, the answer to this question is a resounding yes.

The mere existence of a class of organisms is enough to seize the imagination of at least a few scientists, irrespective of whether such organisms might fulfil any biologically useful role or not. And when one recognizes that there might be in excess of 3,000 species of earthworm, about only 5% of which number is there any significant scientific knowledge, then the scientific challenge is obvious (Lee, 1985). But earthworms do much more than just exist – and for at least three quite different but very important reasons the scientific community is becoming increasingly interested in them: namely, relative to soil formation and fertility, relative to environmental monitoring and relative to the study of evolution.

First, what of soil formation and fertility? Since the time of Charles Darwin a hundred years ago it has been recognized that many species of earthworm play a large and perhaps critical role in the development and maintenance of soil fertility (Satchell, 1983). The recycling of dead organic material derived from surface vegetation and its incorporation into and through the upper soil profile, together with modifications to soil physical properties, are all an earthworm contribution. With developing interest in various forms of no-tillage or minimum-tillage farming, with and without chemical herbicides, and with increasing concern for all factors relevant to soil erosion, the role of earthworms in both the biological and physical aspects of pedogenesis is assuming greater significance (Stockdill, 1982).

Related to this is a second scientific task recently allotted to earthworms, *i.e.*, what might be loosely termed, environmental monitoring. Worms can accumulate, often in high concentration, various environmental pollutants – including especially certain heavy metals and industrial chemicals – as well as various biocides and other agricultural chemicals and their residues (Ireland, 1983; Martinucci *et al.*, 1983). Given this, plus the worm's recognized place in the food chains of many and diverse higher organisms – fish, birds, carnivorous insects and small mammals (Macdonald, 1983) (and even other earthworms, Lavelle, 1983) – there is increasing interest in how, and how much, extraneous material different earthworms can accumulate, where did it come from? where is it going? what damage is likely along the way?

Earthworms in waste and environmental management
edited by C.A. Edwards and E.F. Neuhauser, pp. 3–7

The third developing area of scientific interest in earthworms is rather different from these first two. It now appears that earthworms represent very early stages in the biochemical evolution of some quite complex macromolecular systems – such as for instance oxygen-carrying proteins (David & Daniel, 1974; Ascoli *et al.*, 1978) and the immune system (Cooper, 1979) – which are of considerable importance in higher organisms, including man. Much understanding of the origin and function of these systems can be gained from knowledge of their structure and operation in the simplified versions found in earthworms.

Thus, as I have said, I believe that an increased scientific interest in earthworms over the next few years, *i.e.*, the further development of earthworm science, is inevitable and likely to be considerable. But what of an earthworm industry? Just as any assessment of the potential for development of the science depends upon a satisfactory answer to a very simple question, so too the potential for development of the industry will depend upon a satisfactory answer to a simple question – or perhaps rather to a set of three simple questions: (i) is there a demonstrable need for the goods and services that might be provided by the vermiculture industry? (ii) can vermiculture meet these needs at a reasonable price? and (iii) can these needs be met by vermiculture in a manner more efficient and more effective than by any alternative means?

The basis for a legitimate vermiculture industry is no different in principle, from that of any other industry, namely the provision of goods and/or services in sufficient quantity and quality and in such a manner that all those who invest in the industry, at whatever level, can expect to make a reasonable return on their time or money or expertise so invested. So, what goods and services can be provided by vermiculture, and are they needed? As I see it the need at least is not only indisputable, but often massive, especially with regards to potential services. And in the context of vermiculture 'services' generally equates with 'waste and resource management'.

Almost wherever we turn our society, particularly our highly urbanized and industrialized society, seems to generate vast mountains or extensive lakes of waste material (*e.g.*, Edwards, 1983; Millis, 1980; Walsingham, 1981) – wastes that are wastes simply because no one has yet devised a suitable (read profitable) use for them – or otherwise of course they would not be wasted, and would indeed most likely be called by-products. All of our agricultural industries, especially our intensive animal industries, and all of our food processing industries, and all of our industries currently collecting and processing human refuse and effluent, all of these generate large quantities of biological waste, waste that presents to us a major two-fold problem. First along with this material there are discarded and lost a whole array of valuable primary nutrients – energy, proteins, minerals – all of which had taken a considerable input of time and effort and money originally to produce. Moreover, the actual physical disposal of this material, *i.e.*, in a manner aesthetically acceptable to our same society, requires yet another considerable input of time and effort and money. Earthworms, in dense culture and in large quantities, can physically handle virtually any biological waste – and can do it potentially at a fraction of the cost of conventional disposal mechanisms (Appelhof, 1981; Hartenstein, 1978). But can they really do it consistently, reliably, and economically? The need is undoubtedly there.

But our society indulges also in yet another quite wasteful activity. Various land utilization practices of ours create large tracts of despoiled countryside – land that is now socially unattractive and productively useless. Strip mining, quarries, rubbish dumps, highway construction and similar activities leave varying areas of variably degraded, but generally unsightly and unproductive, scars upon the landscape. For millenia earthworms have been preparing soil for the introduction and development of higher vegetation under conditions often originally no less harsh than we are now creating. Can they be put to work for us again, for a need is certainly there?

So much for the services provided by vermiculture, what about the goods? – for while earthworms are busily devouring our biological wastes, and thus we hope decreasing our disposal problems, they are also and concurrently manufacturing two new products – earthworm compost and more earthworms. Are these materials any good in the marketplace?

For the answer to this question also to be in the affirmative, particular attention must be paid to the notion of quality. Basically of course quality in these goods, as for any other agricultural commodity, will depend upon the nature of the starting material originally provided for the worms to eat and the biological dynamics of the culture system. Vermicompost, for instance, if it is high in required plant nutrients, if these nutrients are in readily available form, if additional growth stimulants might be present, and if physical texture, odour and water-holding capabilities are satisfactory, and if it contains no undesirable pollutants nor pollutant residues, then undoubtedly it will be great news for home gardeners and market gardeners, for horticulturalists, floriculturalists and so on. If all this applies, and if the worms can produce it economically, then vermicompost has a great commerical future.

And the excess worms – do they too have a great commercial future and are they valuable? Certainly, both individually and collectively, earthworms are valuable – but the real question is, how valuable? An estimated 50 million recreational anglers in North America, and presumably comparable numbers elsewhere (Tomlin, 1983), are looking for bait – and are mostly prepared to pay for the luxury of not having to find it themselves. Collecting worms to satisfy such a market is already a significant industry, *i.e.*, collecting worms, and collecting the right worms. Can a culture industry produce the worms, the right worms, more efficiently and more effectively?

But worms are more than just bait for a hook – they are also protein, protein that is high in biological value and that can appropriately substitute for other protein sources, expensive protein sources, in various livestock rations (Sabine, 1983). Pigs and poultry, and I presume dairy cattle, thrive on rations in which the major added protein comes as vermimeal. Exotic species too – birds and fish and frogs and shellfish – may also thrive, as could household pets, on diets containing significant amounts of earthworms, either fresh or processed. But, and it is a very large but, whereas an individual angler might cheerfully pay several cents or more for one commercial bait worm, to avoid the time and effort of finding his own, that same worm as protein is worth perhaps one hundreth to one thousandth of this amount – depending upon quantity and quality, and the price of alternative protein sources. But protein is protein, and as such earthworms are valuable. Can the vermiculture industry produce this worm protein, free from chemical and biological contamination, more economically than any alternative?

Two final thoughts for the future remain – in the form of two popular but overworked and often misunderstood phrases: 'low technology' and 'high technology'. Vermiculture as a low-tech industry (Appelhof, 1982), at individual or household or village level, for improved husbanding of scarce biological resources and reduction of biological losses in less-well-developed countries, or for many dispossessed, underprivileged, handicapped citizens of even well-developed countries, has much to recommend it. Not a major commercial enterprise, certainly, but nevertheless still of significant social and economic consequence.

Vermiculture as a high-tech industry, however, is quite another matter. High technology implies high levels of financial input, high levels of scientific interest and, significantly, high levels of risk. But all of this would readily be justified if there were any real likelihood of equally high returns for all who invest their money or their time or their knowledge, at whatever level of involvement. Is there such a likelihood?

Again, high technology implies high science. Of the several thousand species of earthworm potentially available, less than a dozen have received serious attention as possible waste controllers, or compost manufacturers, or protein producers. Is the incentive real enough for the scientific community seriously to examine the untried species, to check for contaminating residues, to develop new culture systems, or even to develop new species altogether specifically designed for specific tasks by the incredible techniques of genetic engineering, possibly the current pinacle of modern high technology?

All these questions are simple questions. Obviously, however, their consequences are far from simple and their answers are going to be complex. We know that the future will be built upon the present. For the present, vermiculture can do a very small job very well. For the future it remains to be seen, and perhaps this conference will show us the way, whether it can do a very big job equally well – or better.

For my part ... bring on that future!

References

Appelhof, M. (Ed.) 1981. Workshop on the Role of Earthworms in the Stabilization of Organic Residues. Beech Leaf Press, Kalamazoo, MI. 315 pp.

Appelhof, M. 1982. Earthworms Eat My Garbage. Flower Press, Kalamazoo, MI. 100 pp.

Ascoli, F., Rossi Fanelli, M.R., Chiacone, E., Vecchini, P. and Antonini, E. 1978. Studies on erthrocruorin: VI Ferric derivatives of earthworm erythrocruorin. J. Mol. Biol. 119: 191–202.

Cooper, E.L. 1979. Earthworms and immunology. Trends Biochem. Sci. 4: 295–296.

David, M.M. and Daniel, E. 1974. Subunit structure of earthworm erythrocruorin. J. Mol. Biol. 87: 89–101.

Edwards, C.A. 1983. Production of earthworm protein for animal feed from potato waste. *In* Upgrading Waste for Feed and Food. pp. 153–162. Edited by D.A. Ledward, A.J. Taylor and R.A. Lawrie. Pub. Butterworth's Larda.

Hartenstein, R. (Ed.) 1978. Proc. Conf. Utilization of Soil Organisms in Sludge Management. State University of New York, Syracuse. 167 pp.

Ireland, M.P. 1983. Heavy metal uptake and tissue distribution in earthworms. *In* Earthworm Ecology: from Darwin to Vermiculture. pp. 247–265. Edited by J.E. Satchell. Chapman and Hall, London.

Lavelle, P. 1983. *Agastrodrilus* Omodeo and Vailland, a genus of carnivorous earthworms from the Ivory Coast. *In* Earthworm Ecology: from Darwin to Vermiculture. pp. 425–429. Edited by J.E. Satchell. Chapman and Hall, London.

Lee, K.E. 1985. Earthworms: Their Ecology and Relationships with Soils and Land Use. Academic Press, Sydney. 411 pp.

Macdonald, D.W. 1983. Predation on earthworms by terrestrial vertebrates. *In* Earthworm Ecology: from Darwin to Vermiculture. pp. 393–414. Edited by J.E. Satchell. Chapman and Hall, London.

Martinucci, G.B., Crespi, P., Omodeo, P., Osella, G. and Traldi, G. 1983. Earthworms and TCDD (2,3,7,8-tetrachlorodibenzo-*p*-dioxin) in Seveso. *In* Earthworm Ecology: from Darwin to Vermiculture.

pp. 275–283. Edited by J.E. Satchell. Chapman and Hall, London.

Millis, N.E. (Ed.) 1980. Proc. Conf. Treating Food Wastes for Profit. Australian Acad. Techn. Sci. Parkville, Vic.

Sabine, J.E. 1983. Earthworms as a source of food and drugs. *In* Earthworm Ecology: from Darwin to Vermiculture. pp. 285–296. Edited by J.E. Satchell. Chapman and Hall, London.

Satchell, J.E. (Ed.) 1983. Earthworm Ecology: from Darwin to Vermiculture. Chapman and Hall, London. 495 pp.

Stockdill, S.M.J. 1982. Effects of introduced earthworms on the productivity of New Zealand pastures. Pedobiologia 24: 29–35.

Tomlin, A.D. 1983. The earthworm bait market in North America. *In* Earthworm Ecology: from Darwin to Vermiculture. pp. 331–338. Edited by J.E. Satchell. Chapman and Hall, London.

Walsingham, J.M. 1981. The use of invertebrates in recycling. *In* Vegetable Productivity. pp. 239–246. Edited by C.R.W. Spedding. Inst. Biol. Symp. 25. Macmillan, London.

THE POTENTIAL OF EARTHWORMS FOR MANAGING SEWAGE SLUDGE

EDWARD F. NEUHAUSER*, RAYMOND C. LOEHR** and MICHAEL R. MALECKI
Department of Agricultural Engineering, Cornell University, Ithaca, NY 14853

Introduction

Increasingly strict water pollution regulations throughout the world have resulted in the generation of increasing amounts of sludge for ultimate disposal. Many alternative sludge management methods have been developed, such as different methods of composting, land application and the use of earthworms to help stabilize sludges (Camp, Dresser and McKee, 1981 and Dinges, 1982).

The use of earthworms in sludge management is known as vermicomposting (Neuhauser *et al.*, 1979) or vermistabilization (Loehr *et al.*, 1984a). Because vermistabilization is a relatively new process, there are many fundamental factors that have to be evaluated to assure the technical and economic success of this process. These factors include: (a) how earthworms affect and are impacted by sludge characteristics, (b) the comparative ability of different earthworm species to grow and reproduce in sludge, and (c) the effect of mixed earthworm species in a vermistabilization system.

Information in this paper resulted from a project to assess the feasibility of using earthworms to stabilize wastewater treatment sludges (Loehr *et al.*, 1984a). Research in this project was conducted along two complementary lines: (a) basic studies to identify fundamental factors that affect the performance of the vermistabilization process and (b) applied studies to determine design and management relationships. The studies reported in this paper describe some of the results of the basic studies. The applied studies have been reported in a subsequent conference paper (Loehr *et al.*, 1988).

Results and discussion

Degree of stabilization

If earthworms are to be useful in stabilizing sludge, they must increase the stabilization rate. This can be shown if the presence of earthworms in sludge causes an increase in the rate of volatile solids reduction. Maximum reduction of volatile solids is a goal of any sludge stabilization system.

Figure 1 shows that the earthworm, *Eisenia fetida*, can increase the sludge volatile solids destruction rate when present in aerobic sludge. This increase in the volatile solids destruction rate reduces the probability of putrefaction occurring in the sludge due to anaerobic conditions. The more rapid degradation of the organic matter was

* *Present address*: Niagara Mohawk Power Corp., 300 Erie Boulevard W, Syracuse, NY 13202
** *Present address*: University of Texas, Austin, TX 78712.

Earthworms in waste and environmental management
edited by C.A. Edwards and E.F. Neuhauser, pp. 9–20

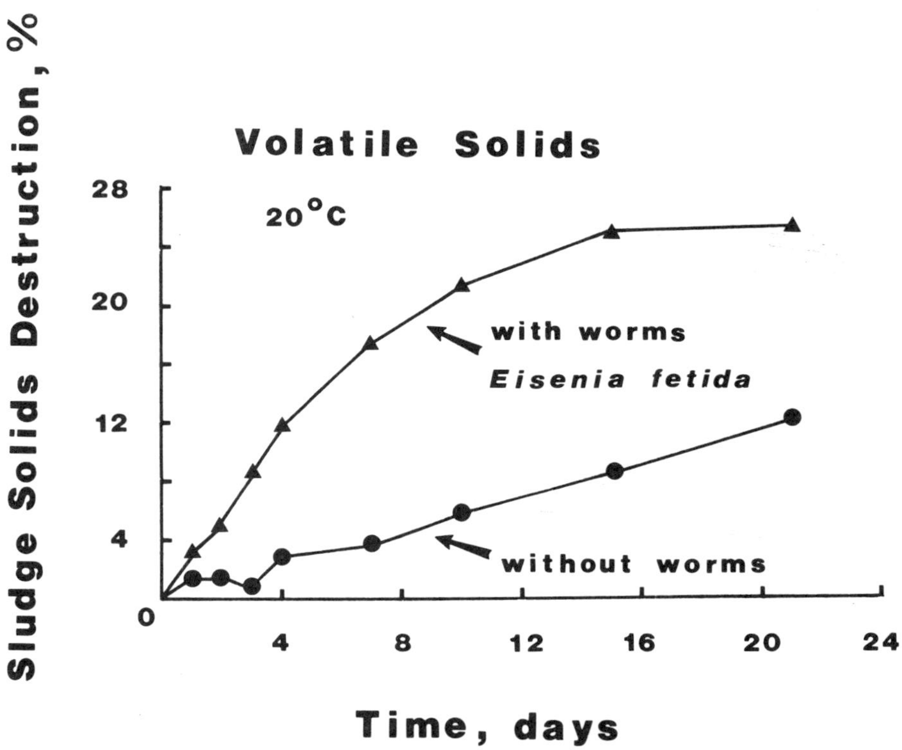

Fig. 1. The effect of earthworm on volatile solids destruction in aerobically digested sludge.

probably due to increased aeration and other factors brought about by the earthworms.

Moisture content

Both excessive and insufficient moisture can adversely impact earthworm growth (Edwards and Lofty, 1977). A series of experiments were conducted to determine the desirable moisture content of the media that will exist in vermistabilization units. Such information is important since it identifies the conditions that may be necessary for successful vermistabilization.

Aerobically digested sludge was dewatered to a moisture content of 75% (total solids = 25%). Dewatering was accomplished without the aid of chemicals. The drier sludge was mixed with a more liquid sludge to vary the moisture content of the sludge used in experiments.

Only one species, *E. fetida*, and one temperature, 25°C, were tested. Seventy grams of sludge were placed in each of five 20 × 100 mm glass petri dishes, with 4 young, *E. fetida* (< 10 mg each) added per dish. Earthworm weight gains were determined after four weeks.

The results of these studies are presented in Fig. 2. Although there was some variation at the various sludge concentrations evaluated, a definite pattern can be ob-

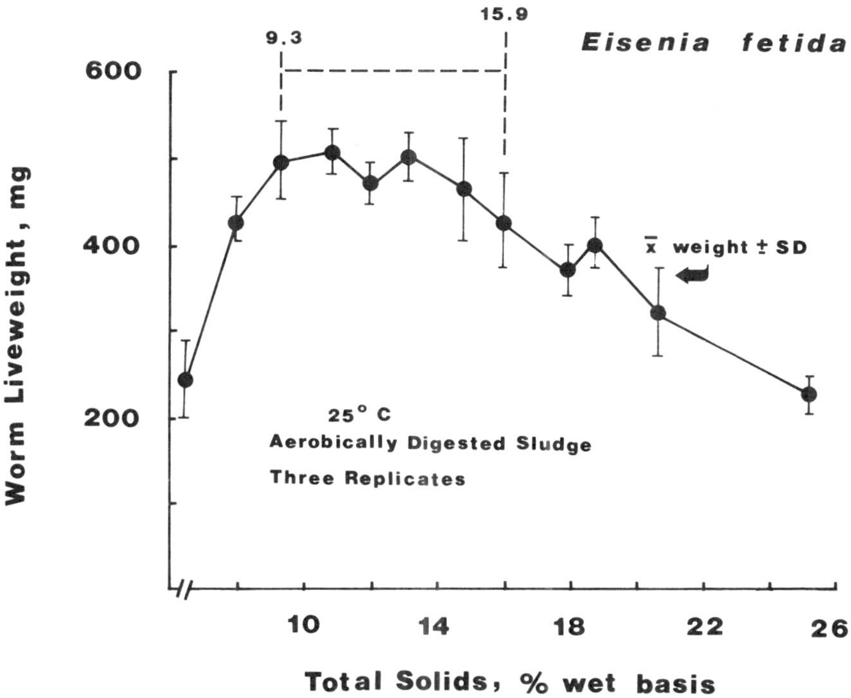

Fig. 2. The total solids content of aerobically digested sludge as a function of earthworm weight at four weeks. N = 5.

served. Worm growth was hindered at the lower and higher solids concentrations.

The data were analyzed statistically to discern differences. Worm growth at the low and high total solids levels (6.3 to 7.9 and 17.9 to 25.1% solids) was statistically different, at the 1% level, from the worm growth at the middle range of solids (9.3 to 15.9% solids). This indicates that optimum worm growth occurred over a range of total solids, wet basis, from about 9 to 16% in the media to which the worms were exposed.

The length of the study did not permit an evaluation of the complete reproductive cycle. However, the study was conducted long enough to permit early cocoon production to be determined. Optimum cocoon production occurred at concentrations of 11.9 to 18.6% solids. No cocoon production occurred at concentrations of from 6.3 to 7.9% and from 20.5 to 25.1% solids. The cocoon production data reinforced the worm growth data in defining the appropriate moisture content range for vermistabilization.

It should be emphasized that the identified appropriate range of total solids (9–16%) (91–84% moisture) is the total solids content to which the worm were exposed. More liquid wastes can be applied to the vermistabilization process as long as the liquid drains from media, the organics are retained, and aerobic conditions are maintained. The desirability of having the total solids content of the media in the range noted does not mean that sludge slurries cannot be used with the vermistabiliza-

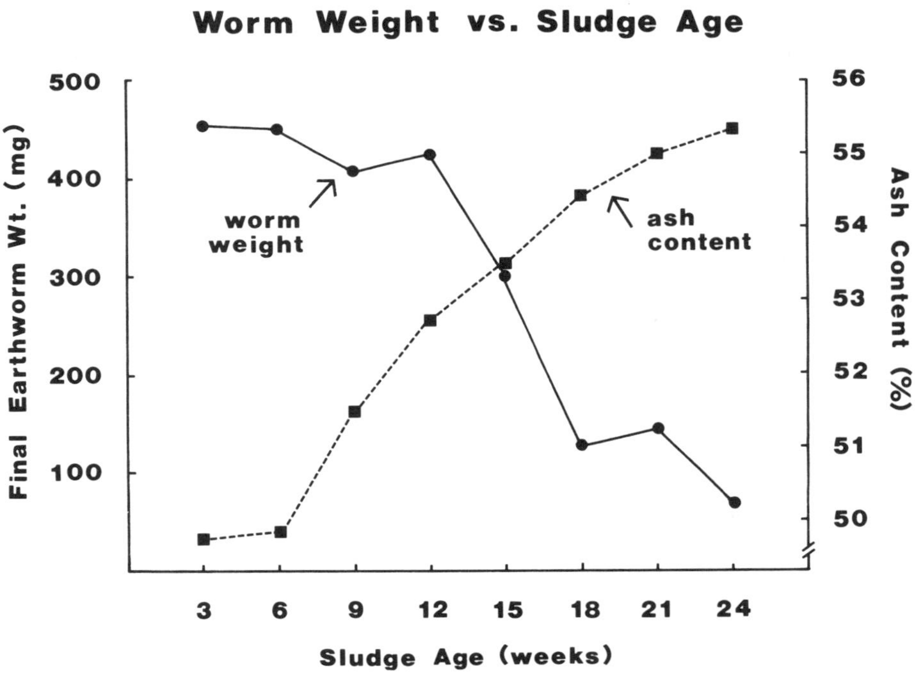

Fig. 3. The effect of aging aerobically digested sludge, after removal from the digester, on the growth of *E. fetida* at four weeks. N = 10.

tion process. Work in our laboratory has demonstrated that liquid sludges can be used with this process (Loehr *et al.*, 1988).

The data indicates that the upper range of the appropriate total solids content for the sludge used was about 16% in the absence of stressed conditions. In an operational sense, it could be necessary to add moisture to a vermistabilization process if the media should become too dry. Such could be the case in hot and/or dry climates.

Sludge age

Since earthworms can help stabilize sludges, it is important to find out when in the stabilization process earthworms can be most useful. An approach for this purpose is to relate earthworm growth to the period of time after a sludge is removed from an aerobic reactor. Sludge age in these experiments refers to the time after the sludge was removed from the reactor and immediately dewatered.

The sludge used was activated sludge which had been digested aerobically. The sludge was removed from the digester, dewatered to 10–12% solids, placed in individual 20 × 100 mm petri dishes and stored at 20°C for aging. Sludge age for this experiment started from the time the sludge was dewatered and continued until the earthworms were placed in the petri dishes. For example, a designated sludge age of three weeks means that the sludge was aged for three weeks before the worms were

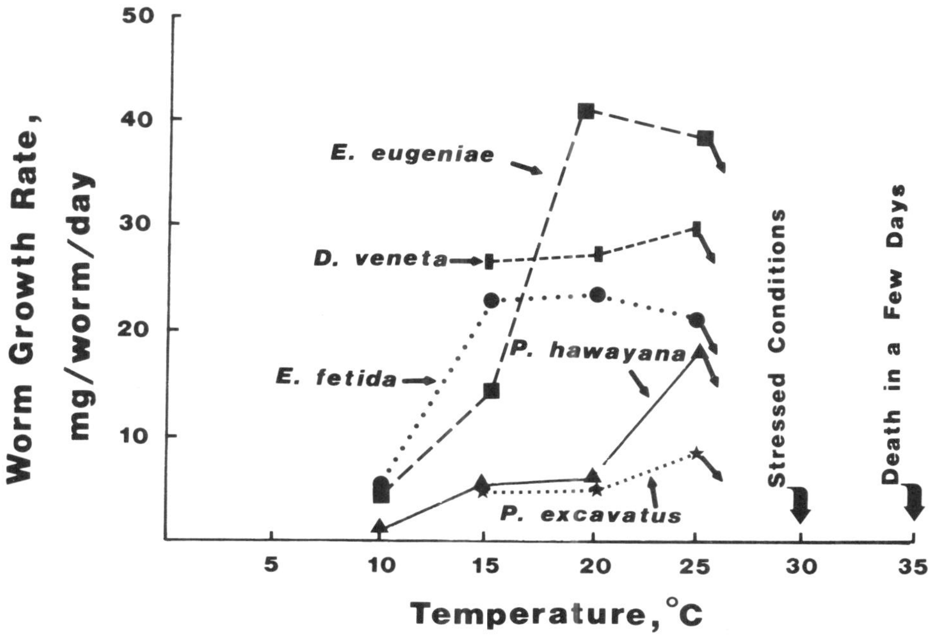

Fig. 4. The growth rate of five species of earthworms on aerobically digested sludge after 70 days, as a function of temperature. N = 4.

added. In these experiments, the worms were then allowed to grow in the sludge for an additional four weeks. Two young worms, < 10 mg each, were placed in each petri dish with 100 mg of sludge (10–12% solids).

Ten replicates were used for each sludge age evaluated. After allowing the worms to grow for four weeks, the final weights of the earthworms was determined.

The effect of sludge age on earthworm growth is shown in Fig. 3. As the sludge aged, its nutritive value to earthworms decreased rapidly after about twelve weeks removal from the digester. The ash content of the sludge increased with time, an indication of sludge stabilization.

Growth and reproduction

There are many earthworm species that have potential to be used in sludge stabilization systems (Neuhauser *et al.*, 1979b; Kaplan *et al.*, 1980a, b). Since earthworm growth and reproductive rates are a way of indicating potential for use in sludge management systems, five species (*Dendrobaena veneta, E. fetida, Eudrilus eugeniae, Perionyx excavatus* and *Pheretima hawayana*) were evaluated. The objective of this study was to determine the optimum temperature for growth and reproduction for these five species.

Four young worms of each species, < 10 mg each, were placed in 300 cc glass dishes and fed 200 g (wet weight) of dewatered (10–12% solids) aerobically digested sludge. Four replicates at six different temperatures were tested. Every two weeks the cocoons were removed and the worms counted, weighted, and refed fresh dewatered

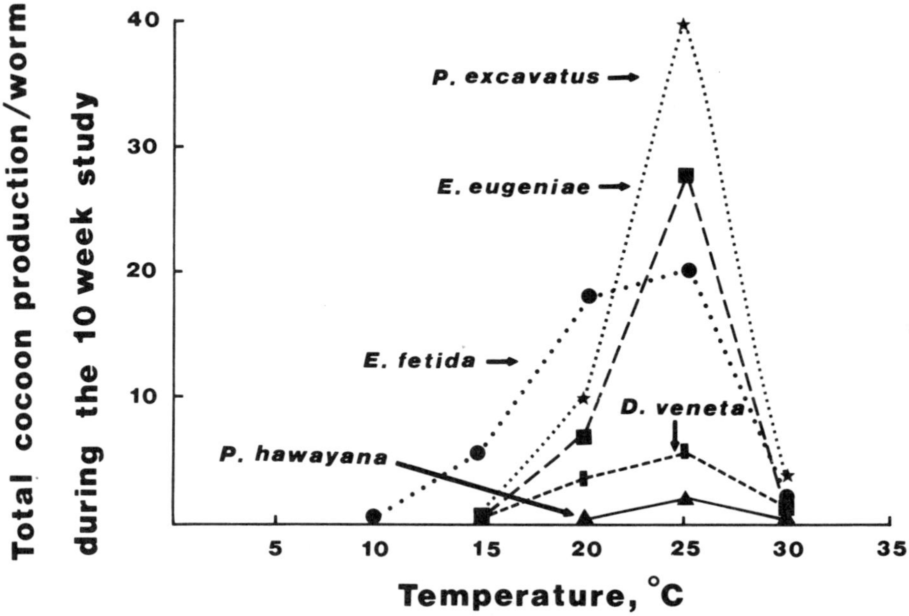

Fig. 5. Cocoon production of five species of earthworms on aerobically digested sludge after 10 weeks, as a function of temperature. N = 4.

sludge. After 70 days, growth rates for the five species were determined as a function of temperature (Fig. 4). Reduced growth rates for all species were observed at 30°C, with death of all species at 35°C. The data indicated that *D. veneta, E. fetida* and *P. excavatus* have a broad temperature optimum. Temperature variations within the range of 15–25°C are not likely to affect the growth of these species. The optimum temperature for growth of *E. eugeniae* was narrower, 20–25°C, while that for *P. hawayana* was 25°C.

Figure 5 indicates the cocoon production over a ten week period, for five species of worms as a function of temperature. Cocoon production was more restricted by temperature than growth. Only *E. fetida* produced cocoons at 15°C and only *P. excavatus* produced cocoons at 30°C. All five species produced the highest number of cocoons at 25°C.

Longer term studies were then conducted to determine how the five species compared in terms of growth and reproductive capabilities at one temperature. Four young worms of each species, < 10 mg each, were placed in 300 cc glass dishes and fed 500 g of dewatered (10–12°C solids) aerobically digested sludge. The four replicates were weighed, cocoons removed and refed fresh sludge every two weeks. All species were kept for twenty weeks at 25°C.

The individual live weight of the five species that occurred over the 20 weeks is shown in Fig. 6. *P. excavatus* had the smallest increase in weight and *E. fetida* the next smallest increase. Both *P. hawayana* and *E. eugeniae* reached their peak weight at about 10–12 weeks. The weight of *P. hawayana* remained constant for the remainder of the experiment. *E. eugeniae* lost weight after 14 weeks. *D. veneta*

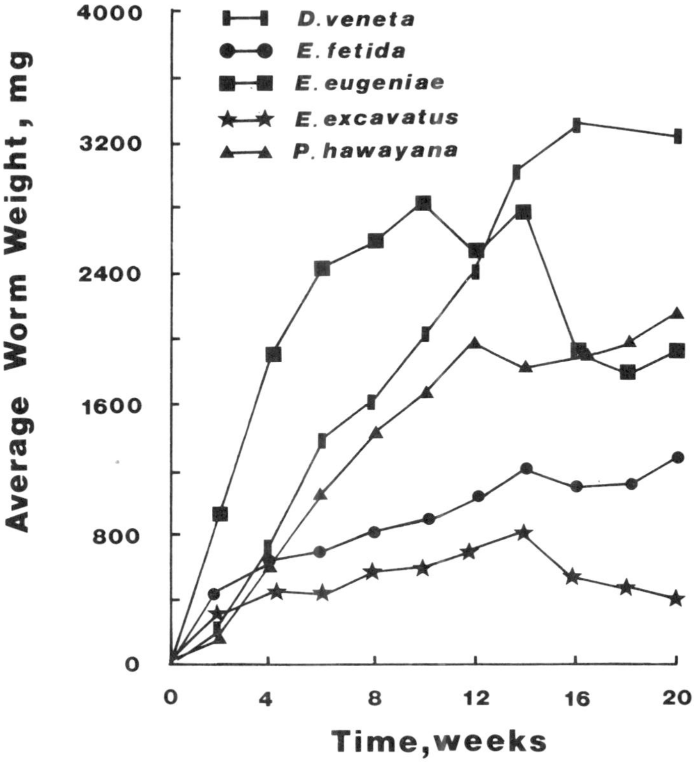

Fig. 6. The growth of five species of earthworms in aerobically digested sludge at 25°C. N = 4.

achieved the largest increase in weight of the five species, but took 16 weeks to achieve maximum weight.

Cocoon production for the five species of earthworms over the twenty weeks is shown in Fig. 7. *P. excavatus* produced the largest number of cocoons, while *D. veneta* produced the lowest number of cocoons and did not start producing cocoons until 10 weeks. The other three species (*E. fetida, E. eugeniae* and *P. hawayana*) all produced a similar number of cocoons, peaking at the middle of experiment (10 weeks) and gradually decreasing toward the end of the experiment.

Since not all earthworm cocoons hatch and it is possible to have more than one young earthworm per cocoon, it is important to evaluate the number of live young that would be obtained from the cocoons of each species. To determine such data, cocoons from five species of worms (*D. veneta, E. fetida, E. eugeniae, P. excavatus* and *P. hawayana*) were collected and allowed to hatch. Individual cocoons were kept under non-stressed conditions, at 25°C, and were checked twice per week to determine the number of cocoons that hatched and the number of worms that were produced per hatched cocoon. The results are noted in Table 1.

On the basis of number of live worms produced per cocoon, *E. fetida* produced more live worms per cocoon (3.8) than any of the other species tested. *E. eugeniae* produced over two live worms per cocoon.

Table 1. Cocoon viability for four earthworm species

-Earthworm Cocoon Viability-

Species	Total No. Cocoons Evaluated	Cocoons Hatched (%)	Young per Cocoon Hatched	Young per Cocoon Evaluated
D. veneta	470	78	1.1	0.86
E. fetida	1410	83	3.8	3.1
E. eugeniae	300	73	2.5	1.8
P. excavatus	100	50	1.1	0.55
P. hawayana	100	79	1.2	0.95

The overall reproductive capability of each species can be estimated by using the total number of cocoons produced over the twenty week breeding study (Table 2). Using this basis, of the five species evaluated, *E. fetida* appears to be an appropriate species to use in vermistabilization studies since it produced the greatest number of young worms per initial parent worm. Although *P. excavatus* produced the largest number of cocoons, the hatchability and live worms produced per cocoon were low (Table 1).

Conversion of sludge volatile solids to earthworm tissue

It is useful to know the ability of different earthworm species to convert the organic

Table 2. Reproductive potential of five earthworm species

-Earthworm Reproduction-

Species	Cocoons per Worm per Week +	Young per Cocoon Evaluated	Young per Worm per Week +	Young per Worm in 20 Weeks
D. veneta	5	0.86	4.3	19
E. fetida	6	3.1	19	223
E. eugeniae	11	1.8	20	173
P. excavatus	24	0.55	13	179
P. hawayana	10	0.95	9.5	72

+maximum values observed

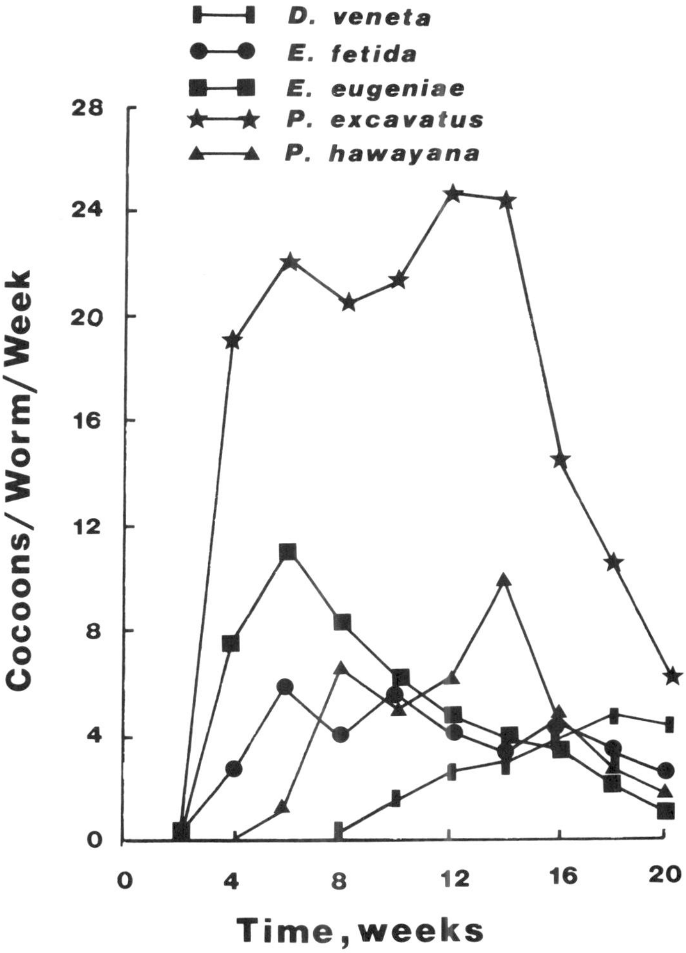

Fig. 7. Cocoon production of five earthworm species in aerobically digested sludge at 25°C. N = 4.

material in sludge into earthworm tissue. Ten young worms (< 10 mg each) of five earthworm species (*D. veneta, E. fetida, E. eugeniae, P. excavatus* and *P. hawayana*) were placed in petri dishes (20 × 100 mm) with 70 g wet weight (12% solids) of aerobically digested sludge. Dishes were incubated at 25°C and after 10 days four dishes were examined every two days for earthworm weight changes until 36 days. This resulted in a total of 56 dishes for each species.

Table 3 shows the time required to achieve maximum earthworm live weight for the five species. The 70 g of dewatered sludge (12% total solids) per dish used as a nutrient media is equivalent to 8.4 g of total solids per dish on a dry weight basis. The dry sludge contained 66% volatile solids, so each dish contained a total of 5.54 g of volatile solids. The five species of earthworms collectively have an average dry weight (total solids) of 18%. The quantity of worm total solids (dry weight) produced per quantity of sludge volatile solids (dry weight) in the dishes (Table 3) shows that

Table 3. The maximum weight obtained by five earthworm species grown in 70 g sludge (12% solids)

-Earthworm Growth-

Species	Max. Wt. Obtained for 10 Worms, mg	Days to Obtain Max. Wt.	Dry Wt. Worm Tissue, mg	Dry Wt. Worm Tissue/V.S. in Sludge, %
D. veneta	3614	34	650	11.8
E. fetida	1985	27	357	6.5
E. eugeniae	3076	22	554	10.0
P. excavatus	1032	31	186	3.4
P. hawayana	1289	28	232	4.2

N = 4.

D. veneta was the most efficient species tested, followed by *E. eugeniae, E. fetida, P. hawayana* and *P. excavatus. E. eugeniae* was the species to reach maximum weight the earliest (22 days), while *D. veneta* required the longest time (34 days) to achieve maximum weight.

These experiments demonstrate that there can be a three-fold difference in the efficiency of producing worm biomass per unit of sludge volatile solids. This points out the differences in feeding efficiencies of the five earthworm species.

Polyculture

In nature, several different earthworm species may exist in the same area of soil, each filling a different niche and using different substrates for food (Edwards and Lofty, 1977). Therefore, it is possible that a combination of worm species in a vermistabili-

Table 4. Statistical analysis of the polyculture experiment using *E. fetida*

Polyculture Study

statistical analysis of data

6 Worms/dish, N=5 at 25°C for 28 days

	Control (no worms)	*Eisenia fetida*	*Eudrilus eugeniae*	*Perionyx excavatus*	*E. fetida & P. excavatus*	*E. fetida & E. eugeniae*	*E. eugeniae & P. excavatus*	*E. fet., E. eug. & P. ex.*
$\bar{x}$ % Volatile Solids Reduced	13.01[a]*	27.95[b]	32.31[d]	27.55[b]	28.05[b]	30.41[c]	30.02[c]	30.03[c]
Total Biomass Produced, g/6 worms	—	2.88[f]	4.89[h]	2.36[e]	2.69[ef]	4.52[h]	3.76[g]	3.70[g]

*Numbers with a common letter are not significantly different, $p < 0.05$.

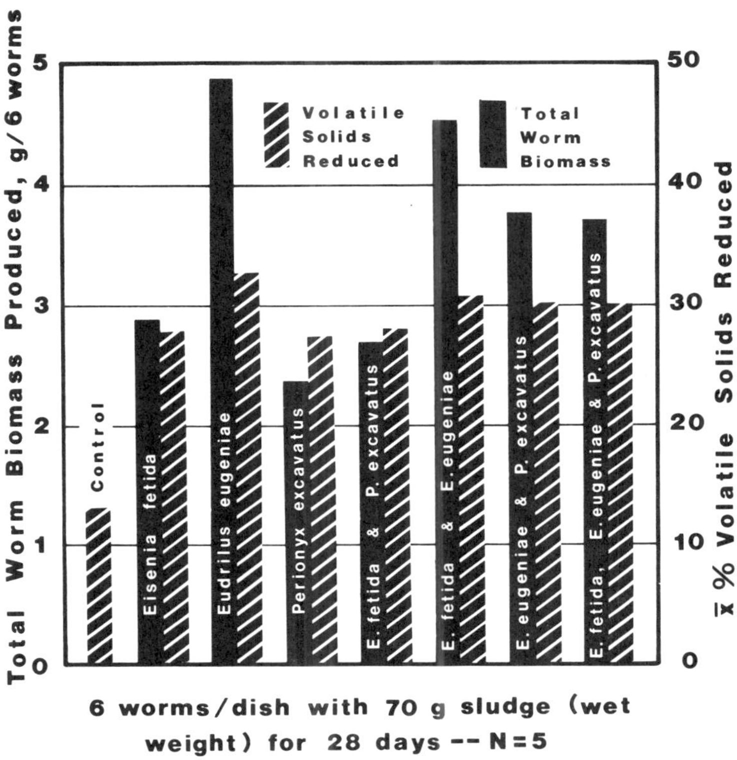

Fig. 8. The effect of combining different species of worms on volatile solids reduction and total worm biomass produced in sludge.

zation process (polyculture) could accomplish greater stabilization than a single species (monoculture) (Loehr *et al.*, 1984a). The objective of these experiments was to evaluate this hypothesis.

Young worms (< 10 mg each) were used and the studies conducted at 25°C with aerobically digested sludge dewatered to 12% solids without chemicals. The worms and the combination that were evaluated were: *E. fetida* alone, *E. eugeniae* alone, *P. excavatus* alone, *E. fetida* and *E. eugeniae* together, *E. fetida* and *P. excavatus* together, *E. eugeniae* and *P. excavatus* together and all three species together. Parameters measured at the end of the four week study were total worm weight and volatile solids reduction.

The results are presented graphically in Fig. 8. In each case, when the worms were included, the volatile solids reduction was greater, generally at least twice as great, as the controls which did not contain worms. This provides additional verification of the data in Fig. 1, *i.e.*, that worms do assist the vermistabilization process. The statistical analysis of the data is shown in Table 4. The results clearly indicate that the presence of the earthworms caused a statistically significant reduction of volatile solids when compared to the controls which did not contain earthworms.

In terms of total weight gain, in every situation in which *E. eugeniae* was present alone or in one of the combinations, the total weight gain was significantly greater than in situations in which *E. eugeniae* was absent. This is consistent with data from the previous study (Fig. 6) which also indicated that *E. eugeniae* had a greater weight gain over a ten-week-period that the other species. Whether this apparent advantage of *E. eugeniae* would be sustained over a long period of time is unclear especially since the weight of *E. eugeniae* appeared to decrease with time (Fig. 6).

In terms of volatile solids reduction, the differences are not as dramatic (Fig. 8) but again, the reductions were greater when *E. eugeniae* was alone or part of a combination. The data also indicate that the weight gain of *P. excavatus* was less than that of the other two species.

The major goal of vermistabilization is the stabilization of organic matter and not the production of earthworm biomass. When the small differences in volatile solids reduction (Table 4) are considered with the changes in worm biomass (Table 4), it is not obvious that polyculture has any advantages over monoculture using either *E. fetida* or *E. eugeniae*.

Conclusions

Several important factors were evaluated to examine possible methods of optimizing the use of earthworms in sludge stabilization (vermistabilization). These studies were done with activated sludge which had been aerobically digested.

Acknowledgements

This report is based on research supported by a grant from the Program on Appropriate Technology, National Science Foundation, NSF Grant No. ISP- 8016764. Any opinions, findings, summaries or conclusions expressed in this report are those of the project investigators and do not necessarily reflect the views of the National Science Foundation. The authors would like to thank the many students who have helped on this project.

References

Camp, Dresser and McKee, Inc. 1981. Engineering Assessment of Vermicomposting Municipal Wastewater Sludges, prepared for the Municipal Environmental Research Laboratory, Office of Research and Development, Environmental Protection Agency, Cincinnati, Ohio, EPA 600/2-81-075.

Dinges, R. 1981. *Natural Systems for Water Pollution Control*. Van Nostrand Reinhold Company, New York.

Edwards, C.A. and Lofty, J.R. 1977. The Biology of Earthworms. Chapman and Hall Ltd., London, 333 pp.

Kaplan, D.L., Hartenstein, R., Neuhauser, E.F. and Malecki, M.R. 1980a. Physio-chemical requirements in the environment of the earthworm *Eisenia foetida*. Soil. Biol. Biochem. 12: 347–352.

Kaplan, D.L., Hartenstein, R., Malecki, M.R. and Neuhauser, E.F. 1980b. Role of the pheretemoid worms, *Amynthas hawayanus* and *A. rodericensis* in soils and recycling. Rev. Ecol. Biol. du Sol. 17: 165–171.

Loehr, R.C., Martin, J.H., Neuhauser, E.F. and Malecki, M.R. 1984a. Waste Management Using Earthworms – Engineering and Scientific Relationships. PB84-193218. NTIS, Springfield, VA. 76 pp.

Loehr, R.C., Martin, J.H. and Neuhauser, E.F. 1988. Stabilization of Liquid Municipal Sludge Using Earthworms (See p. 95).

Neuhauser, E.F., Hartenstein, R. and Kaplan, D.L. 1979a. A second progress report on the potential use of earthworms in sludge management. *In* Proceedings of the Conference on Sludge Composting, Nov. 14–17. Sponsored by Information Transfer, Inc., Silver Spring, MD.

Neuhauser, E.F., Kaplan, D.L. and Hartenstein, R. 1979b. Life history of the earthworm *Eudrilus eugenae*. Rev. Ecol. Biol. du Sol. 16: 525–534.

BREAKDOWN OF ANIMAL, VEGETABLE AND INDUSTRIAL ORGANIC WASTES BY EARTHWORMS

CLIVE A. EDWARDS
*The Ohio State University, Columbus, Ohio 43210**

Introduction

Extremely large amounts of organic wastes are produced by domesticated animals all over the world. With increasingly specialized agricultural production in developed countries, it is often very difficult to dispose of these wastes on nearby land. Similarly, large amounts of vegetable wastes are produced in the horticultural and food processing industries. Finally, there are a number of industries, including the brewery, processed potato and mushroom industries, that produce large quantities of organic wastes. The amounts of organic wastes produced from different sources in the U.K. are summarized in Tables 1 and 2. These wastes are often expensive to dispose of and may have to be kept in lagoons or land disposal sites before they can be put on to agricultural land. Moreover, they cause odour problems and are a potent source of pollution of groundwater by nitrogen and a source of odour. Breakdown of organic wastes by earthworms can turn such a negative asset into a profit, by producing useful materials, and at the same time minimizing environmental pollution (Edwards *et al.*, 1985).

It has been known for many years that earthworms can breakdown a wide range of organic wastes and this is demonstrated by their common presence and activity in compost heaps. Worms are also bred commercially on a large scale in organic wastes for fish bait. The species most commonly bred in this way is the 'brandling worm', *Eisenia foetida* (Savigny). Work in the United States, at the State University of New York, has shown that earthworms can breakdown activated sludges very effectively (Hartenstein *et al.*, 1979a; Neuhauser *et al.*, 1979). Much of the basic biology of *E. foetida* was worked out by these workers (Hartenstein *et al.*, 1979b; Neuhauser *et al.*, 1980b; Kaplan *et al.*, 1980; Hartenstein, 1982) on a laboratory scale using sewage sludge as the principal growth medium. Major findings from this research are summarized in the chapter previous to this one by Neuhauser *et al.*.

Research began at Rothamsted in 1980, on using earthworms to break down animal manures, such as pig and cattle solids and slurries, wastes from laying chickens, broilers, turkeys and ducks, horse manure and rabbit droppings (Edwards, 1983). This was extended later to studies using worms to break down vegetable wastes including those from the mushroom industry, processed potato industry, brewery industry and paper pulp. This work began as fundamental research on a laboratory scale but was extended to development research on a field scale and was eventually commercialized with the formation of a government-aided company, British Earthworm Technology.

The research at Rothamsted had two main aims: (i) to turn animal and other agricultural wastes into useful materials that could be added to agricultural land to

* Formerly; Senior Principal Scientific Officer, Rothamsted Experimental Station, Harpenden, Herts, U.K.

Earthworms in waste and environmental management
edited by C.A. Edwards and E.F. Neuhauser, pp. 21–31

Table 1. Excreta from farm livestock in England and Wales (1980)

Livestock	Numbers (millions)	Excreta output (kg per animal per day)	Excreta from housed animals (million tonnes per year)	Total excreta all livestock (million tonnes per year)
Cattle	9.44	19–41	43.06	83.17
Pigs	6.59	1.5–8.0	9.11	9.11
Poultry	109.79	0.115	4.60	4.60
Sheep	22.49	1.5–4.0	—	22.69

Adapted from Neilsen (1982)

improve soil structure and fertility, and also have considerable potential in horticulture as a plant growth medium or component of commercial potting composts; (ii) to harvest worms from the worked waste and process them into a high nutritious protein feed supplement for fish, poultry and pigs. To accomplish these aims, a complex network of collaborative research was set up involving six research stations, six university or college departments and eight commercial organizations with a research program coordinated by an interdisciplinary research committee. The following areas of research and development were studied:

a. A laboratory screening programme into the suitability of five different earthworm species and 10 different waste materials to assess their biological and economic potentials, and study the biology and ecology of these worms in organic wastes. This research included studies of the population dynamics, life tables and the influence of environmental factors on the growth, reproduction and survival of these different species.
b. Microbiological studies to assess the source of nutrition of these earthworms that live on organic wastes, the relative importance of bacteria, protozoa, fungi and nematodes in their diet, and whether particular microorganisms were essential to their survival (see p. 71 *et seq*).

Table 2. Plant and vegetable wastes in U.K.

Source of material	Quantity (millions of tonnes tonnes of dry matter)	Metabolisable energy (MJ × 10^6
Straw	9.30	60450
Arable crops	0.85	8460
Sugar beet	0.50	6000
Brewery	0.24	2640
Paper	0.25	2700
Potato	0.21	1800
Spent mushroom compost	0.20	1300
Vegetable	0.05	178

(from various sources)

Table 3. Maximum reproduction rate of earthworms in animal and vegetable wastes

Species	No. of cocoons per week	% Hatch	No. of hatchlings	Net reproductive rate per week
Eisenia foetida	3.8	83.2	3.3	10.4
Eudrilus eugeniae	3.6	81.0	2.3	6.7
Perionyx excavatus	19.5	90.7	1.1	19.4
Dendrobaena veneta	1.6	81.2	1.1	1.4

c. Evaluation of productivity of worm biomass and processed waste in relation to type of waste, worm stocking rate and environmental factors (Edwards *et al.*, 1985).
d. Development of systems of production of worm protein and processed waste (see p. 111 *et seq*).
e. Harvesting and processing of worms into animal feed protein, animal feeding trials and toxicological tests (Edwards, 1985), (see p. 169 *et seq*).
f. Production and preparation of plant growth media processed by worms from organic wastes, plant growth trials and amendment of media to produce maximum plant growth (Edwards, 1983; Edwards *et al.*, 1985) (see p. 211 *et seq*).

Biology of earthworm species that breakdown organic wastes

Five earthworm species were identified as candidate species for breakdown of organic wastes. Three species, *Eisenia foetida, Dendrobaena veneta* and *Lumbricus rubellus*, were of temperate climatic origin and two, *Eudrilus eugeniae* and *Perionyx excavatus* came from the tropics. The survival, growth and reproduction of these species was studied in the laboratory in pig, cattle, duck, turkey, poultry, potato, brewery and paper wastes and compared with that in activated sewage sludge.

The reproductive rates of four of these species are summarized in Table 3, the length of the life cycles in Table 4, and the biomass production in Table 5. All grew satisfactorily and survived well in a wide range of organic wastes but some were very prolific, some grew rapidly, some attained a large biomass; all characters contributing to the practical usefulness of the worms.

Table 4. Productivity of earthworms in animal and vegetable waste – length of life cycle

Species	Time for cocoons to hatch (days)	Time to sexual maturity (days)	Time egg to maturity (days)
Eisenia foetida	32–73	53–76	85–149
Endrilus eugeniae	13–27	32–95	43–122
Perionyx excavatus	16–21	28–56	44–71
Dendrobaena veneta	40–126	57–86	97–214

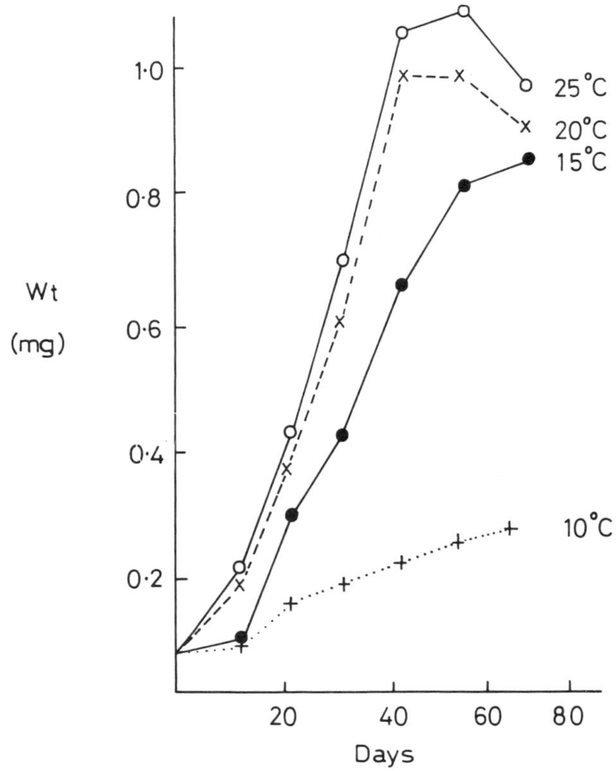

Fig. 1. Growth of *Eisenia foetida* at different temperatures.

Influence of environmental and other factors on growth and survival of earthworms

The growth of four species of earthworms in different animal and vegetable wastes was studied in relation to a range of environmental factors. All species of earthworms required the waste in which they lived to be aerobic, and as soon as anaerobicity developed the worms moved out of the waste. Each of the four species differed very considerably in terms of their response to and tolerance of different temperatures. The optimum temperature for *E. foetida* was 25°C (Fig. 1), and it had a tolerance of temperatures from 0–35°C. *D. veneta* had a rather lower temperature optimum and rather less tolerance of extreme temperatures. The optimum temperatures for *E. eugeniae* and *P. excavatus* were about 25°C, but they died at temperatures below 9°C and above 30°C. Optimum temperatures for cocoon production were much lower than for growth for all species.

Species also differed in their optimum moisture requirements from those of *E. foetida* (Fig. 2). Some, such as *Dendrobaena veneta*, were able to withstand a much wider range of moisture than others, such as *Perionyx excavatus*.

Earthworms are very sensitive to ammonia and will not survive in organic wastes containing much ammonia, *e.g.*, fresh poultry litter (Fig. 3). They also die in wastes with large amounts of inorganic salts (Fig. 4). Both ammonia and organic salts have

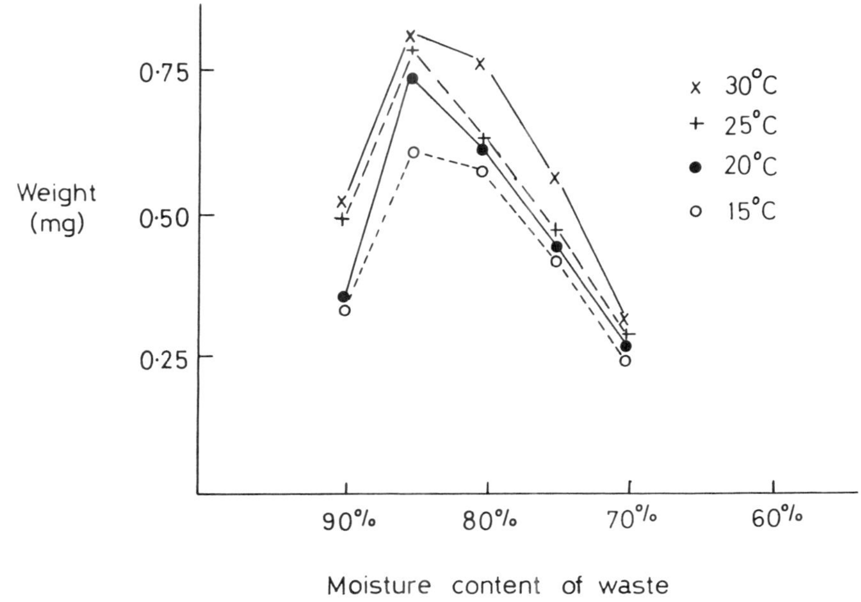

Fig. 2. Growth of *Eisenia foetida* at different moisture contents.

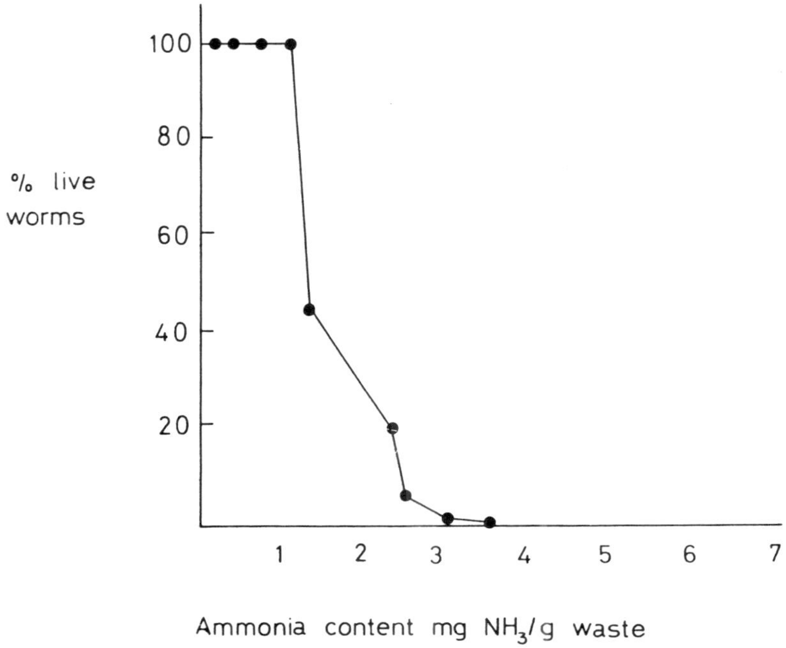

Fig. 3. Survival of *Eisenia foetida* in different ammonia contents.

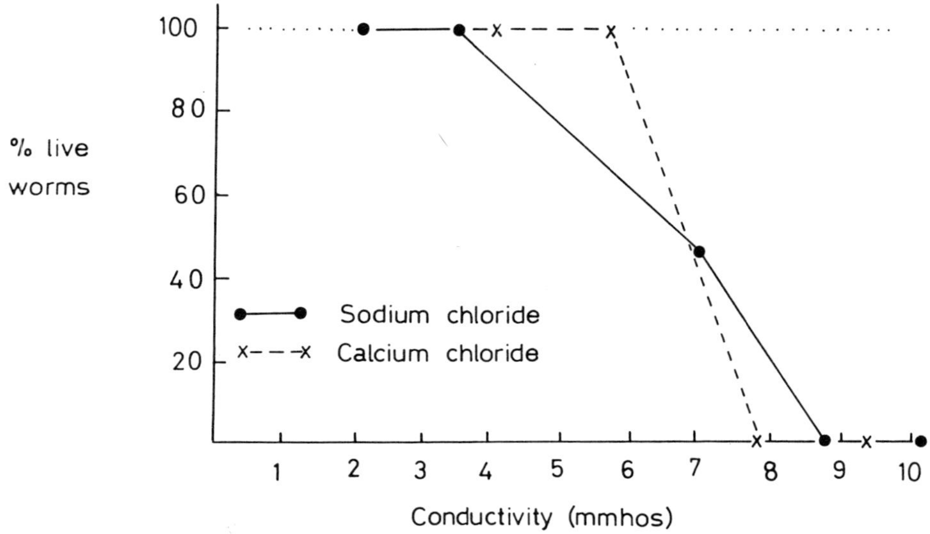

Fig. 4. Survival of *Eisenia foetida* at different salt contents.

Table 5. Productivity of earthworms in animal and vegetable wastes – biomass production

Species	Maximum net reproductive rate/worm/week	Mean mature weight (g)	Minimum time to maturity (weeks)	Biomass production per worm per week (g)
Eisenia foetida	10.4	0.55	8.4	0.68
Eudrilus eugeniae	6.7	4.3	5.0	5.76
Perionyx excavatus	19.4	1.3	4.0	6.3
Dendrobaena veneta	1.4	0.92	8.1	0.16

very sharp cut off points between toxic and non-toxic. Wastes that have too much ammonia became acceptable after this is removed by a period of composting or both excessive ammonia and salts can be washed out of the waste. They are relatively tolerant with regard to pH, but when given a choice in a pH gradient, they moved towards the more acid material with a preference of pH = 5.0 (Fig. 5). The optimal conditions

Table 6. Optimal conditions for breeding *E. foetida* in animal and vegetable wastes

Condition	Requirements
Temperature	15–20°C (Limits 4–30°C)
Moisture content	80–90% (Limits 60–90%)
Oxygen requirement	Aerobicity
Ammonia content of waste	Low: < 0.5 mg/g
Salt content of waste	Low: < 0.5%
pH	> 5 and < 9

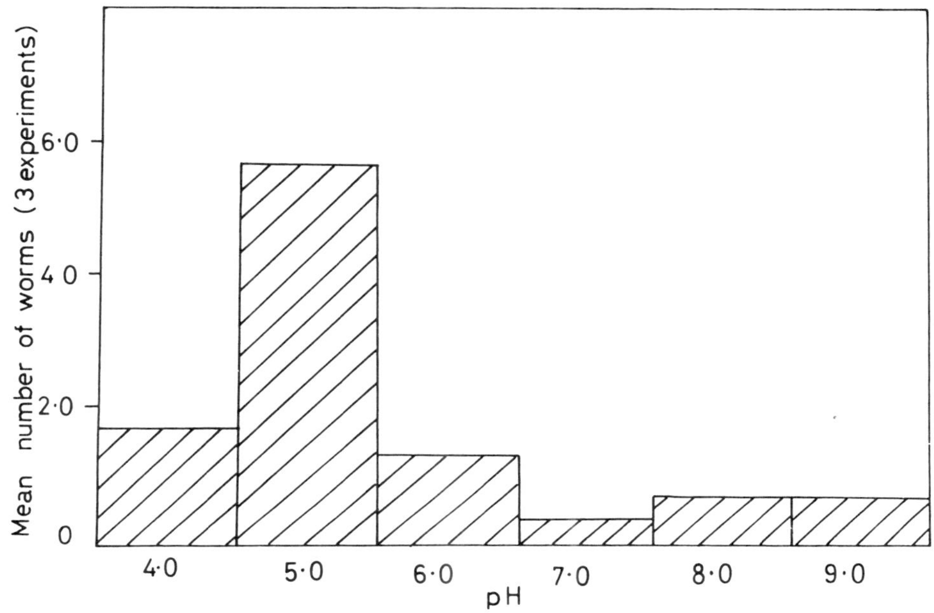

Fig. 5. pH preference of *Eisenia foetida.*

for breeding *E. foetida* are summarized in Table 6. These do not differ too much from those suitable for the other species.

Evaluation of species for breakdown of wastes

The species most commonly used for degrading organic wastes is *E. foetida*. There are a number of reasons why this species is preferred. It is ubiquitous, and many organic wastes became naturally colonized by this species. It has a wide temperature tolerance and can live in wastes with a good range of moisture contents. It is a tough worm, readily handled and in mixed cultures usually becomes dominant, so that even when field systems begin with other species they often end up with a large proportion of *E. foetida*.

Eudrilus eugeniae is a large worm that grows extremely rapidly and is reasonably prolific. Under optimum conditions it would be ideal for protein production. Its main disadvantages are poor temperature tolerance and poor handling capabilities so that it can be difficult to harvest. It is a good species for tropical conditions.

Perionyx excavatus is an extremely prolific earthworm, almost as easy to handle as *E. foetida*, and very easy to harvest. Its main drawback under temperate conditions is its inability to withstand adverse temperature conditions. For tropical conditions it is an ideal species.

Dendrobaena veneta is a large worm but is not very prolific and does not grow very fast. It is probably the least suitable for organic matter breakdown of the four species studied in detail.

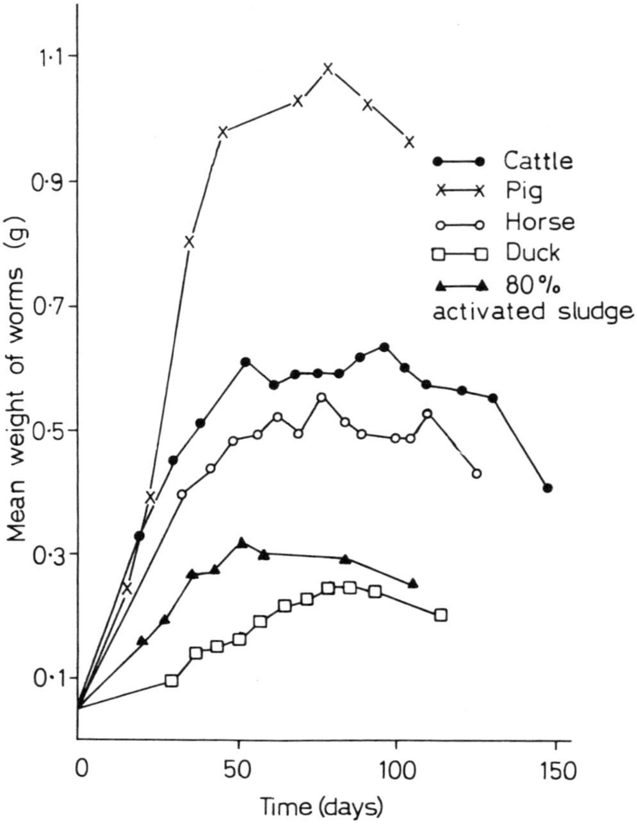

Fig. 6. Growth of *Eisenia foetida* in different animal wastes.

Suitability of types of animal and vegetable wastes for breakdown by earthworms

Most organic wastes can be broken down by earthworms, but some have to be made acceptable and not all grow worms equally well. The growth of *E. foetida* in five different animal wastes is illustrated in Fig. 6.

Pig solids are probably the most productive waste for growing worms. If the waste is in the form of a slurry, the solids must be separated mechanically or by sedimentation. Pig waste tends to have some ammonia and inorganic salts and may have to be composted for about two weeks prior to inoculation with worms. Pig wastes tend to have a content of heavy metals, particularly copper.

Cattle solids are the easiest animal waste in which to grow worms. They usually contain no materials that are unfavorable to growing worms. Solids have to be separated from slurries as with pig wastes, before they can be used to grow worms.

Horse manure is an excellent medium for growing worms and needs no modification.

Poultry wastes including chicken, duck and turkey manures, contain significant amounts of inorganic salts and ammonia which may kill worms in fresh waste. After

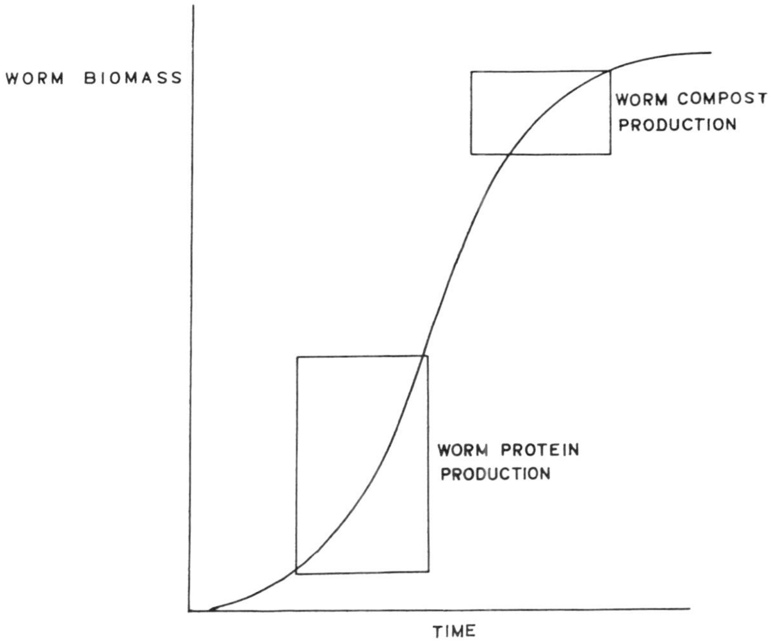

Fig. 7. Worm populations for protein production and compost production.

removal of these materials, worms grow well in them and the compost produced is high in nutrients.

Potato waste in the form of peel from the processed potato industry is an ideal growth medium for earthworms needing no modification in terms of moisture content.

Paper pulp solids produced by mechanical separation or sedimentation of solids from the press washings is an excellent material for growth of worms.

Brewery waste needs no modification in terms of moisture content to grow worms. Worms process it very quickly.

Spent mushroom compost is a good medium for growing earthworms, which are able to breakdown the straw it contains.

Productivity of earthworms in protein production and waste processing

For protein production, it is important to inoculate waste with cocoons or young worms rather than fully grown or mature worms. This maintains maximum biomass productivity (Fig. 7). The stocking rate in terms of ratio of biomass of worms to biomass of waste is important for maximum productivity, a stocking rate of about 1:10 giving maximum biomass productivity in the majority of organic wastes (Fig. 8). For production of plant growth media, the fastest processing of the waste can be achieved by inoculation with nearly fully-grown earthworms which achieve maximum throughput of materials (Fig. 7).

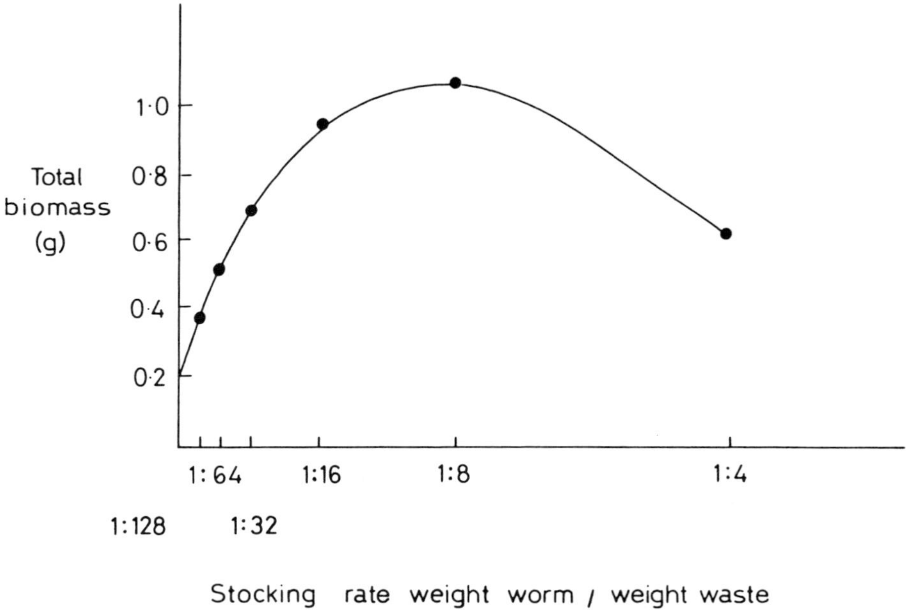

Fig. 8. Optimum stocking rates with *E. foetida* for maximum biomass production.

Table 7. Systems for breeding earthworms

Method	Comments
Windrows and beds	Labour intensive. Needs relatively large area of land. Seasonal unless covered
Crates or boxes	Handling of units and watering difficult
Moving belt system	High initial cost
Continuous flow suspended beds	High initial cost, but economic to run due to low labour costs
Trickling filter systems	Has excellent potential for liquid wastes. Needs further development research.

Systems of earthworm production

Systems of growing earthworms range from low technology simple systems in windrows, heaps or boxes to complex continuous breeding systems (Table 7). The engineering details of such systems are dealt with in a later section, but the basic principle of breeding systems is to add wastes in small regular batches and allow the earthworms to process successive aerobic layers of wastes (Fig. 9). Many of the operations can be mechanized, a suitable balance being needed between costs of mechanization and savings in labor. The key to maximum productivity is to maintain aerobicity and optimal moisture and temperature conditions. Hence, for all the year round

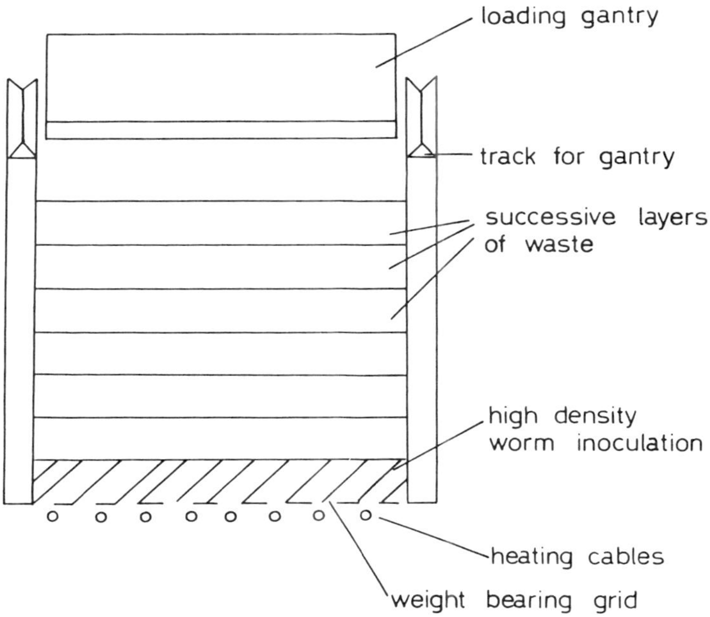

Fig. 9. Design of automatic loading earthworm bed.

production, the processing must be done under cover, although heating is not usually necessary, because the decomposing waste maintains warm conditions.

References

Edwards, C.A. 1982. Production of earthworm protein for animal feed from potato waste. *In* Upgrading Waste for Feed and Food. pp. 153–162. Edited by D.A. Ledward, A.J. Taylor and R.A. Lawrie. Publisher: Butterworth's, London. 321 pp.

Edwards, C.A. 1985. Production of feed protein from animal wastes by earthworms. Phil. Trans. Royal Soc. Londen. B310: 299–307.

Edwards, C.A., Burrows, I., Fletcher, K.E. and Jones, B.A. 1985. The use of earthworms for composting farm wastes. *In* Composting of Agricultural and Other Wastes. pp. 229–242. Edited by J.K.R. Gasser. Elsevier Publ. Co., Amsterdam.

Edwards, C.A. 1983. Earthworms, organic waste and food. Span. Shell Chem. Co. 26(3): 106–108.

Hartenstein, R., Neuhauser, E.F. and Kaplan, D.L. 1979a. A Progress Report on the Potential Use of Earthworms in Sludge Management. *In* Proceedings of the 8th National Sludge Conference, Florida, 1979. Information Transfer Inc., Silver Spring, MD.

Hartenstein, R., Neuhauser, E.F. and Kaplan, D. 1979b. Reproductive potential of the earthworm *Eisenia foetida*. Oecologia 43: 329–340.

Hartenstein, R. 1982. Metabolic parameters of the earthworm *Eisenia foetida* in relation to temperature. Biotechn. and Bioengin. 24: 1803–1811.

Kaplan, D.L., Hartenstein, R., Neuhauser, E.F. and Malecki, M.R. 1980. Physiochemical requirements in the environment of the earthworms *Eisenia foetida*. Soil Biol. Biochem. 12: 347–352.

Neuhauser, E.F., Hartenstein, R. and Kaplan, D.L. 1980b. Growth of the earthworm *Eisenia foetida* in relation to population density and food rationing. Oikos 35: 93–98.

THE USE OF ANAEROBICALLY DIGESTED CATTLE SOLIDS FOR VERMICULTURE

JIM FREDERICKSON and DAVID KNIGHT*
U.K. Technology Faculty, Open University, Walton Hall, Milton Keynes, Herts, U.K.

Introduction

Anaerobic digestion of animal wastes in the United Kingdom has been shown to be a technically feasible process, primarily as a means of producing a net excess of energy in the form of methane gas but also as a method of reducing the total solids, BOD and odour of wastes. (Pain *et al.*, 1984; Friman, 1984) With intensification of livestock housing, increasing slurry management problems and the prospect of escalating energy costs it is possible that the anaerobic digestion process will be adopted to a greater degree on medium to large scale farms. Despite the obvious benefits of the process, anaerobic digestion is not thought to be economically attractive at present (Anon, 1982; James & Campbell, 1985). Economic assessments of digestion plants, however, tend to focus primarily on methane production and utilization with little value being ascribed to the other benefits of the process or to the digested residue. Yet, an important method of enhancing the overall economic performance of anaerobic digestion is to increase the value of the digested residue above its gross fertilizer value when landspread.

This paper explores the extent to which vermiculture may be used to upgrade the value of mechanically separated digested cattle solids to produce worm protein and a worm-worked material of value as a horticultural growing medium. Little published information has addressed the growth of worms in digested animal wastes, while some reports have suggested that sewage sludge, having undergone anaerobic digestion, is toxic to worms. (Hartenstein, 1981a; Mitchell *et al.*, 1977). To enable the effect of digestion on worm growth and the nature of worm-worked material to be directly assessed, cattle waste from two sources was selected, before and after digestion. Chemical analyses of the wastes were undertaken to detect major physico-chemical changes brought about by digestion. Worm growth trials using *Eisenia foetida* were conducted on the mechanically-separated undigested and digested solids and the resulting worm-worked material analyzed for major plant growth characteristics. Plant growth trials were then undertaken. Having demonstrated that digested solids could support worm growth successfully and that the resulting worm-worked material appeared to have potential as a growing medium when further processed, a preliminary study was undertaken to assess the potential improvement in the economic performance of a digester plant arising out of the incorporation of a vermiculture unit producing worm-worked material.

The anaerobic digestion process

Anaerobic digestion may be seen as the microbial conversion of complex organic materials, in the absence of air, to methane gas and a range of inorganic end products including carbon dioxide and ammonia. In cattle waste, the organic materials are

* *Present address*: Department of Biological Sciences, University of Exeter, U.K.

Earthworms in waste and environmental management
edited by C.A. Edwards and E.F. Neuhauser, pp. 33–47

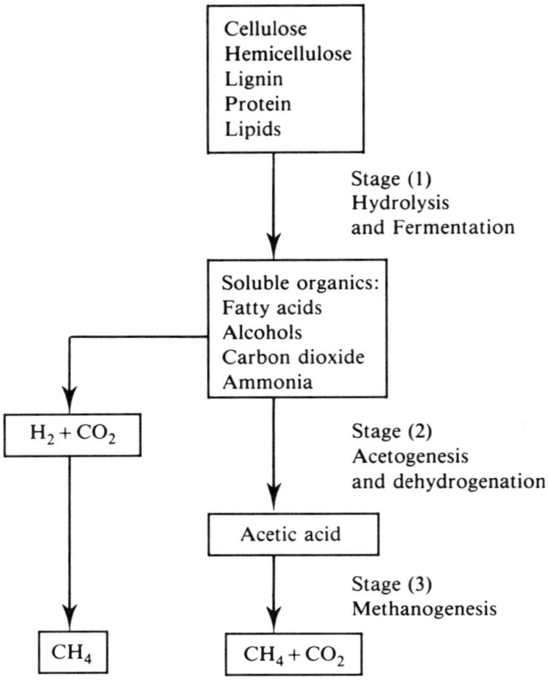

Fig. 1 Anaerobic digestion of organic materials.

mainly cellulose, hemicellulose and lignin with smaller fractions of protein and lipids. The digestion or fermentation process is usually considered to take place in stages by three physiologically distinct groups of bacteria (Pfeffer, 1980).

Figure 1 illustrates the stages of fermentation. It can be seen that the first group of hydrolyzing and fermenting bacteria convert complex organic polymers such as carbohydrates, proteins and lipids to fatty acids, alcohols, carbon dioxide and ammonia. The second group of hydrogen producing acetogenic bacteria convert the products of the first group into hydrogen, carbon dioxide and acetic acid. The third group of methanogenic bacteria in turn converts these products to a mixture of methane and carbon dioxide.

Effect on waste characteristics

It can be seen that the digestion process involves the destruction of the more readily degradable solid organic materials, thereby significantly reducing total and volatile solids content. Since the only major losses from the system are from methane and carbon dioxide, valuable inorganic nutrients including nitrogen mainly in the form of ammonia, are conserved. The extent of the conversion of biodegradable materials to soluble organics and then to CH_4 and CO_2 depends greatly on the nature of the waste, the efficiency of the digester, operating conditions and residence time.

Table 1. Chemical changes in waste during digestion.

Component	Influent	Effluent
Dry matter (%)	8.00	5.34
	% of dry matter	
Cell wall (NDF)	50.66	45.33
ADF	32.34	32.22
Cellulose	24.18	21.03
Hemicellulose	18.32	13.11
Lignin	8.16	11.19
Ash	11.43	16.18

Martin *et al.*, 1983.

Table 1 illustrates some of the major changes in the chemical composition of the waste brought about by digestion. In general, it can be seen that the loss in dry matter or total solids on passage through the digester can be accounted for by losses in the cellulose and hemicellulose fractions of the waste while lignin and ash are virtually undegradable. Other common physico-chemical effects are small increases in the ammonia content and ionic conductivity of the digested waste while the loss of total solids enhances the crude protein content on a dry matter basis (Hayes *et al.*, 1980; Martin *et al.*, 1983).

Earthworm growth trials

The purpose of this section is to investigate the extent to which anaerobically digested cattle solids may support the growth of *E. foetida* compared to the undigested solids from the same source, and in so doing assess the effect of digestion on worm growth. As illustrated in the section on anaerobic digestion, both undigested and digested solids comprise a complex mix of hemicellulose, cellulose, lignin and microbial biomass with varying degrees of biodegradability and physico-chemical characteristics depending on diet, slurry handling practices, type of bedding material used, digester efficiency, etc.

Factors relating to worm growth may be usefully subdivided into physico-chemical and nutrient characteristics. Broad categories of physico-chemical requirements have been identified for *E. foetida* grown on organic substrates (see materials and methods) but the nutritional element remains unclear. However, it is known that the microbial element of agricultural wastes is essential to earthworm growth (Hartenstein, 1981b; Neuhauser *et al.*, 1980). Day (1950) showed a decrease in certain strains of bacilli passing through the gut of *Lumbricus terrestris*, suggesting their breakdown as food, and Dash *et al.* (1979) list eight species of fungi involved in the feeding of tropical earthworms. Miles (1963) showed that specific mobile protozoa were an essential part of the diet of *E. foetida* if it is to reach sexual maturity. Parle (1963) indicated the presence of an earthworm cellulase enzyme in *L. terrestris*, indicating the importance of cellulose in earthworm nutrition.

Hartenstein (1981b) suggested that *E. foetida* may derive a considerable amount of energy from cellulose and that it may selectively remove a larger proportion of microbes from the ambient cellulosic matrix. In contrast, however, he suggested that wood did not lend itself to production of *E. foetida* mainly because of the ligneous components but also as a result of possible natural fungicidal material or preservatives in the wood.

To enable interpretation of the results from the comparative worm growth trials to be undertaken, detailed chemical analyses were carried out to determine possible differences between undigested and digested solids in relation to biodegradable carbon, total nitrogen content and physico-chemical characteristics. Worm growth trials were carried out as described in the next section and the resulting worm-worked material re-analyzed.

Materials and methods

Mechanically-separated fresh and digested cattle solids were obtained from two dairy cattle sources, Bore Place Farm and the National Institute for Research in Dairying (NIRD). The Bore Place cattle were fed a silage-based complete diet, and bedded on wood chips. The digester was of the plug flow type operating on a retention time of approximately 15–19 days, receiving slurry with a 13% TS content. A continuous belt press separator was used. The NIRD cattle were fed a maize silage-based diet and bedded on sawdust. The digesters were of a mixed tank design with a mean residence time of 20 days receiving slurry with an 8% TS content. A brush and roller fixed screen separator was used.

Prior to worm growth trials the cattle solids were allowed to age aerobically until acceptable to worms. Table 3 illustrates the physico-chemical changes which took place on ageing. Digested and undigested solids (60 g) were placed in pots (10.5 cm diameter) at 85% moisture content and 4 *E. foetida* added. Ten replicates were used and the pots were incubated at 20°C ± 1. The pots were sampled weekly. The physico-chemical characteristics of the wastes conformed to limits suitable for worm growth for pH (5–9) moisture content (43–90%), temperature optimum (20–25°C), ionic conductivity (less than 15,000 μmhos) and ammonia (less than 1,000 mg g^{-1}) (Hartenstein, 1982; Kaplan *et al.*, 1980; Edwards *et al.*, 1984).

The solids were analyzed for pH, total solids, volatile solids, ionic conductivity, ammonium ion concentration, Kjeldahl nitrogen and fiber content. Total solids were determined by oven drying at 105°C for 18 hours, volatile solids by ashing at 600°C for two hours. Free ammonia and ammonium ions were determined directly using an electrode in an aqueous suspension (Byrne and Power, 1974). Conductivity was measured using a conductivity cell on the wet extract. The Van Soest method for forage fiber analysis (Goering and Van Soest, 1975) was used to determine NDF, ADF, hemicellulose, cellulose lignin and ash. NDF is the insoluble fraction remaining when a neutral detergent solution has removed the very soluble cell content fraction. ADF is the insoluble fraction remaining after application of acid detergent solution and corresponds to the cellulose, lignin and the residue is ashed. Hemicellulose was determined as NDF minus ADF. Total nitrogen and crude protein were measured by the macro-Kjeldahl method (Osborne and Voogt, 1978).

Results of chemical determinations are given as the means of three samples. Results from the comparative worm growth trials were analyzed statistically using a one and two tailed Student's *t*-test (Bishop, 1980).

Results

Substrate characteristics

The mechanical separation process concentrates coarse fiber from the slurry with the resulting loss of much soluble and colloidal nitrogen to the liquid fraction. Moreover, much of the solids are derived from the bedding material of the cattle, in this case sawdust and woodchips typically forming up to 20% of the original slurry on a dry weight basis and comprise mainly highly undegradable lignocellulose material (Hills and Roberts, 1981). Overall, the substrate is characterized by high levels of carbon compared to relatively low levels of nitrogen.

Prior to separation the anaerobic digestion process has the effect of removing volatile solids from the slurry, converting a proportion of plant protein to NH_3 and NH_4^+ and incorporating a fraction into microbial biomass, whilst increasing ionic conductivity levels. The digested slurry also contains much undegraded lignocellulose material. The subsequent separation passes this recalcitrant material into the solid fraction whilst much of the nitrogenous products of plant protein breakdown is lost to the liquid fraction. Clearly, the separation of digested slurry compared to undigested leads to an overall reduction of solids yield because of the breakdown of volatile solids.

Both digested and undigested material must be aged aerobically to reduce the initially high levels of NH_3 and NH_4^+ which are toxic to earthworms (Hayes *et al.*, 1980). However, ageing would appear to be more important in laboratory scale compared to field scale trials.

Tables 2 and 3 show the major differences in physico-chemical characteristics and chemical composition of undigested and digested mechanically separated cattle solids from NIRD and Bore Place.

For the NIRD solids little difference in physico-chemical characteristics for undigested and digested solids could be detected, once acceptable to *E. foetida*. Fiber analysis, however, indicated higher proportions of easily soluble material mainly

Table 2. Composition of raw solids

	NIRD		Bore Place	
% Total solids	Undigested	Digested	Undigested	Digested
N.D.F.	78	81	74	78
Neutral detergent soluble material	22	19	26	22
A.D.F.	49	59	48	55
Hemicellulose	29	22	26	23
Cellulose	32	30	28	32
Lignin	15	28	19	22
Ash	2	1	1	1
Crude protein (TKN × 6.25)	7.0	6.8	9.6	8.8
Total Kjeldahl nitrogen	1.2	1.09	1.53	1.40

Table 3. Physico-chemical characteristics of fresh and acceptable separated solids

	N.I.R.D.				Bore Place			
	Undigested		Digested		Undigested		Digested	
	Fresh	Aged	Fresh	Aged	Fresh	Aged	Fresh	Aged
Total solids (%)	14.8	18.0	16.0	20.3	17.6	20.9	18.6	23.6
Volatile solids (% of T.S.)	93.4	91.9	86.4	83.2	87.3	84.3	89.0	87.6
pH	7.9	8.2	8.3	8.1	8.0	8.1	8.4	8.4
Ionic conductivity (μmhos)	—	6093	—	6785	—	7928	—	10423
NH_3–N (mg/L)	63	< 1	364	< 1	84	< 1	210	< 1
NH_3–N + NH_4^+–N	560	6	1120	3	1330	5	1680	19
(mg/L)	560	1120	1330	1680				

comprising cell contents (proteins, lipids, etc.) and non-lignin biodegradable carbon in the form of hemicellulose and cellulose in the undigested solids. Total nitrogen levels were found to be low but slightly higher in undigested solids.

For Bore Place solids, once acceptable, physico-chemical characteristics were found to be similar with the exception of ionic conductivity for the digested material which remained high. Fiber analysis indicated the undigested solids contained a greater proportion of easily soluble material and hemicellulose, but the digested solids contained a higher proportion of cellulose. It is probable that much of the cellulose in Bore Place material is shielded by lignin rendering it highly undegradable to anaerobic digestion, as shown by its increased proportion in the digested fraction (Hills and Roberts, 1981). Undigested solids showed a higher total nitrogen content than digested solids.

Earthworm growth trials

Figure 2 shows the growth of *E. foetida* on digested and undigested NIRD solids. It can be seen that the patterns of growth are similar for both materials for approximately 42 days. Thereafter, the mean weights of earthworms on digested material began to fall compared to undigested solids. However, after 49 days there was no significant difference at the 5% level between the two weights. The conversion ratios on a dry weight basis, assuming an earthworm solids content of 20% (Sabine, 1983) were 3.0% for undigested solids, and 2.9% for digested solids taken from the highest point of each curve.

Figure 3 shows the growth of *E. foetida* on digested and undigested Bore Place solids. After 14 days, the mean weight of earthworms grown in undigested solids was significantly greater at the 5% level, and this trend continued for the duration of the experiment. After 42 days, the mean weight of earthworms grown in digested solids began to drop whereas the weights in undigested solids remained constant. The conversion ratios were 2.8% and 2.3% for undigested and digested solids respectively.

The conversion ratios quoted in this study were based on a stocking density of 4 earthworms per 60 g of waste. This was considered appropriate given the comparative nature of the study. However, it has been shown that higher stocking densities

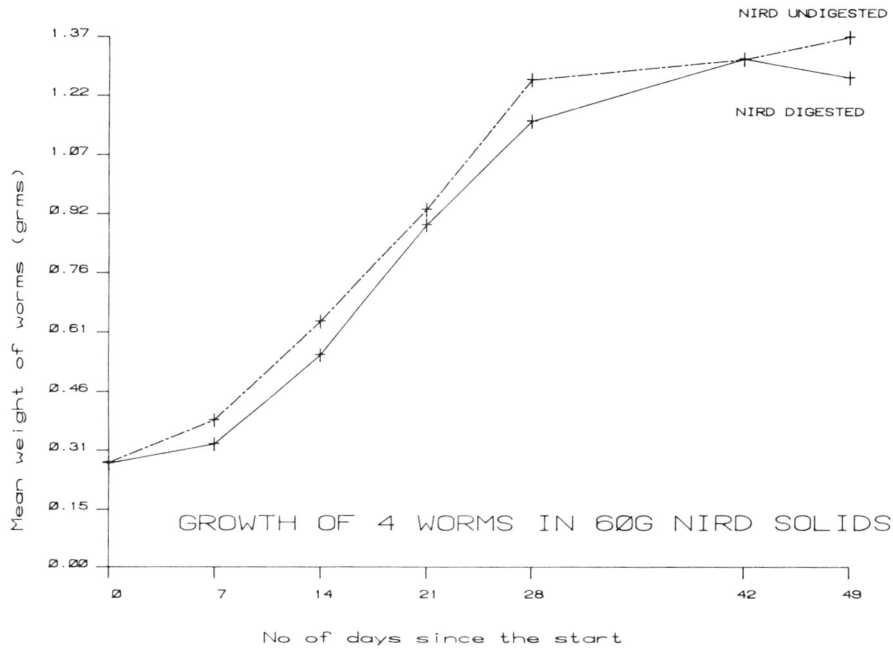

Fig. 2. Growth of worms in NIRD cattle solids

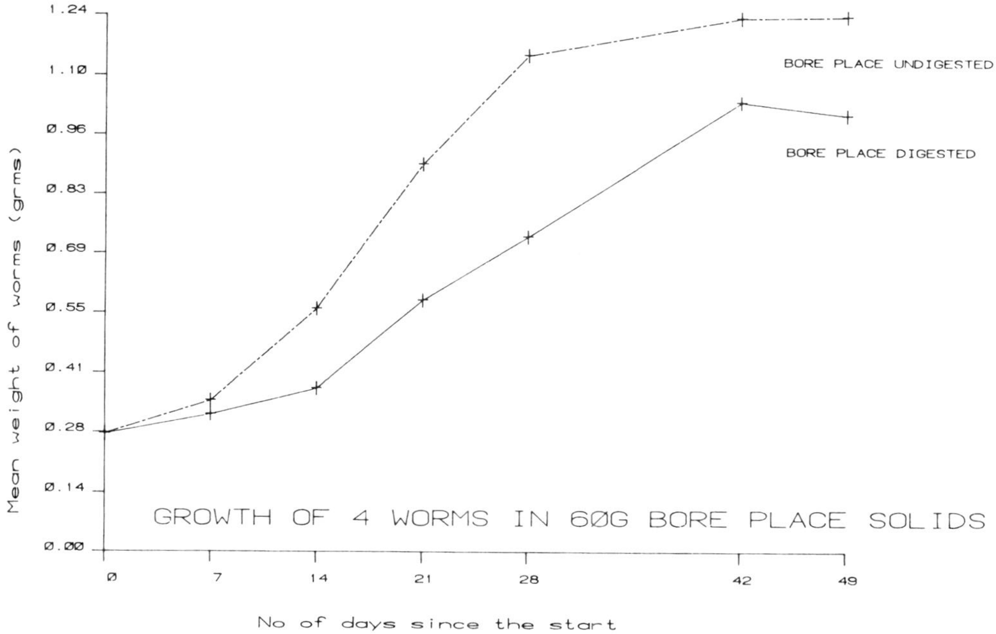

Fig. 3. Growth of worms in Bore Place cattle solids

have produced greater conversion ratios in other organic wastes (Edwards, 1982).

Discussion

For NIRD solids, chemical analyses indicated a slightly higher total nitrogen level for undigested solids but little difference in physico-chemical characteristics. Fiber analysis showed the undigested solids to contain a greater biodegradable and available carbon component. From these results it may be expected that the maximum theoretical conversion of solids to earthworm biomass would be marginally greater for the undigested material. The earthworm growth trials produced conversion ratios for undigested and digested solids of 3.0% and 2.9% respectively, the variance not being statistically significant (5% level). After 42 days, however, the weight of *E. foetida* grown on digested material was seen to decrease, indicating that the nutritional content of the waste was becoming limiting.

For the Bore Place solids the digested material was found to have a much greater ionic conductivity, higher pH, less total nitrogen and a reduced biodegradable and available carbon component. The growth trial results showed a significantly greater (5% level) earthworm biomass production from undigested solids compared to digested solids which was reflected in conversion ratios of 2.8% and 2.3% respectively. As with NIRD solids the earthworm biomass from digested Bore Place solids was seen to decrease at 42 days. The differences in conversion ratios would appear to be the result of the difference in physico-chemical factors as well as nutritional value.

In general, earthworm growth would be expected to reflect differences in the nutritional value and physico-chemical characteristics of substrates under test. The NIRD results demonstrated that for undigested and digested solids with similar physico-chemical characteristics, earthworm growth rates could be expected to be broadly comparable over the period when available carbon and nitrogen were not limiting. The nutritional value of the digested material should become limiting more quickly than for the undigested material. The Bore Place results indicated that over the period when nutritional value was not limiting the higher levels of ionic conductivity and pH appeared to adversely affect growth in the digested solids.

Digested worm worked material as a plant growth medium

Introduction

Digested solids have been shown to support earthworm growth to produce a worm worked material. This section deals with the relative value of digested worm worked material compared to its undigested counterpart, with respect to its potential as a plant growth medium for horticultural uses similar to commercially available peat-based loamless composts with added nutrients. Plant nutrient analyses and preliminary plant growth trials were carried out to this end.

Table 4. Plant nutrient analysis of raw and worm-worked material

Variable	Raw NIRD undigested solids	Raw NIRD digested solids	Raw Bore Place undigested solids	Raw Bore Place digested solids	Worm-worked NIRD undigested solids	Worm-worked NIRD digested solids	Worm-worked Bore Place undigested solids	Worm-worked Bore Place digested solids
pH	7.4	8.1	7.7	8.6	7.7	7.7	7.8	7.7
Ionic conductivity (μmhos)	1024	1070	1517	1851	832	788	1280	1448
P (mg/L)	97	180	202	208	112	109	252	252
K (mg/L)	864	1080	1656	1746	1382	1353	2045	2419
Mg (mg/L)	92	56	110	30	53	43	110	98
NO_3^- (mg/L)	13	13	19	14	26	24	28	28
NH_4^+ (mg/L)	282	402	348	1080	6	6	6	6

Materials and methods

Plant nutrient analyses were carried out using ADAS/MAFF recommended methods for peat-based soils on samples of worm-worked digested and undigested solids from Bore Place and compared with Levington multipurpose compost used as a control in plant growth trials as described below. Further plant nutrient and fiber analyses were also carried out on other batches of solids from Bore Place and NIRD in order to quantify more fully, the plant nutrient characteristics of digested and undigested solids.

The plant growth trials were carried out as follows; F_1 (Sonato) tomato seeds were germinated in trays and transplanted into 4 inch and later 6 inch plastic pots before being transferred to polyethylene growing tunnels. Plants were watered on demand, and when the first truss had set a propriety liquid fertilizer (NPK 4–4.5–8) was fed at every other watering. The tops of plants were picked out after the 5th truss had set. Sixteen replicates for each material were used.

Results

Plant growth characteristics

Table 4 and comparison between Tables 2 and 5 show that worm-working has contributed to a reduction in ionic conductivity and an increase in nitrate-nitrogen levels when compared to the fresh solids. There was a breakdown of degradable carbon in the form of hemicellulose and cellulose and a concomitant increase in the readily soluble component. This stabilization process was accompanied by a reduction in fiber particle size. There was little change in P, K and Mg levels on worm-working. Table 6 also shows the similarities between pH and plant nutrient characteristics for digested and undigested worm-worked material from the two sources.

Table 6 shows the comparison between the digested and undigested worm-worked material used in the plant growth trials and the compost control. The plant nutrient levels of P, K and Mg were similar for digested and undigested worm-worked material, but were quite high when compared to the compost control, as was ionic conductivity. NO_3^- levels were low for all materials, and NH_4^+ levels were higher in

Table 5. Composition of worm-worked solids

	NIRD		Bore Place	
% Total solids	Undigested	Digested	Undigested	Digested
N.D.F.	66	72	65	69
Neutral detergent soluble material	34	28	35	31
A.D.F.	54	60	52	54
Hemicellulose	11	12	13	15
Cellulose	20	19	18	20
Lignin	33	39	33	33
Ash	1	2	1	1

Table 6.

	Compost control	Worm-worked digested solids	Worm-worked undigested solids
pH	5.8	8.0	7.3
Ionic conductivity (μmhos)	750	1593	1930
P (mg/L)	184	144	320
K (mg/L)	720	2376	2959
Mg (mg/L)	100	138	178
NO_3^- (mg/L)	<10	>10	>10
NH_4^+ (mg/L)	348	174	720

undigested solids although the difference was not important with respect to plant growth.

The overall results show that, although there has been a breakdown of fiber on anaerobic digestion to release energy in the form of methane, plant nutrients have passed through relatively unchanged, and this similarity is reflected in the worm-worked material.

Table 6 shows the comparison between the worm-worked solids used in the plant growth trial, and the compost control. The worm-worked material has a higher pH, conductivity and general nutrient level than the control, but as in Table 4 there are few differences between the digested and undigested worm-worked material with respect to plant nutrients.

The development of tomatoes grown in the above media can be seen in Table 7.

Table 7.

	Control		Worm-worked digested solids		Worm-worked undigested solids	
	95 days	109 days	95 days	109 days	95 days	109 days
Mean height (cm)	111	125	25	41	42	58
σ_{n-1}	12	20	6	14	7	14
Number of leaves	17	20	7	14	7	14
Number of trusses	1	4	0	1	0	1

The data shows a reduced rate of development for tomatoes grown in worm-worked material. No data are available for fruit yields, but initial tomato harvesting suggested lower yields for worm-wormed material, in line with the slower rate of plant development.

Discussion

Although reduced development rates were observed for plants in worm-worked material, there was no sign of any trace element deficiency on the vegetative parts of plants. The reduced growth would therefore seem to be due to the high pH conductivity and nutrient levels in the worm-worked material.

Table 8. ADAS recommended nutrient levels for peat based tomatoes

pH	P (mg/L)	K (mg/L)	Mg (mg/L)	NO_3-N (mg/L)	Conductivity (μmhos)
5.5–6.8	29–40	176–250	26–35	81–200	500–600

This is confirmed if the pH and nutrient values from Table 6 are compared with the ADAS recommended nutrient requirements for peat grown tomatoes seen in Table 8 (MAFF, 1981b).

There was a high incidence of Blossom End Rot (BER) on set fruit from worm-worked material. In digested solids, 16% of fruit was affected, compared with 52% for undigested solids, and none on control plants. This symptom is considered to be caused by an induced calcium deficiency in the fruit because of reduced water transport of calcium ions from the growing medium (MAFF, 1981). This can be caused by high conductivity levels in the material, especially potassium ions. In this experiment the digested worm-worked material had lower conductivity and potassium levels, possibly accounting for the reduced incidence of BER in this material.

In conclusion, high conductivity, pH and nutrient levels in the worm-worked material have in general reduced the development rates of plants grown in worm-worked material, and in particular induced BER. For the culture of tomato plants the worm-worked material requires amendment to reduce the pH and conductivity levels and dilute the high nutrient content. This can be achieved easily by bulking with an acidic peat, or leaching with an appropriate quantity of water to reach the desired nutrient levels. Plant growth trials in amended materials are reported elsewhere in the proceedings.

None of the data presented here suggests that digested worm-worked material is any less able to support plant growth than worm-worked undigested solids.

Economics of digestion and vermiculture

Many economic assessments of anaerobic digestion plants have been undertaken (Anon, 1982; James and Campbell, 1987; Hashimoto and Chen, 1981; West, 1985) and while it is clear that the viability of individual units depends to a large extent on the specific characteristics of each farm, a feature common to most appraisals is the emphasis on gas production and utilization as prime determinants of economic viability. Other benefits of the digestion process such as odour reduction and improvements in slurry handling tend to be ascribed little economic value while the digested residue is often valued only for its gross fertilizer potential. Hence the economic viability of anaerobic digestion is seen as being marginal on the basis of gas production alone. Taking full account of the additional benefits of the process or upgrading the digested residue can often enhance the overall economic performance of the plant. Some studies have concluded that, although gas production only partially offset total production costs, utilizing the digester residue as an animal feedstock greatly improved the economic performance (West, 1983; Martin *et al.*, 1983).

In the previous sections it has been demonstrated that separated digested cattle solids can be used to support worm growth successfully and that the resulting worm-worked material appears to have potential as a horticultural plant growth medium,

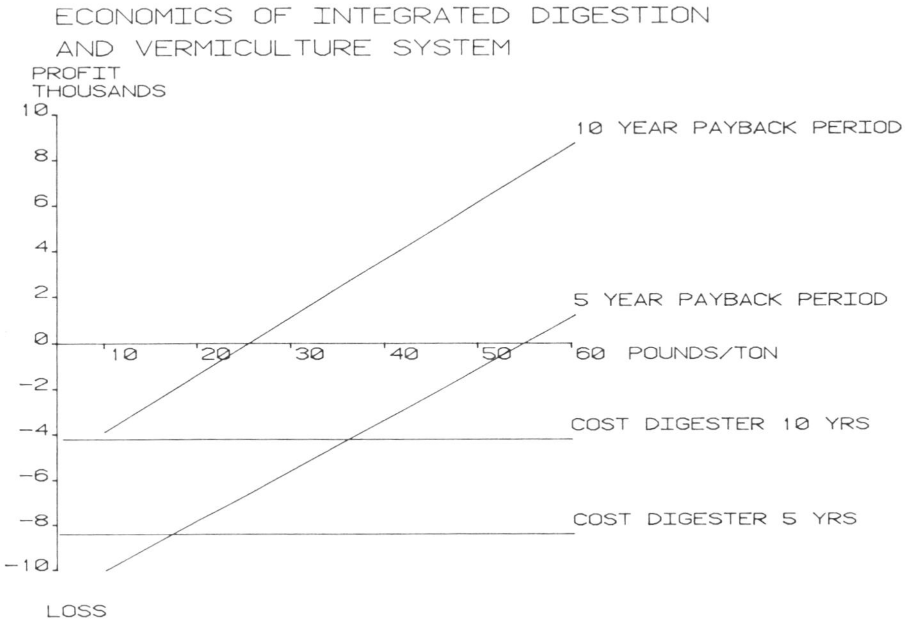

Fig. 4.

when further processed. A preliminary study was undertaken to assess the effect on the economics of the digestion plant arising out of the processing of the digested residue to worm-worked material. The value of the worm protein produced during the processing of the waste has not been included in the analysis. The study was based on data relating to the experimental digesters at NIRD and the economic model developed by NIAE.

Operating the digester in conventional mode and using existing data on capital and running costs, gas production etc., the model showed that for 10-year- and 5-year-payback-periods, the annual net costs of digester operation approximated £4,000 and £8,500 respectively. See Fig. 4. These figures were based on a 200 cow herd housed for 180 days.

The economic feasibility of incorporating a vermiculture unit to enhance the economic performance of the digester as assessed by calculating the required off-farm selling price of the unsterilized worm-worked material, to allow the combined system to break even based on 10-year- and 5-year-payback-periods.

Estimated costs of installing a vermiculture unit comprising a separator (£8,800) and an appropriately sized, covered, heated bed system plus storage pad (£5,200) were included in the value of capital equipment. Energy for bed heating during the winter months was assumed to be available from the digester at minimal cost. Labor charges have not been included at this stage but may be considered by adjusting the final selling price of the worm-worked material. The weight of worm-worked material produced was taken to be 50% of the original weight of solids in the slurry to account for losses during digestion and worm working.

Figure 4 illustrates that the combined digester and vermiculture unit would breakeven on 10-year- and 5-year-payback-periods if the unsterilized worm worked material could command off-farm prices of £25/tonne and £55/tonne.

This preliminary case study should not be viewed as a definitive economic analysis of the value of vermiculture but rather as an illustration of how innovative processes may interact for mutual benefit. In this case by using vermiculture to upgrade the waste from anaerobic digestion which otherwise would be seen as having only minimal value.

Acknowledgements

The authors would like to thank Dr. B.F. Pain of NIRD and Mr. S. Crocker and Mr. C. Nichols of Bore Place Farm for their permission and help in collecting samples of separated solids, Ms. P. Furniss and Dr. N. Watson for their help and advice in carrying out laboratory analyses, and Dr. J. Towriss for his assistance in computer programming.

References

Anon, 1982. An Economic Assessment of Anaerobic Digestion. Discussion document of an ADAS/BABA working party. London: MAFF.

Bishop, O.N. 1980. Statistics for Biology. Longmans, London. 182 pp.

Byrne, E. and Power, T. 1974. Determination of ammonium nitrogen in animal slurries by an ammonium electrode. Comm. Soil Sci. and Plant Anal. 5(1): 51–54.

Dash, M.C., Mishra, P.C. and Behera, N. 1979. Fungal feeding by a tropical earthworm. Tropical Ecology 20(1): 9–12.

Day, G.M. 1950. The influence of earthworms on soil micro-organisms. Soil Science 69: 175–184.

Edwards, C.A. 1982. Production of earthworm protein for animal feed from potato waste. *In* Upgrading Waste for Feed and Food. pp. 153–162. Edited by D.E. Ledward, A.J. Taylor and R.A. Lawrie. Butterworths, London.

Edwards, C.A., Burrows, I., Fletcher, K.E. and Jones, B.A. 1984. The use of earthworms for composting farm wastes. Report of Rothamsted Experimental Station.

Friman, R. 1984. Monitoring anaerobic digesters on farms. J. Agri. Engin. Res. 29: 357–365.

Goering, H.D. and Van Soest, P.J. 1975. Forage Fibre Analysis. US Dept. of Agriculture, Agricultural Research Service, Washington DC, USA. 20 pp.

James, P.J. and Campbell, R.J. 1985. The economics of anaerobic digestion on farms. *In* Anaerobic Digestion of Livestock Wastes. pp. 73–80 Reading: NIRD. Technical Bulletin No. 2.

Hartenstein, R. 1981a. Assimilation by the earthworm *Eisenia foetida*. *In* Earthworm Ecology from Darwin to Vermiculture. pp. 297–308. Edited by J.E. Satchell. Chapman and Hall, London.

Hartenstein, R. 1981b. Use of *Eisenia foetida* in organic recycling based on laboratory experiments. *In* Workshop on the Role of Earthworms in the Stabilisation of Organic Residues. pp. 155–165. Vol. 1, Proceedings. Compiled by M. Appelhof. Beech Leaf Press, Kalamazoo, Michigan, USA.

Hartenstein, R. 1982. Metabolic parameters of the earthworm *Eisenia foetida* in relation to temperature. Biotechn. and Bioengin. 14: 1803–1811.

Hashimoto, A.G. and Chen, Y.R. 1981. The overall economics of anaerobic digestion. *In* Anaerobic Digestion. pp. 449–466. Edited by D.A. Stafford, B.I. Wheatley and D.E. Hughes. Applies Science Publishers, London.

Hayes, T.D., Jewell, W.J., Dell'Orto, S., Fanfoni, K.J., Leuschner, A.P. and Sherman, D.F. 1980.

Anaerobic digestion of cattle manure. *In* Anaerobic Digestion. pp. 255–288. Edited by D.A. Stafford, B.I. Wheatley and D.E. Hughes. Applied Science Publishers, London.

Hills, D.J. and Roberts, D.W. 1981. Anaerobic digestion of dairy manure and field crop residues. Agricult. Wastes 3: 179–189.

Kaplan, D.L., Hartenstein, R., Neuhauser, E.F. and Malecki, M.R. 1980. Physicochemical requirements in the environment of the earthworm *Eisenia foetida*. Soil Biol. and Biochem. 12: 347–352.

MAFF 1981a. The Analysis of Agricultural Materials, 2nd Edition. London, HMSO. 248 pp.

MAFF 1981b. Tomato Production, Part 3 Growing Media and Nutrition. ADAS Booklet 2246. 11 pp.

Martin Jr., J.H., Loehr, R.C. and Pilbeam, T.E. 1981. Animal manures as feedstuffs: nutrient characteristics. Agricult. Wastes 6: 131–166.

Martin Jr., J.H., Loehr, R.C. and Pilbeam, T.E. 1983. Animal manure feeding trials. Agricult. Wastes 7: 81–110.

Miles, H.B. 1963. Soil protozoa and earthworm nutrition. Soil Science 95: 407–409.

Mitchell, M.J., Mulligan, R.M., Hartenstein, R. and Neuhauser, E.F. 1977. Conversion of sludges into 'topsoils' by earthworms. Compost Science 18(4): 28–32.

Neuhauser, E.F., Kaplan, D.L., Malecki, M.R. and Hartenstein, R. 1980. Materials supporting weight gain by the earthworm *Eisenia foetida* in waste conversion systems. Agricult. Wastes 2: 43–60.

Osborne, D.R. and Voogt, P. 1978. The Analysis of Nutrients in Foods. Academic Press, London. 251 pp.

Pain, B.F., West, R., Oliver, B. Hawkes, D.L. 1984. Mesophilic anaerobic digestion of dairy cow slurry on a farm scale: first comparisons between digestion before and after solids separation. J. Agri. Engin. Res. 29 (3): 249–256.

Parle, J.N. 1963. Micro-organisms in the intestines of earthworms. J. Gen. Microbiol. 31: 1–11.

Pfeffer, J.T. 1980. Anaerobic digestion processes. *In* Anaerobic Digestion. pp. 15–36. Edited by D.A. Stafford, B.I. Wheatly and D.E. Hughes. Applied Science Publishers, London.

Plaskett, L.G. 1978. The Generation of Gases from Biomass by Anaerobic Digestion and their Practical Utilisation a Fuel. A systems study undertaken for ETSU, Harwell and DoE by Biotechnical Processes Ltd.

Sabine, J.R. 1983. Earthworms as a source of food and drugs. *In* Earthworm Ecology from Darwin to Vermiculture. pp. 285–296. Edited by J.E. Satchell. Chapman and Hall, London.

West, R. 1985. Matching energy supply and demand in anaerobic digestion of farm waste. pp. 55–71. NIRD Technical Bulletin No.7.

THE VERMICOMPOSTING OF COW SLURRY

P. HAND[1], W.A. HAYES[2], J.E. SATCHELL[3] and J.C. FRANKLAND[3]
[1]Glasshouse Crops Research Institute, Littlehampton, W. Sussex; [2]Aston University, Birmingham; [3]I.T.E. Merlewood, Grange Over-Sands, Cumbria

In traditional farming systems, the application of animal manures to croplands is an integral part of mixed farming practice. The recent adoption of intensive livestock rearing in many countries has led to problems of disposal of the large volumes of animal wastes produced.

Indiscriminate landspreading of intensively-produced animal wastes damages soil fertility, causes water pollution and may present a health risk. Some form of treatment of animal wastes is therefore often necessary to make them suitable for safe disposal into the environment. Present treatment methods are based upon those employed in the treatment of sewage (Loehr, 1971; Vanderholm, 1979). In recent years, the application of low-cost, low technology systems to the treatment of animal waste has been considered. Several workers have investigated the utilization of natural decomposition processes involving earthworms in organic waste disposal (Mitchell *et al.*, 1977; Neuhauser *et al.*, 1980; Pincince *et al.*, 1980). The species *Eisenia fetida* is considered suitable for such application, because of its rapid growth rate, reproductive potential, and occurrence in rich organic substrates in nature. It has been suggested that some earthworms utilize micro-organisms in their substrate as a food source and are able to selectively digest certain micro-organisms (Cooke & Luxton, 1980; Day, 1980; Dash *et al.*, 1979). The efficiency of vermicomposting systems may therefore depend upon the number and types of micro-organisms present in the substrate.

The objectives of this study were to assess the suitability of cow slurry as a substrate for vermicomposting, to investigate the changes brought about in slurry by *E. fetida*, and to examine the specific relationships between *E. fetida* and micro-organisms isolated from slurry.

Methods

Vermicomposting systems investigated were the mixing of slurry with bedding materials, and the application of slurry to the surface of earthworm beds. In the first system, peat and paper tissue waste were wetted to 60% moisture content and neutralized. Fresh cow slurry was then mixed with these bedding materials in different proportions. Controls were watered with distilled water only. The mixtures were aged for three days in the laboratory, then placed in 100 g amounts in deep Petri dishes. Five juvenile *E. fetida* were then added to each dish, and the dishes were incubated at 25°C for five weeks. The earthworms were weighed every seven days, and regular samples were taken from a tissue waste/cow slurry mixture for chemical and microbiological analysis. Earthworm reproduction was assessed by counting the number of cocoons present in each dish after five weeks.

In the second system, cow slurry was applied at different rates to the surface of

Earthworms in waste and environmental management
edited by C.A. Edwards and E.F. Neuhauser, pp. 49–63

pots containing earthworms and bedding materials. Fresh slurry was added as required over a five-week-period. The earthworms were weighed every seven days, and the number of cocoons in each pot was counted at the end of the experiment.

The changes occurring in fresh cow slurry applied to the surface of earthworm beds which make it acceptable to *E. fetida* as a food source were investigated. Surface and 50 mm depth samples were taken from tissue waste vermiculture beds at zero, two, five and seven days after application of cow slurry. The pH, electrical conductivity, ammonia volatilization and ammonium nitrogen content of the samples and total numbers of bacteria and fungi were determined.

Interactions between *E. fetida* and micro-organisms isolated from cow slurry and vermiculture beds were assessed initially by screening experiments. Glass tubes containing moist sand/α-cellulose mixture (5:1 v/v) were sterilized by autoclaving. Earthworms were treated by 24-hour immersion in an antibiotic mixture (Cycloheximide, 100 $\mu g\ ml^{-1}$; Aureomycin, 6 $\mu g\ ml^{-1}$; Nisin, 50 $\mu g\ ml^{-1}$; Novobiocin, 50 $\mu g\ ml^{-1}$) to reduce microbial contamination, then rinsed in sterile deionized water, weighed and added to the sterile tubes. Washed liquid cultures of chosen micro-organisms were added to the surface of the tubes as live or heat-killed cells. Controls had added deionized water only. Experiments were run for seven days at 25°C with re-feeding of the tubes over 48 hours. At the end of the experiment, the earthworms were retrieved and weighed. The interaction of *E. fetida* with micro-organisms was assessed as the mean weight change of fed worms compared with unfed controls. In a second series of feeding experiments, the interaction of micro-organisms with axenic *E. fetida* was investigated. Axenic earthworms were prepared by prolonged immersion in an antibiotic mixture (Aureomycin, 6 $\mu g\ ml^{-1}$; Cycloheximide, 50 $\mu g\ ml^{-1}$; Nisin, 50 $\mu g\ ml^{-1}$; Novobiocin, 50 $\mu g\ ml^{-1}$), with regular transfer of the earthworms to fresh solution. When microbial contamination of the earthworms could no longer be detected, they were utilized in feeding experiments. Suspensions of live bacteria in 0.1% peptone water were added to large Pyrex boiling tubes containing 40 g of sterilized paper tissue waste, with 10 replicates of each treatment. Control tubes were watered with peptone water only. The tubes were stored at room temperature for three days, then individual axenic *E. fetida* were weighed aseptically and transferred to the tubes. The experiment was run for 21 days at 25°C, with re-feeding with bacterial suspensions every seven days. At the end of the experiment earthworms were retrieved and weighed. Mean weight change of earthworms fed with different bacteria was compared with control worms.

In a further experiment, a potential vermicomposting system was investigated, in which paper tissue waste containing *E. fetida* was seeded with a suspension of bacteria at regular intervals. The earthworms were weighed every seven days for three weeks, and the percentage of the substrate which had been converted to worm casts were estimated visually each week.

Results

The weight changes of *E. fetida* in different peat/slurry mixtures is shown in Fig. 1. Peat alone provided little or no nutrients to *E. fetida*. In the 2:1 peat/slurry mixture, the slurry fraction was quickly consumed, and the earthworms were unable to obtain

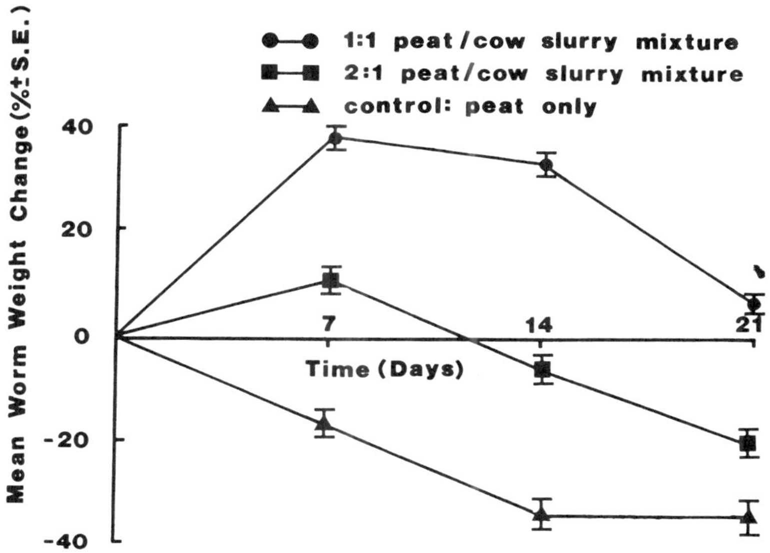

Fig. 1. % Weight changes of *E. fetida* with time in different peat/cow slurry mixtures

further nutrients from the substrate, and so lost weight. The 1:1 peat/slurry mixture supported significantly greater earthworm weight gain than the control, with maintenance of weight for a time, but once more, rapid earthworm weight loss occurred when the slurry fraction was consumed. The use of paper tissue waste in mixtures with slurry as a substrate is shown in Fig. 2. The addition of slurry in different proportions produced similar large earthworm weight increases. Table 1 shows cocoon production by *E. fetida* in peat/slurry and tissue waste/slurry mixtures, indicating the greater nutritional value of this substrate to *E. fetida*. When slurry was applied to the surface of tissue waste vermiculture beds at two feeding rates (Fig. 3), earthworms in pots with 20% (w/v) slurry additions showed much lower weight gains than those fed at the 10% (w/v) level. In the latter treatment earthworms rapidly consumed applied slurry throughout the experiment, but lost weight from day 21 onwards. Table 2 shows the cocoon production of *E. fetida* in worm beds top-fed with cow slurry. Cocoon production was higher in tissue waste with added slurry than in soil pots with added slurry. The earthworms in tissue waste with slurry added at the 10% level produced more cocoons than those with slurry added at the 20% level.

Table 3 shows the chemical changes which occurred in a 2:1 (w/w) tissue waste/slurry mixture with time, in the presence and absence of *E. fetida*. Most of the initial ammonium nitrogen contained in the slurry mixture was lost by day 14, probably largely due to volatilization of ammonia. However, some conversion of ammonium-N to nitrate-N was observed. In the treatment containing earthworms, nitrate-N was conserved throughout the experiment, while in the mixture without worms, a smaller initial increase in nitrate occurred, and the level of this constituent subsequently fell to a level below that at the beginning of the experiment. Organic-N declined at a steady rate in the presence of earthworms, while a small overall decrease occurred in the control. Organic carbon content of the mixture containing worms declined rapidly at the beginning of the experiment, but stabilized after day 21. At

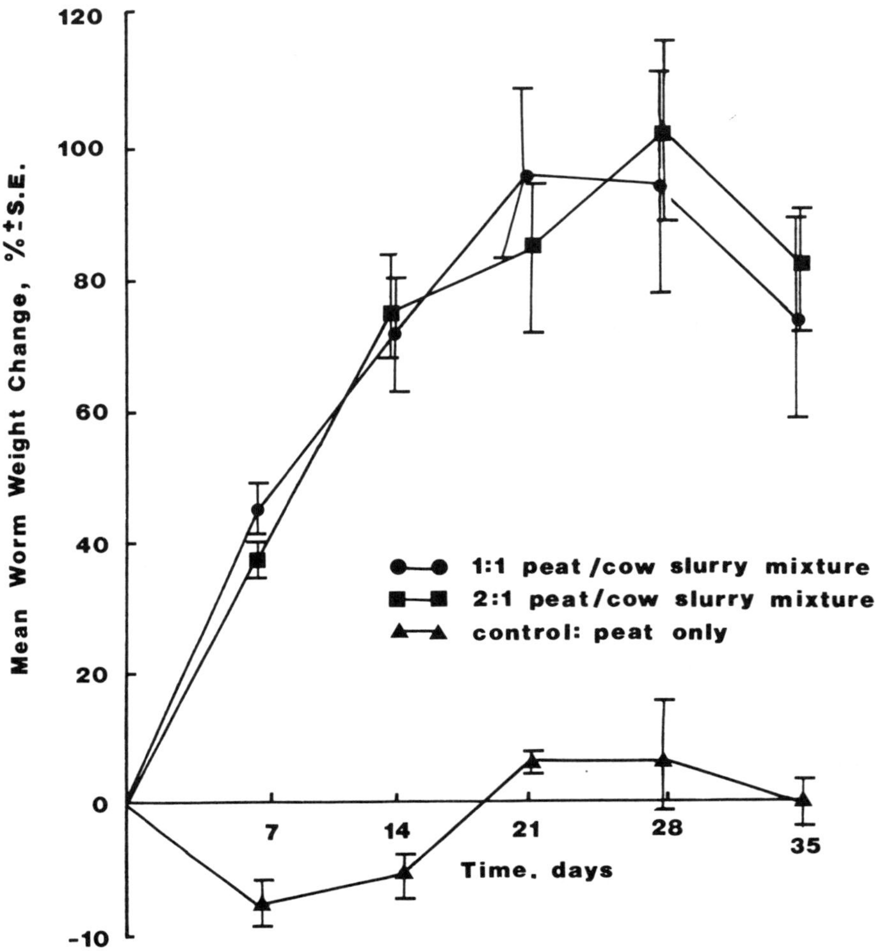

Fig. 2. % Weight changes of *E. fetida* with time in different paper tissue waste/cow slurry mixtures.

Table 1. Reproduction of *E. fetida* in cow slurry mixed with two bedding materials. Mean number of cocoons after 35 days

Treatment	Mean no cocoons per worm (n = 3)	S.E.
Peat control	0	—
2:1 peat/slurry	0.4	± 0.12
1:1 peat/slurry	0.7	± 0.17
Tissue waste control	1.8	± 0.18
2:1 T.W./slurry	4.2*	± 0.28
1:1 T.W./slurry	6.1*	± 0.37

S.E. = Standard error of the mean.
* = Treatment significantly different from control: $p < 0.05$.

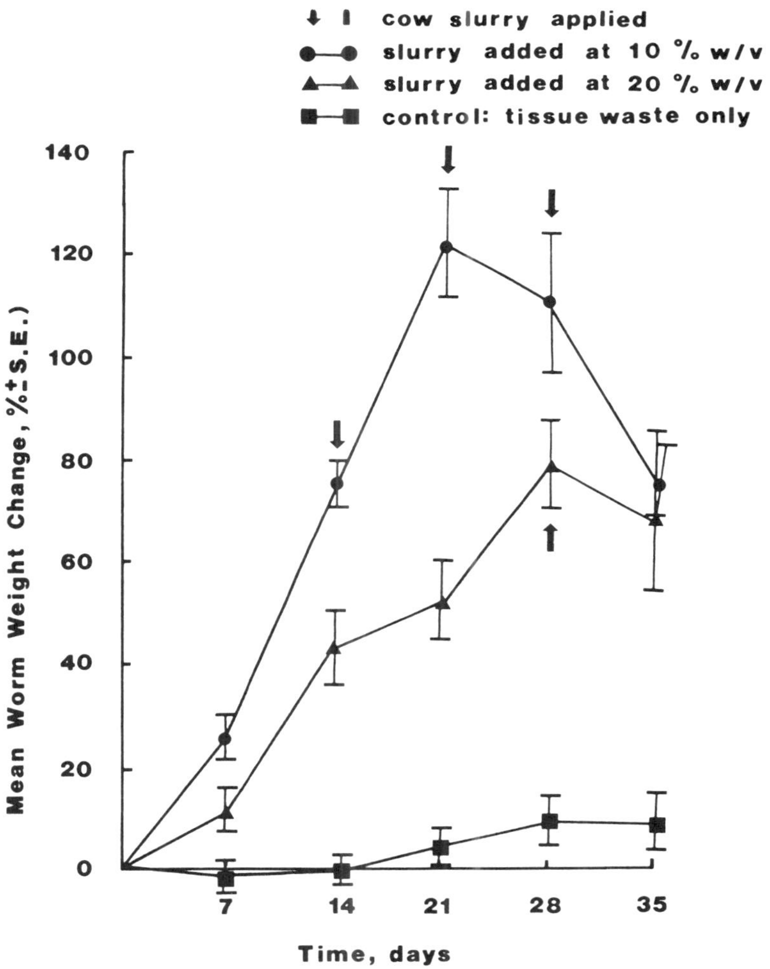

Fig. 3. % Weight changes of *E. fetida* in paper tissue waste top-fed with cow slurry (10 and 20% w/v).

the end of the experiment the organic carbon content of this mixture was lower than that of the mixture without worms. The C/N ratio of the mixture containing earthworms decreased rapidly initially as organic C was lost from the substrate. The ratio then stabilized and finally rose slightly due to continued loss of organic N. At the end of the experiment the control mixture had a slightly lower C:N ratio than the mixture containing earthworms.

The cellulose content of the mixture containing earthworms showed no change during the experiment. However, in the control mixture, some evidence of cellulose decomposition was observed over six weeks. The proportion of lignin increased slightly with time in the mixture containing earthworms, but showed a such greater increase in the control mixture. Total numbers of bacteria in the two mixtures were similar for six weeks, showing the same trend of a slow decline (Fig. 4). Between

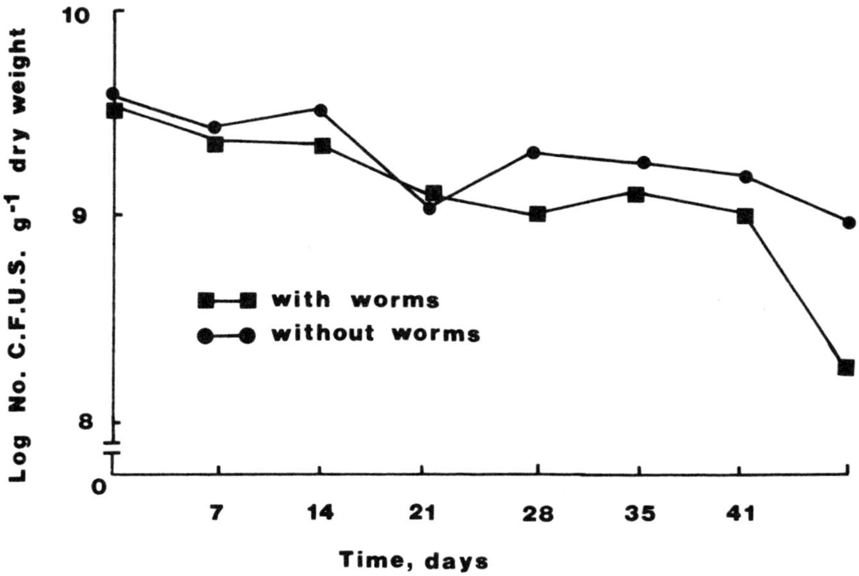

Fig. 4. Changes in total numbers of aerobic bacteria in a 2:1 (w/w) paper tissue waste/cow slurry mixture.

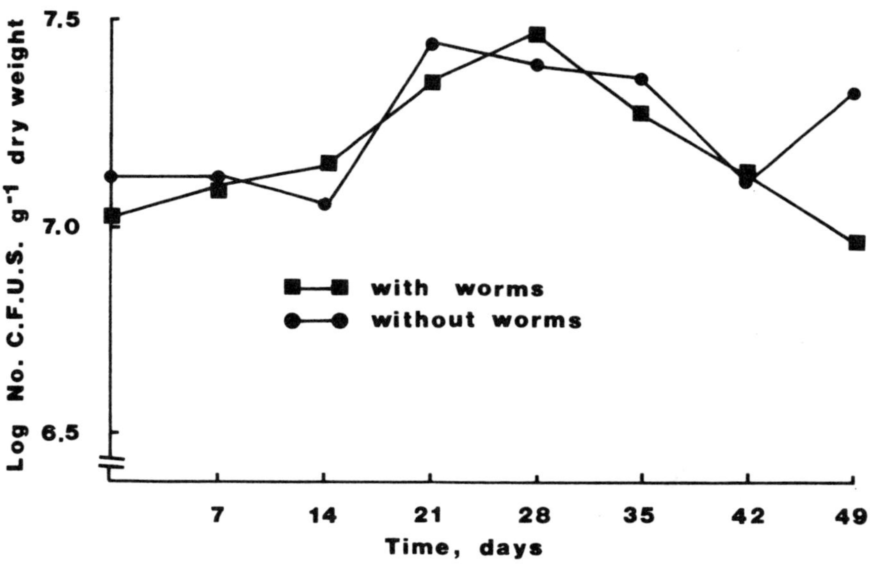

Fig. 5. Changes in total numbers of fungi in a 2:1 (w/w) paper tissue waste/cow slurry mixture.

weeks six and seven, however, the number of bacteria in the mixture containing earthworms fell sharply compared with the control.

Figure 5 shows the changes in the numbers of fungi present with time. Again the pattern of change in numbers was similar between the two treatments, with an initial

Table 2. Reproduction of *E. fetida* in worm beds top-fed with cow slurry. Mean number of cocoons after 35 days

Treatment	Mean no. cocoons per worm (n = 5)	S.E.
Soil control	0	—
Soil + 10% (w/w) cow slurry	0.8	± 0.22
Tissue waste control	1.8	± 0.16
Tissue waste + 10% (w/w) slurry	6.3*	± 0.44
Tissue waste + 20% (w/w) slurry	4.3*	± 0.48

S.E. = standard error of the mean.
* = Treatment significantly different from control: $p < 0.05$.

Table 3. Changes in a 2:1 (w/w) paper tissue waste/cow slurry mixture with time (Upper figure = with *E. fetida*, lower figure = without *E. fetida*)

Constituent	Time (weeks)			
	0	2	3	6
Ammonium-N (mg 100 g^{-1})	37	0.8	1.8	1.6
	37	1.5	0.6	0.6
Nitrate-N (mg 100 g^{-1})	3.0	23	22.5	22.5
	3.0	8.0	1.5	1.0
Organic-N (%)	1.65	1.6	1.6	1.5
	1.65	ND	ND	1.6
Organic-C (%)	56	48	47	47
	56	ND	ND	49
C/N ratio	34	30	30	31
	34	ND	ND	30
Cellulose (%)	5.3	5.3	5.3	5.3
	5.3	ND	ND	4.8
Lignin (%)	25.5	26.0	26.0	27.0
	25.5	ND	ND	32.0

Table 4. Physicochemical changes at the surface of earthworm beds containing paper tissue waste (Upper figure = slurry added; lower figure = controls without slurry)

	Time (days			
	0	2	5	7
pH	9.1	9.1	9.2	9.0
	6.3	5.6	5.6	5.8
Conductivity	5.0	4.0	3.0	3.0
(mmhos cm^{-1})	2.4	2.5	3.0	2.9
Ammonia	10	10	5	3
Volatilization (ppm)	0	0	0	0
Ammonium-N	13.0	12.0	9.0	6.0
(mg 100 g^{-1})	ND	ND	ND	ND

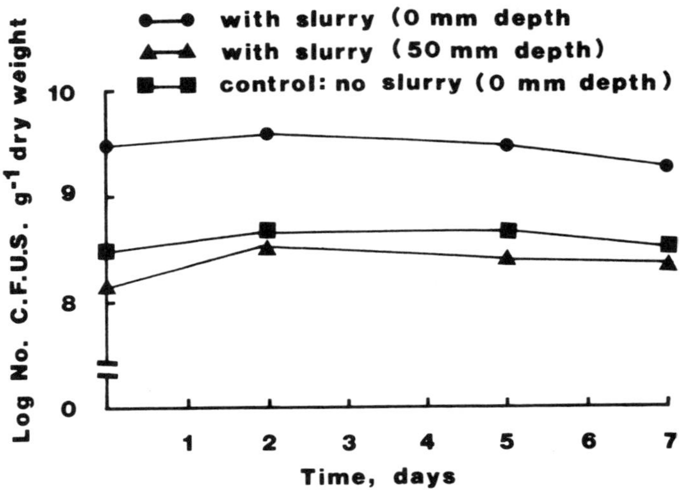

Fig. 6. Total numbers of aerobic bacteria in paper tissue waste beds following surface application of cow slurry.

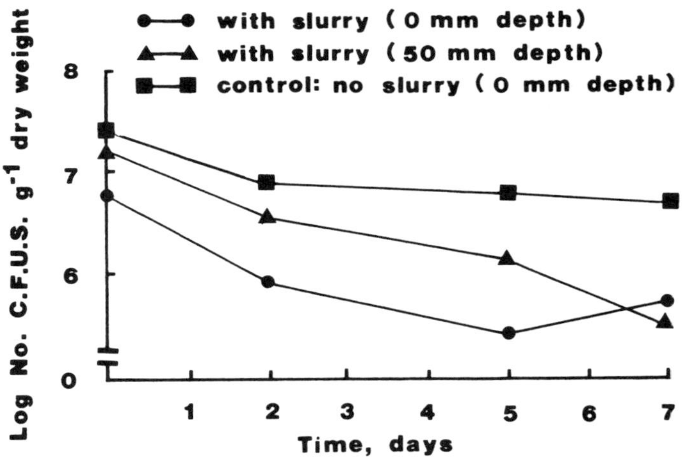

Fig. 7. Total numbers of fungi in paper tissue waste vermiculture beds following surface application of cow slurry.

rise followed by a gradual decrease. Once more, numbers in the mixture containing earthworms fell between weeks six and seven compared with controls.

Table 4 shows the physical and chemical changes with occurred at the surface of earthworm beds containing tissue waste following the application of cow slurry. The surface pH of beds with slurry applied was higher than that of controls throughout the experiment and did not change significantly during this time. The initial high electrical conductivity at the surface of beds with applied slurry quickly declined, becoming equal to that in control beds by day five. Ammonia volatilization at the surface of slurry pots was constant for two days, then declined over the remainder of the ex-

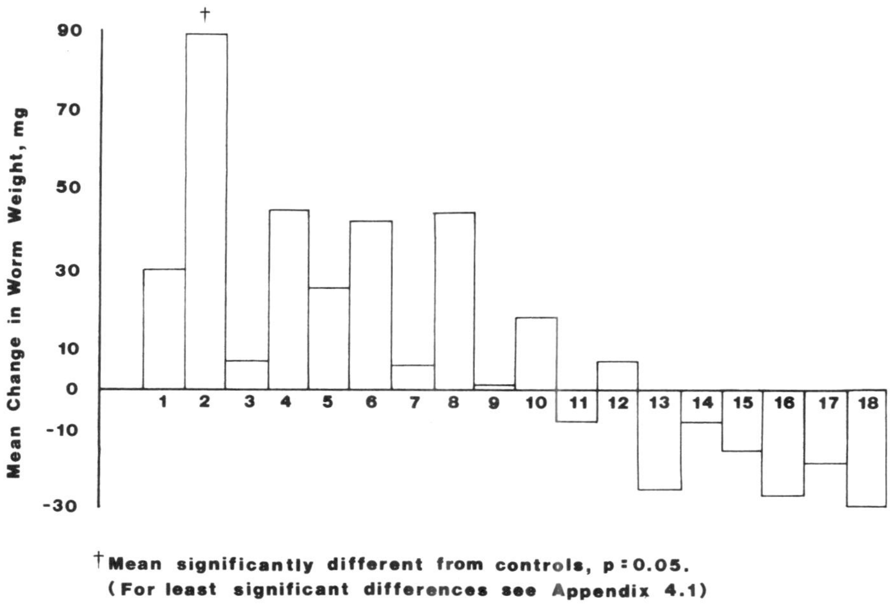

Fig. 8. Effects of feeding individual species of bacteria and actinomycetes to *E. fetida*. 1 = *Acinetobacter lwoffi* (L); 2 = *Ac. lwoffi* (H)*; 3 = *Ac. citroalkaligenes* (L); 4 = *Ac. citroalkaligenes* (H); 5 = *Alkaligenes faecalis* (L); 6 = *Alk. faecalis* (H); 7 = *Nocardia salmonicolor* (L); 8 = *N. salmonicolor* (H); 9 = *Enterobacter clocae* (L); 10 = *E. cloacae*; 11 = *Micrococcus varians* (L); 12 = *M. varians* (H); 13 = *Arthrobacter simplex* (L); 14 = *A. simplex* (H); 15 = *Micrococcus luteus* (L); 16 = *M. luteus* (H); 17 = *Artrobacter tumescens* (L); 18 = *A. tumescens* (H). * = Mean significantly different from control, p = 0.05; L = live cells; H = heat killed cells.

periment. The ammonium nitrogen content of slurry at the surface of the experimental beds declined slowly throughout the experiment. Figures 6 and 7 show the total numbers of aerobic bacteria and of fungi at the surface and at 50 mm depth in tissue waste beds with applied slurry, compared with surface samples of controls without slurry. The results show that no gross microbial changes occurred in the surface-applied slurry during this period. During the initial screening of micro-organisms as a food source for *E. fetida*, four species were found to be toxic to earthworms: *Flavobacterium lutescens*; *Pseudomonas fluorescens*; *Ps. putida*; and a *Streptomyces* species. High earthworm mortality occurred within seven days when these species were presented as both live and heat-killed cells. Figure 8 shows the effects of feeding single species of other bacteria to *E. fetida*. Both control and experimental earthworms lost weight during the experiment. The results are expressed as the mean worm weight change compared with unfed controls. Positive, negative and indifferent effects of micro-organisms on earthworm nutrition were observed. Heat-killing of the cells tended to enhance both the positive and negative effects of micro-organisms. Heat-killed *Acinetobacter lwoffi* cells had a significant positive effect upon earthworm nutrition.

Figure 9 shows the results of feeding single species of fungi to *E. fetida*. Again both control and experimental worms generally lost weight during the experiment. Most of the species of fungi tested had a more positive effect upon earthworm nutrition.

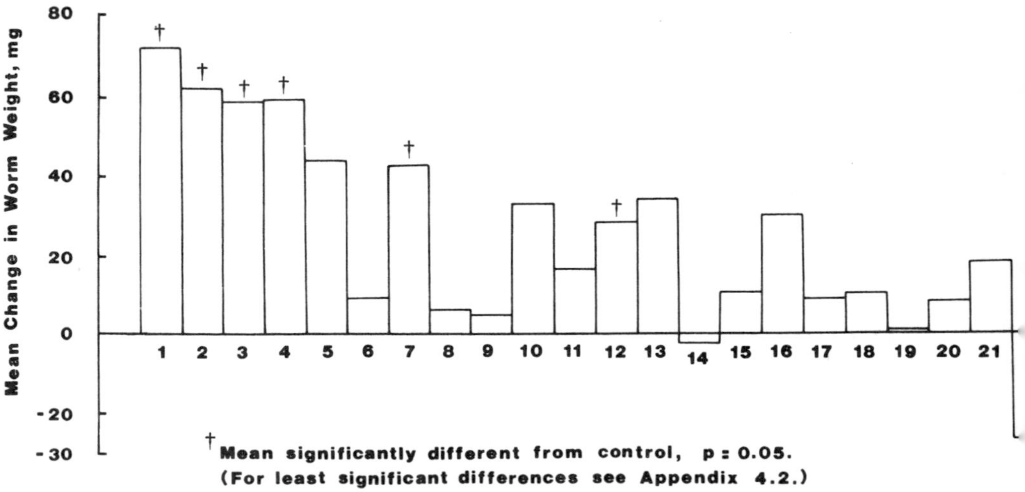

Fig. 9. Effects of feeding individual species of fungi to *E. fetida*.
1 = *Trichoderma harzianum* (L)*; 2 = *T. harzianum* (H)*; 3 = *Mucor plumbeus* (L)*; 4 = *M. plumbeus* (H)*; 5 = *Gliomastix murorum* var. *felina* (L); 6 = *G. murorum* var. *felina* (H); 7 = Trichoderma hamatum (L)*; 8 = *T. hamatum* (H); 9 = *Scopulariopsis carbonaria* (L); 10 = *S. carbonaria* (H); 11 = *Rhodotorula* species (L); 12 = *Rhodotorula* species (H)*; 13 = *Gliocladium deliquescens* (L); 14 = *G. deliquescens* (H); 15 = *Aurebasidium pullulans* (L); 16 = *A. pullulans* (H); 17 = *Aspergillus fumigatus* (L); 18 = *A. fumigatus* (H); 19 = *Penicillium purpurogenum* (L); 20 = *P. purpurogenum* (H); 21 = *Fusarium oxysporum* (L); 22 = *F. oxysporum* (H)*. * = Mean significantly different from control, p = 0.05; L = live cells; H = heat killed cells.

Nutritional value of the fungi to *E. fetida* was less marked than that observed with bacteria. However, *Fusarium oxysporum* produced a significant negative effect upon earthworm growth when fed as heat-killed cells. Significant positive growth effects upon *E. fetida* were produced by *Trichoderma harzianum* (live and heat-killed); *Mucor plumbeus* (live and heat-killed; *T. hamatum* (live cells); and *Rhodotorula* sp. (heat-killed). The feeding of bacteria to axenic earthworms in sterile culture was investigated to determine if the results of the initial screening experiments represented true interactions between *E. fetida* and test micro-organisms. Figure 10 shows the results of this experiment. Both unfed controls and earthworms fed with live bacteria lost weight during the experiment, indicating that *E. fetida* was unable to utilize the species tested as a sole source of nutrition. The observed effects were similar to those obtained in the initial feeding experiments.*Acinetobacter lwoffi* produced a significant weight change compared with controls, indicating the importance of this species in earthworms nutrition. The effects of *Alkaligenes faecalis* and *Enterobacter cloacae* on earthworm weight change were very similar to those produced in the initial screening experiments, suggesting that these experiments provided an accurate indication of the types of interactions occurring between *E. fetida* and the species tested. The toxicity of *Pseudomonas fluorescens* to earthworms was reduced slightly in this experiment, but surviving earthworms lost considerably more weight than control earthworms, confirming the adverse effects of this species upon *E. fetida*.

The significant effect of certain micro-organisms upon earthworm growth suggested that a vermicomposting system could be developed in which substrates could be

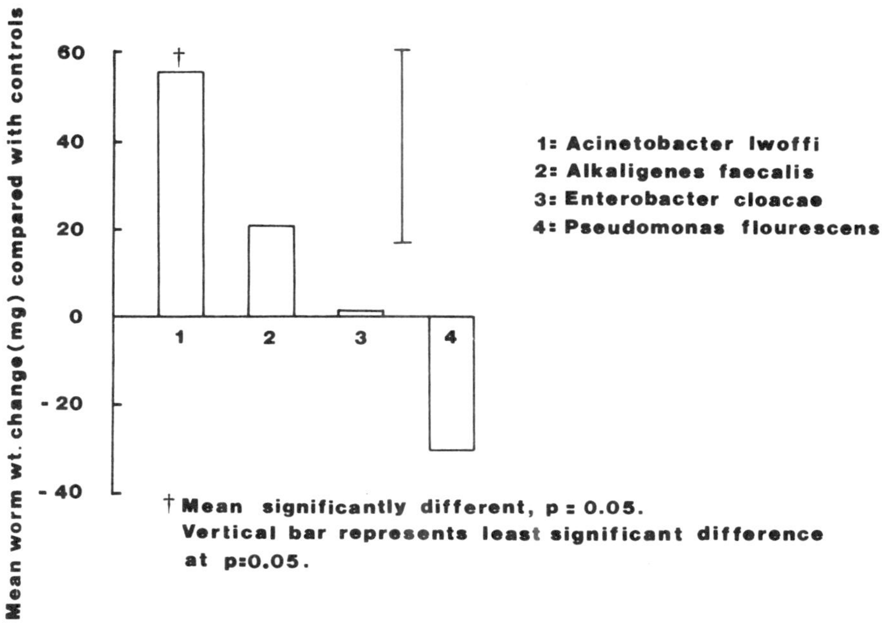

Fig. 10. Effects of feeding bacteria to axenic *E. fetida*.

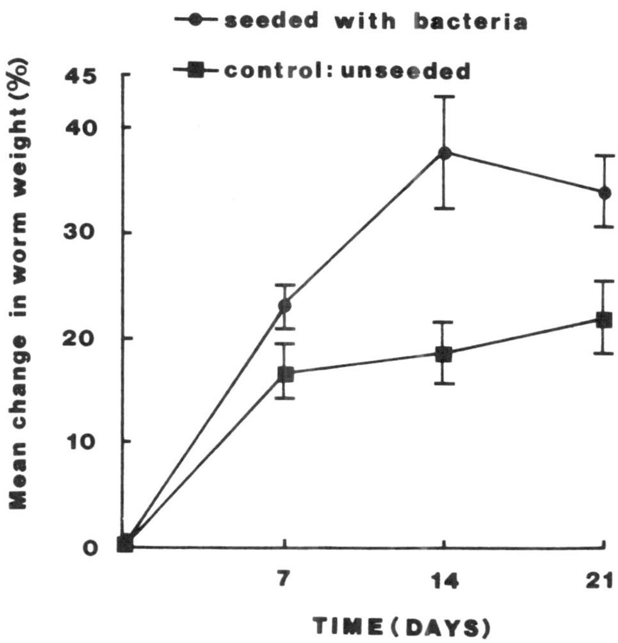

Fig. 11. Effect of adding *Acinetobacter lwoffi* to earthworm beds on the growth of *E. fetida*.

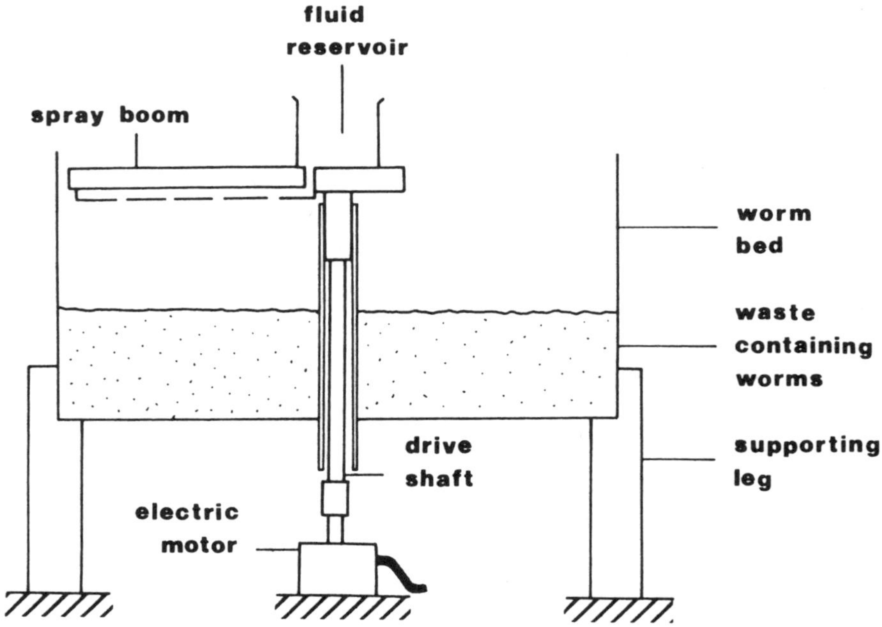

Fig. 12. Laboratory-scale system for vermicomposting wastes by seeding with pure cultures of micro-organisms.

seeded with suitable micro-organisms to enhance the efficiency of vermicomposting. Figure 11 shows the results of an experiment in which *Acinetobacter lwoffi* was added to tissue waste in vermiculture beds. Weight gain of earthworms in beds with added bacteria was significantly greater than that of controls feeding on unamended tissue waste. After 14 days, approximately 70% of the tissue waste was subjectively estimated to have been consumed by the earthworms feeding upon seeded tissue waste, compared with approximately 30% of the control material. After 21 days, 90–95% of the tissue waste with added bacteria was consumed, compared with approximately 50% of the tissue waste in control beds. Similar results were obtained in a larger-scale laboratory system which utilized a method comparable to that employed in sewage treatment trickling filters to apply a suspension of micro-organisms to the substrate (Fig. 12).

Discussion

These results show that cow slurry is a suitable substrate for vermicomposting, both when mixed with solid materials and when applied to the surface of bedding materials containing earthworms. Mixtures of slurry with paper tissue waste produced greater earthworm growth and cocoon production per unit of slurry consumed than mixtures with peat. Although the proportion of slurry mixed with tissue waste did not affect the growth rate of *E. fetida*, the number of cocoons produced was proportional to the amount of slurry present. It is therefore possible that earthworm growth rate was maximal in the 2:1 tissue waste/slurry mixture, and increased amounts of slurry in the substrate were utilized by the earthworms for reproduction. Similar results were reported by Neuhauser *et al.* (1980) when sewage sludge was added to soil containing *E. fetida*: little difference in growth rate was observed on the addition of 10–100% sludge.

The lower weight gain of *E. fetida* in the beds with the higher loading rate of top-fed slurry may have been due to slower de-toxification of slurry when a larger volume was applied, with a corresponding greater loss of available nutrients before feeding by earthworms could begin. The greater rate of cocoon production at the lower rate of slurry application also suggests that slurry fed at this rate had a greater nutritional value to *E. fetida*.

The presence of earthworms had a marked effect upon nitrogen transformations in a tissue waste/cow slurry mixture. Nitrogen mineralization was greater in the presence of earthworms, and this mineral nitrogen was retained in the nitrate form. This suggests that *E. fetida* produced conditions which favored nitrification. Earthworms assimilate organic nitrogen and excrete approximately equal amounts of nitrogen as ammonium and muco-proteins (Needham, 1957). This is a possible explanation for the reduction in organic nitrogen in the substrate containing worms. The fact that ammonium-N levels were not elevated in the presence of earthworms suggests that excreted ammonium-N was rapidly converted to nitrate. This effect has been noted by several workers to occur in soils (Syers *et al.*, 1979; Parle, 1963), and following application of slurries to land (Debry *et al.*, 1982). The fluctuations in nitrate content of the substrate without earthworms may have been due to immobilization of nitrogen by the microflora, once ammonium-N became limiting. Some denitrification may also have occurred, if conditions of low oxygen tension developed in large particles of the substrate without worms. The presence of *E. fetida* appeared to inhibit cellulose decomposition. This may have been due to disturbance of the substrate by earthworm activity. The greater increase in the proportion of lignin in the absence of earthworms may reflect the greater decomposition of cellulose in this substrate. Little change was observed in the numbers of micro-organisms in the presence and absence of earthworms. Satchell (1967) pointed out that the effect of earthworms in readily degradable organic matter, which contains a high population of micro-organisms, is likely to be less significant than in soils. Study of changes following surface application of slurry to vermiculture beds showed that the main chemical change was a reduction in ammonium-N levels, accompanied by a fall in electrical conductivity. The high level of ammonia volatilization measured suggests that the nitrogen was lost as ammonia gas. These results suggest that ammonium ion concentration is a major toxic factor in fresh slurry, but that this toxicity is quickly reduced to a level which *E. fetida* can tolerate.

In this study, definite nutritional interactions were found to exist between *E. fetida* and micro-organisms. These interactions appeared to be very species-specific, supporting the findings of Day (1950), and Rouelle (1983). The types of interaction observed with bacteria appeared to be correlated with their cell wall structure, Gram-negative bacteria generally had positive effects upon earthworm nutrition, while Gram-positive species generally produced indifferent or negative effects. The fact that heat-killing of bacteria tended to increase their nutritional value to *E. fetida* suggests that the earthworms were feeding directly upon the bacteria, rather than upon products of the living cells. Of the species found to be toxic to earthworms, the *Pseudomonas* and *Streptomyces* species are known to produce antimicrobial compounds, and it is possible that these were responsible for the observed toxic effects. Toxicity of *Pseudomonas* species and other bacteria to earthworms has been reported by Rao *et al.* (1983).

Fungal species appeared to provide a more readily-available food source than bacteria, as the majority of species tested produced a positive effect upon earthworm nutrition. The negative growth effect of *Fusarium oxysporum* when presented as heat-killed cells may have been due to the production or release of some toxic factor on autoclaving.

The results of feeding experiments using axenic earthworms suggest that secondary interactions between micro-organisms in the initial screening experiments were not significant. This is in agreement with the accepted view that earthworms do not possess a specific gut microflora which has an important nutritional role (Edwards & Lofty, 1977). The enhanced growth of earthworms in vermiculture beds seeded with *Acinetobacter lwoffi* may have been in part, due to interactions between this organism and the established microflora of the tissue waste. The increased rate of consumption of the substrate when seeded with this bacterium may have been due to enhanced palatability of the tissue waste: a similar effect was observed by Wright (1972).

The technique of seeding wastes with beneficial micro-organisms has a potential application in the vermicomposting of many agricultural and industrial wastes which are naturally poor food sources for *E. fetida*. A system could easily be developed whereby wastes were seeded with food micro-organisms before or during vermicomposting in order to enhance the speed and efficiency of the process. Such a system would allow the upgrading and safe utilization of many waste materials which currently present a disposal problem.

Acknowledgements

This work was performed while the senior author was in receipt of a N.E.R.C. CASE award.

References

Cooke, A. and Luxton, M. 1980. Effect of microbes on food selection by *L. terrestris*. Rev. Ecol. Biol. Sol. 17: 365–370.

Dash, M.C., Mishra, P.C. and Behera, N. 1979. Fungal feeding by a tropical earthworm. Tropical Ecology 20: 9–12.

Day, G.M. 1950. The influence of earthworms on soil microorganisms. Soil Sci. 69: 175–184.

Debry, J.M.; Houssain, M., Lemasson-Florenville, M., Wauthy, G. and Lebrun, Ph. 1982. Impact of introduced earthworm populations on pH and nitrogen dynamics in soils treated with pig slurry. Pedobiologia 23: 157–171.

Edwards, C.A. and Lofty, J.R. 1977. Biology of Earthworms. Chapman and Hall, London. 333 pp.

Loehr, R.C. 1971. Alternatives for treatment and disposal of animal wastes. J. Wat. Polln. Control Fedn. 43: 668–678.

Mitchell, M.J., Mulligan, R.M., Hartenstein, R. and Neuhauser, E.F. 1977. Conversion of sludges into topsoils by earthworms. Compost Sci. 18: 28–32.

Needham, A.E. 1957. Components of nitrogenous excreta in the earthworm *L. terrestris* and *E. foetida*. J. Exp. Biol. 34: 425–446.

Neuhauser, E.F., Kaplan, D.L., Malecki, M.R. and Hartenstein, R. 1980. Materials supporting weight gain by the earthworm *E. foetida* in waste conversion systems. Agric. Wastes 2: 43–60.

Parle, J.N. 1963. Microorganisms in the intestines of earthworms. J. Gen. Microbiol. 31: 1–11.

Pincince, A.B., Donovan, J.F. and Bates, J.E. 1980. Vermicomposting municipal sludge: an economical stabilization alternative. Sludge 3: 26–30.

Rao, B.R., Karuna, I. and Bhat, J.V. 1983. Enterobacter *aerogenes* infection of *Hoplochaetella suctoria*. *In* Earthworm Ecology from Darwin to Vermiculture. pp. 383–391. Edited by J.E. Satchell. Chapman and Hall, London.

Rouelle, J. 1983. Introduction of amoebae and *Rhizobium japonicum* into the gut of *E. fetida* (Savigny) and *L. terrestris* (Lge.) IBID: 375–381.

Satchell, J.E. 1967. Lumbricidae. *In* Soil Biology. pp. 259–322. Edited by A. Burges and F. Raw. Academic Press, London.

Syers, J.K., Sharpley, A.N., and Keeney, D.R. 1979. Cycling of nitrogen by surface casting earthworms in a pasture ecosystems. Soil Biol. Biochem. 11: 181–185.

Vanderholm, D.H. 1979. Handling of manure from different livestock and management systems. J. Anim. Sci. 48: 113–120.

Wright, M.A. 1972. Factors concerning ingestion by the earthworm *L. terrestris*, with special reference to apple leaves. Ann. App. Biol. 70: 175–188.

REPRODUCTION AND BIOMASS OF *EISENIA FOETIDA* IN DOMESTIC WASTE

VEIKKO HUHTA and JARI HAIMI
Department of Biology, University of Jyväskylä, Finland

Introduction

The role of earthworms in mull formation and breakdown of plant litter in natural soils has been known since the time of Darwin, but the potential use of earthworms in stabilization of organic residues produced by human community was realized only recently (Watanabe and Tsukamoto, 1976; Mitchell *et al.*, 1977). After that, much research has been focused on 'vermicomposting', including reproductive capacity, physicochemical requirements and effects of worms on their substrate (Hartenstein *et al.*, 1979; Neuhauser *et al.*, 1980a, b; Kaplan *et al.*, 1980). Aerobic sewage sludge and domestic animal manure have been shown to be good worm growth media and thus possible materials for vermicomposting (Hartenstein *et al.*, 1979; Neuhauser *et al.*, 1980b; Tomlin and Miller, 1980). When feeding on these wastes, earthworms increase their overall rate of decomposition (Mitchell *et al.*, 1977, 1980; Hornor and Mitchell, 1981). They also decrease the proportion of anaerobic to aerobic decomposition, which results in a decrease of methane and volatile sulphur compounds (Mitchell *et al.*, 1980; Waugh and Mitchell, 1981).

Pathogenic bacteria are destroyed more rapidly in the presence of worms (Brown and Mitchell, 1981). Composting on the scale of one or a few families is probably not considered big business and hence the study of domestic waste composting has been almost neglected. There is, however, increasing interest in on-site treatment of wastes. Composting toilets are under development, ecofarmers are in want of unpolluted soil amendments. What is the potential of earthworms in small-scale composting systems? Does organic household waste support an adequate population of earthworms, and how do they survive and reproduce in conditions provided by domestic composts? Does the population exert a beneficial influence on the composting processes? These questions were studied by the authors in compost experiments and laboratory tests.

Description of the experiments

The first compost experiment was started on 7 Oct. 1979 to test the survival and reproduction of worms in winter conditions, when there is a risk of freezing. Wastes were deposited in double-walled plastic containers, the inner ones measuring 80 cm in each dimension. Three separate plates of 'styrox' 3 cm thick each, were inserted between the inner and outer walls for insulation. Further plates were placed under the lid. Holes were drilled in the walls and lids for ventilation. The compost was provided with 2200 specimens of *E. fetida*, most of them full-grown. All the decomposable waste of a family of five persons: garbage garden residues, and contents of a composting toilet using crushed bark as mixture material were deposited regularly in

Earthworms in waste and environmental management
edited by C.A. Edwards and E.F. Neuhauser, pp. 65–69

the container. The compost was half full at the beginning, and became full at the end of December 1979. After that it was watered at times. More waste was added in winter when the surface had sunk. The content was first inspected in May 1980 but then allowed to ripen till August 1980.

The objective of the second experiment was to follow the development of worm population in a compost that had its active period in summer and ripening phase in winter. In this case, there is a danger that the whole compost mass can become too hot for worms, especially if the thermophilic phase coincides with warm weather. Indeed, measurements revealed that temperature exceeded 35°C in a great part of the compost for considerable periods. This is the thermal tolerance of *E. fetida* (Mitchell, 1979).

This experiment was started on 17 May 1980. A double walled compost bin was constructed from the same containers used in the first experiment. These were placed tightly side by side, all but one plastic wall was cut off from between the containers, and the remaining wall perforated to allow free entrance for worms from one compartment to the other. To improve ventilation, commercial drainage tubes were inserted through one side wall of each container. Kitchen and garden wastes of one family were deposited into the compost. As substitute for toilet waste, faeces of four dogs were regularly collected from two courtyards. Crushed softwood bark was added to damp odors and to provide a proper C:N ratio. A total of ca. 6000 specimens of *Eisenia fetida* were introduced. Fresh material was deposited into the first compartment until 18 September, when the filling of the second one was started. The contents were inspected in August 1980 and May 1981.

A laboratory experiment utilized mixed garbage, dogs' faeces and garden residues. The waste was embedded portion by portion in crushed pine bark in a small scale 'bed compost'. The dimensions of the bed were 0.4 × 0.2 × 2.0 m. The whole amount of waste was mixed at the same time and divided into equal portions of 3.6 kg. These were stored deep-frozen and added to the bed every two weeks.

At the beginning of the experiment, 0.2 kg earthworms were added to the first portion of waste. This equals about 800 earthworms. The contents were inspected four weeks after the last waste addition.

The number of worms in a compost is difficult to estimate by taking a sample, because they are distributed very unevenly and withdraw readily from a disturbed spot. A more reliable estimate is obtained by extracting the worms from the whole compost mass. This is not as laborious as could be expected. Compost material is spread on a plastic cloth and agitated on the surface, which causes the earthworms to escape deeper so that layer after layer can be removed until a solid mass of worms remains. The total mass is weighed and worms are then counted from a weighed subsample. We divided the worms into 'small', 'medium-sized' and 'full-grown', and made an approximate correction for specimens not found; 10% for large, 20% for medium-sized and 40% for small ones. On 17 May 1981 another method was used: a 4 × 1.2 l sample was rapidly taken from the topmost 10 cm of compost, and the whole layer was immediately removed from the container. The same was repeated until the container was empty. Sampling by layers was also used to assess the numbers of cocoons. These were sorted out by washing on 2 mm sieves.

Table 1. Estimated number of earthworms and their cocoons, and live biomass of worms per 0.5 m^3 of compost mass in the two experiments

	Exp. 1 from 7 Oct. 1979		Exp. 2 from 17 May 1980	
	3 May 1980	12 Aug. 1980	16 Sept. 1980	18 May 1981
Number of worms	6000*	16000	58000	9200
% Fullgrown	67	11	29	34
Biomass, g	—	670	8200	1900
Full cocoons	17500	14500	51300	8800

* 2000 removed.

Development of the earthworm populations

The survival and reproduction of earthworms was good in the composts both in winter and in summer conditions. Even during winter, the initial population increased threefold, and the number of cocoons produced was enough to produce a considerable population (Table 1). Development in summer was even more rapid. The population present in September is potentially capable of processing the whole of the compost content (volumic weight ca. 0.5) in two months, if we assume an ingestion rate of 50% of body weight daily (Mitchell, 1979). There was a continuous emigration of worms from the composts; worms could be found in the surroundings especially on wet weather, and they easily invaded the other container or compartment. When the material ages, emigration and mortality exceed reproduction, resulting in gradual population decline (Table 1). However, new worms continue to hatch, so there is a considerable density even in a material of one year in age.

Also in the bed compost, the growth and reproduction of the earthworms were good. Almost all the waste was eaten during the experiment. The total earthworm biomass at the end of experiment was 0.66 kg, two thirds of which were found in the latest portion (Table 2). The biomass in this portion equals 36 kg live weight m^{-3}. There were plenty of cocoons in every portion of the bed compost, quite many still in the oldest portion. On an average, 64000 cocoons were found per m^3 of material.

Discussion

Our experiments have shown that *Eisenia fetida* thrives well and reproduces in household composts both in summer and in winter conditions. In a few months a sufficient worm density was reached to have the potential to process the compost material in a moderate time. When properly insulated, a compost with worms remains active through the winter even during serious frosts. In a 'compartment compost', maximum population is present as soon as environmental requirements of the species are met in the material. According to several studies published in recent years, an adequate density of earthworms in a compost has potential to accelerate stabilization, remove odors, and promote disappearance of pathogenic organisms. The latter aspect is particularly important in small domestic composts, where there are no

Table 2. The quantity of earthworms and cocoons in each portion of the compost at the end of experiment. The initial biomass was 200 g

Age of waste (weeks)	16	14	11.5	9	6.5	4	Total
Biomass of earthworms (g)	50.40	49.47	54.42	36.36	34.46	435.24	660.35
Number of full cocoons	449	924	879	714	597	1017	4580

means to guarantee that all parts of the compost material does attain temperature high enough to kill harmful microbes. On the other hand, during the thermophilic phase, the stabilization proceeds rather long in the absence of worms. In order to gain maximum benefit from the earthworm population, high temperatures should be minimized by good aeration and mixing with nutrient-poor material, which in turn will reduce thermal hygienization. Earthworms also thrive best in such composts in which heating never happens (as in the bed compost described above).

Unless value is put on production of worms for fish bait, the best strategy in household composting may well be just to confirm the presence of earthworms. Under appropriate conditions they will multiply and complete the process without further care.

Acknowledgements

The authors wish to thank Dr. Mirja Salkinoja-Salonen for keeping and taking care of the compost in the first experiment.

References

Brown, B.A. and Mitchell, M.J. 1981. Role of the earthworm, *Eisenia foetida*, in affecting survival of *Salmonella enteritidis ser. typhimurium*. Pedobiologia 22: 434–438.

Hartenstein, R., Neuhauser, E.F. and Kaplan, D.L. 1979. Reproductive potential of the earthworm *Eisenia foetida*. Oecologia (Berlin) 43: 329–340.

Hornor, S.G. and Mitchell, M.J. 1981. Effect of the earthworm, *Eisenia foetida* (Oligochaeta), on fluxes of volatile carbon and sulfur compounds from sewage sludge. Soil Biol. Biochem. 13: 367–372.

Kaplan, D.L., Hartenstein, R., Neuhauser, E.F. and Malecki, M.K. 1980. Physicochemical requirements in the environment of the earthworm *Eisenia foetida*. Soil Biol. Biochem. 12: 347–352.

Mitchell, M.J. 1979. Functional relationships of macroinvertebrates in heterotrophic systems with emphasis on sewage sludge decomposition. Ecology 60: 1270–1283.

Mitchell, M.J., Mulligan, R.M., Hartenstein, R. and Neuhauser, E.F. 1977. Conversion of sludges into 'topsoils' by earthworms. Compost Science 18(4): 28–32.

Mitchell, M.J., Hornor, S.G. and Abrams, B.L. 1980. Decomposition of sewage sludge in drying beds and the potential role of the earthworm, *Eisenia foetida*. J. Environ. Qual. 9: 373–378.

Neuhauser, E.F., Hartenstein, R. and Kaplan, D.L. 1980a. Growth of the earthworm *Eisenia foetida* in relation to population density and food rationing. Oikos 35: 93–98.

Neuhauser, E.F., Kaplan, D.L., Malecki, M.R. and Hartenstein, R. 1980b. Materials supporting weight gain by the earthworm *Eisenia foetida* in waste conversion systems. Agricult. Wastes 2: 43–60.

Tomlin, A.D. and Miller, J.J. 1980. Development and fecundity of the manure worm, *Eisenia foetida*

(Annelida: Lumbricidae) under laboratory conditions. Proceedings of VII-International Colloquium of Soil Zoology. Ed. D. Dindal. pp. 673–378.

Watanabe, H. and Tsukumoto, J. 1976. Seasonal change in size class and stage structure of Lumbricid *Eisenia foetida* population in a field compost and its practical application as the decomposer of organic waste matter. Rev. Ecol. Biol. Sol. 13: 141–146.

Waugh, J.M. and Mitchell, M.J. 1981. Effect of the earthworm, *Eisenia foetida*, on sulfur speciation and decomposition in sewage sludge. Pedobiologia 22: 268–275.

THE ROLE OF MICRO-ORGANISMS IN THE NUTRITION OF *EISENIA FOETIDA*

M.H. MORGAN*
Rothamsted Experimental Station, Harpenden, U.K. AL5 2JQ

Introduction

It is generally agreed that the alimentary canal of earthworms contains the same species of micro-organisms as those found in the environments in which the worms are growing. This may be expected since worms ingest large quantities of material which is rich in micro-organisms. However, which, if any, of the micro-organisms present constitute a major food source is an unresolved problem.

Bassalik (1913) compared the microflora of the alimentary canal and the excreted casts of *Lumbricus terrestris* with that of soil from the worms' habitat. Although he isolated more than fifty species he found no differences between the range of species present in each environment. Khambata and Bhatt (1957), Parle (1963) and Piearce (1978) also studied gut microflora and found only species that were common in soil or plant remains. In recent years, work has focussed on feeding worms on pure cultures of micro-organisms which had been layered over the surface of worm cultures in soil, organic wastes or synthetic bedding material. (Neuhauser *et al.*, 1980; Cooke *et al.*, 1980). However, most of the available literature has not taken into account the presence of micro-organisms other than those added as possible food sources. This omission must affect the interpretation of the results since in a natural habitat (either soil or organic waste) the microbial population is in a 'steady' or equilibrated state, each species occupying its own niche to capacity. If, however, an outside organism is introduced, then although this may swamp the resident microflora initially, it will soon be overcome and included at a level in the population compatible with the other organisms present. Thus, by initially adding micro-organisms to a worm culture in either soil or organic waste material and leaving the culture for a few weeks it is likely that any weight increase observed is not due to the added organism but to the presence of a mixed resident microflora.

The experiments described in this paper were designed to establish if the earthworm *Eisenia foetida* (Savigny) depends on the microflora in organic wastes for nutrition. The effect of adding worms on the population of components of the resident microflora of cattle waste was investigated, as well as assessing growth of worms fed on simple nutrients and pure cultures of microorganisms in natural and synthetic environments.

To obtain an accurate knowledge of the nutritional requirements of *E. foetida* it was necessary to produce worms which were free of micro-organisms both internally and externally as well as placing them in a sterile (*i.e.*, microbe-free) environment. The methods employed enabled direct assessments of the nutritional requirements to be made; rather than an indirect assessment which occurs in a worm/waste system, due to the presence of a mixed population of micro-organisms and their metabolites.

* *Present address:* University of Reading, Berks, U.K.

Earthworms in waste and environmental management
edited by C.A. Edwards and E.F. Neuhauser, pp. 71–82

Materials and methods

Relationship between worms and micro-organisms

Samples of cattle waste (1 g) capable of supporting worm growth (acceptable) were placed in 10 ml sterile distilled water and shaken. Serial dilutions were made from the samples and 0.1 ml of each dilution spread on various media. Nutrient agar (OXOID), malt extract agar (OXOID), cellulose agar and actinomycete agar were used to establish the normal microflora (both numbers and species) in the absence of worms. Some of the cattle waste was heated to 120°C for 60 minutes in order to decrease the micro-organisms present. Samples of this waste were tested at regular intervals to establish the microflora present over a period of eight weeks. Protozoa were counted using a haemocytometer and nematodes by using a counting grid.

Acceptable cattle waste was placed in small glass crystallizing dishes (7 cm × 4 cm). Each dish contained 30 g fresh weight of waste and between 0 and 30 worms weighing between 40 and 50 mg each. The dishes were then incubated at 23°C ± 1°C for six weeks, during which time the worms were weighed at regular intervals and samples taken for microbial analysis. Plate counts were made to establish numbers of bacteria and fungi per g. dry weight. Heated waste was substituted for acceptable waste in some experiments.

Feeding on simple nutrients

Samples of cattle waste (30 g) were placed in small dishes (7 cm × 4 cm) with five equal-sized hatchling worms (mean weight 20 mg). Simple nutrients, such as carbohydrate and protein sources, were added fortnightly at concentrations ranging from 0–5% (w/w) and worm weights monitored regularly. Some of these nutrients were also added at lower concentrations to cellulose (see below) containing two worms and worm weights again monitored.

Production of worms and environments free from micro-organisms

Many media were tested for their ability to be sterilized and remain suitable for worm survival, *i.e.*, sand, diatomaceous silica, organic wastes and cellulose. Of these, cellulose was found to be most suitable since the worms maintained a steady weight for long periods of time and also ingested the cellulose making it suitable as a carrier medium for antibiotics. Carbenicillin was found to be the most suitable antibiotic for worm sterilization. This was incorporated into cellulose and once sterilized, the worms were stored in autoclaved cellulose until required.

Production and harvesting of micro-organisms

Fungi and bacteria were grown in liquid media at 23°C ± 1°C in a shaking water bath, whilst protozoa were grown in static conditions at the same temperature. Bacterial and protozoal cultures were harvested by centrifugation followed by three washes in sterile distilled water. The final pellet was resuspended in the volume of sterile water which, when added to cellulose media, gave a population of micro-

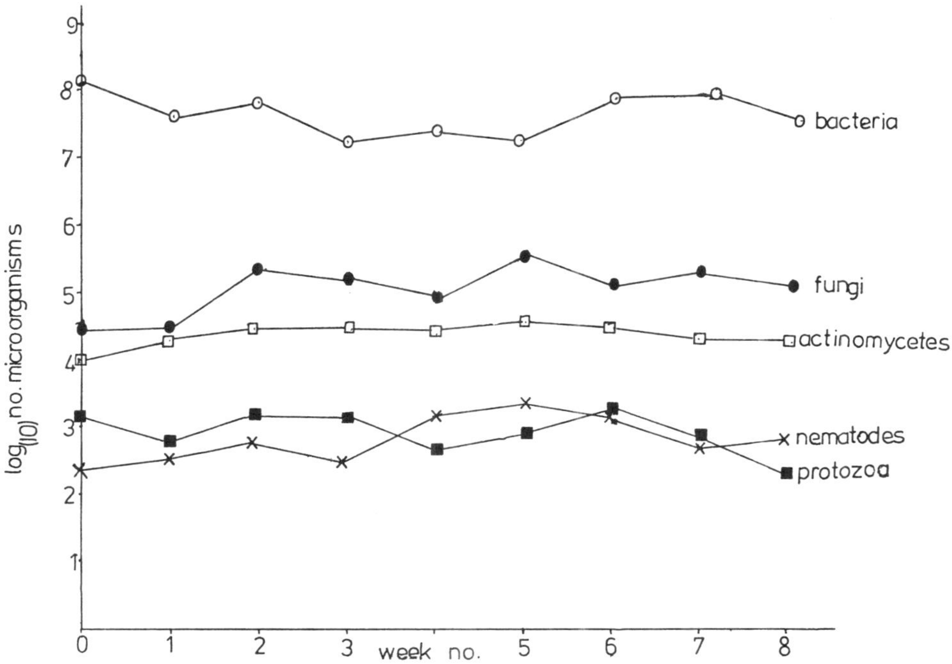

Fig. 1. Nos. of micro-organisms per gm. wet wgt. in acceptable cattle waste

organisms equivalent to that in the waste. Fungi were harvested through coarse tissue paper followed by three washes in sterile distilled water and resuspension at a concentration equivalent to 100 mg dry wgt fungi per cellulose dish.

Results and discussion

Spread plate counts over eight weeks showed that the microflora of cattle waste remained steady over that period (Fig. 1). On further study, it was found that the majority of the bacteria were Enterobacteriaceae, which was expected, due to the origin of the material. Nematode numbers were equivalent to protozoan numbers throughout the experiment. The microflora changed when the waste was heated (Fig. 2). The heating was inadequate to sterilize the waste and initial plate counts showed the presence of bacteria although the fungi, actinomycetes, nematodes and protozoa had been removed. The heat failed to penetrate the centre of the waste because it was heated in 2 kg amounts in glass containers. At week one a microbial 'bloom' was observed, presumably because the inadequate heating process removed only vegetative cells, leaving the heat resistant spores and cysts to germinate, and also due to the growth of airborne contaminants. Heating altered the structure of the waste and it became dense and waterlogged. Presumably, many nutrients were solubilized in the waste during heating providing an increased food supply for the micro-organisms, encouraging rapid growth. Most of the bacteria isolated at this time were *Pseudomonas* sp. and a normal mixed population was not obtained until week four. The

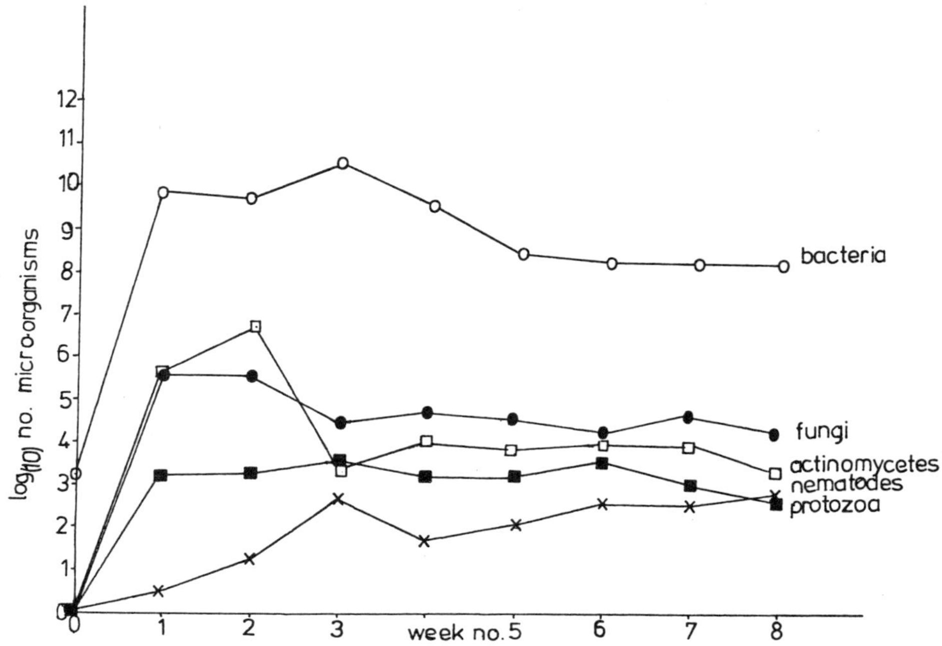

Fig. 2. Nos. of micro-organisms per gm. wet wgt. in autoclaved cattle waste

numbers of organisms present varied widely over the first four weeks after which they stabilized at a population similar in species and numbers to that found in acceptable, untreated waste.

The addition of worms to the waste altered the numbers of micro organisms present. In an enclosed non-draining system, waste without worms showed a decline in numbers of bacteria and fungi over a period of six weeks. By adding worms to the waste, the decline was accelerated, in general becoming greater the more worms added (Figs. 3 and 4). Numbers of bacteria increased over the first two weeks in waste containing 0 and 5 worms. This increase was due to the aeration of the waste during the setting up of the experiment. All the waste used in the experiment came from the same storage pile. Some of the waste (3 kg) was placed in a bucket and stored in the laboratory. It was then divided into 30 g samples for the experiment. Transfer and division of the waste into small amounts and subsequent mixing resulted in increased aeration and a greater surface area, following an increase in microbial activity.

When the numbers of worms were increased, numbers of bacteria declined. This could have been because the worms were feeding directly off the population, preventing it from increasing in numbers, or possibly the worms were feeding off another nutrient source also required by the micro-organisms, so increased microbial growth was prevented due to competition for food. By week six, the numbers of bacteria were the same in each treatment.

Similar patterns were seen with the numbers of fungi, although accurate fungal counts are more difficult than bacterial counts since both hyphae and spores can produce colonies on plates. In this case, the control followed a linear decline over the

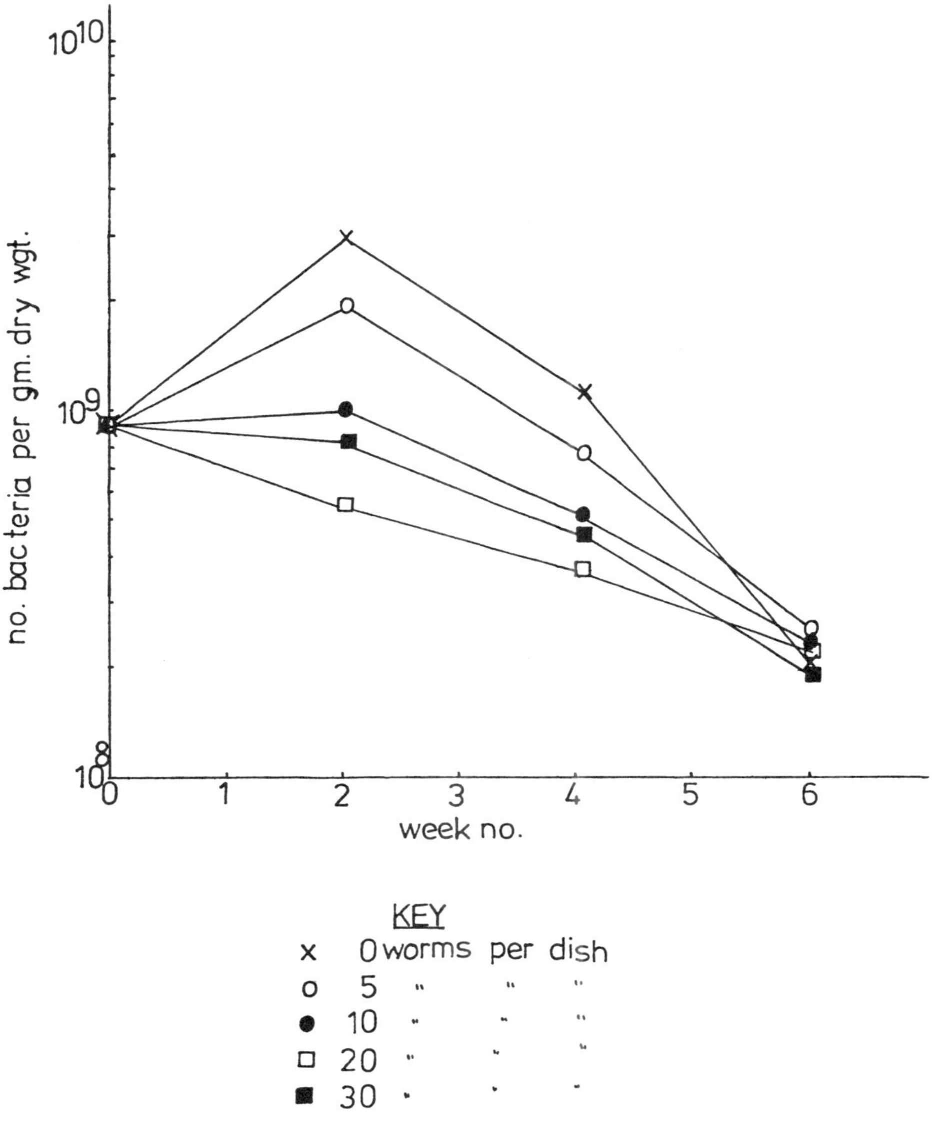

Fig. 3. Relationship between bacterial numbers and worm numbers in an endosed environment (30 gms cattle waste)

first four weeks this pattern being mimicked when there were 5 or 10 worms per dish. Fungal numbers tended to fluctuate when 20 or 30 worms were present.

These changes in microbial numbers appear to be related to worm growth (Fig. 5). Over the first two weeks when numbers of bacteria and fungi were highest, the worms in each dish showed their greatest rate of weight increase. Worms present in low

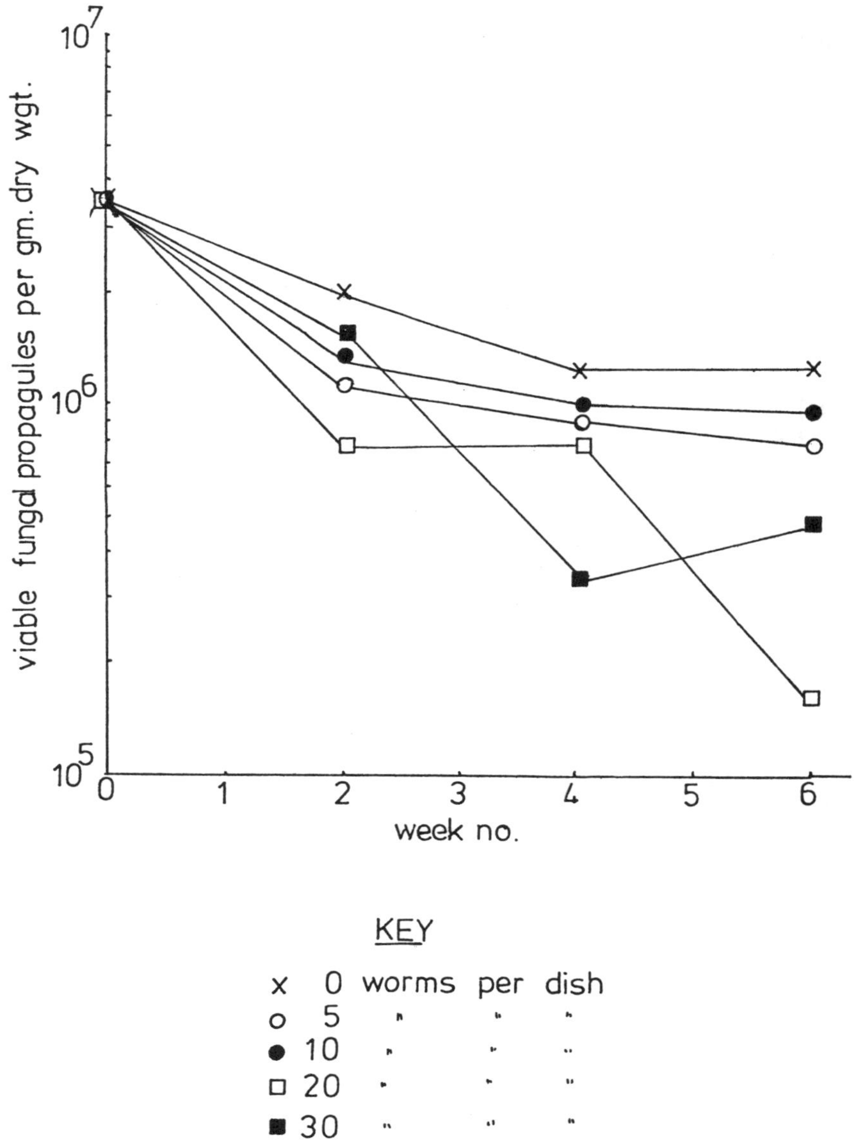

Fig. 4. Relationship between fungi and worm numbers in an enclosed environment (30 gms. cattle waste)

numbers gained weight faster and reached greater weights than when high numbers of worms were present in each dish. As the microflora decreased, the worms started to lose weight and by week six the worms were becoming the same weight regardless of the numbers of worms per dish. The rapid weight loss seen after week four may have been due to a build-up toxic metabolites in the dishes.

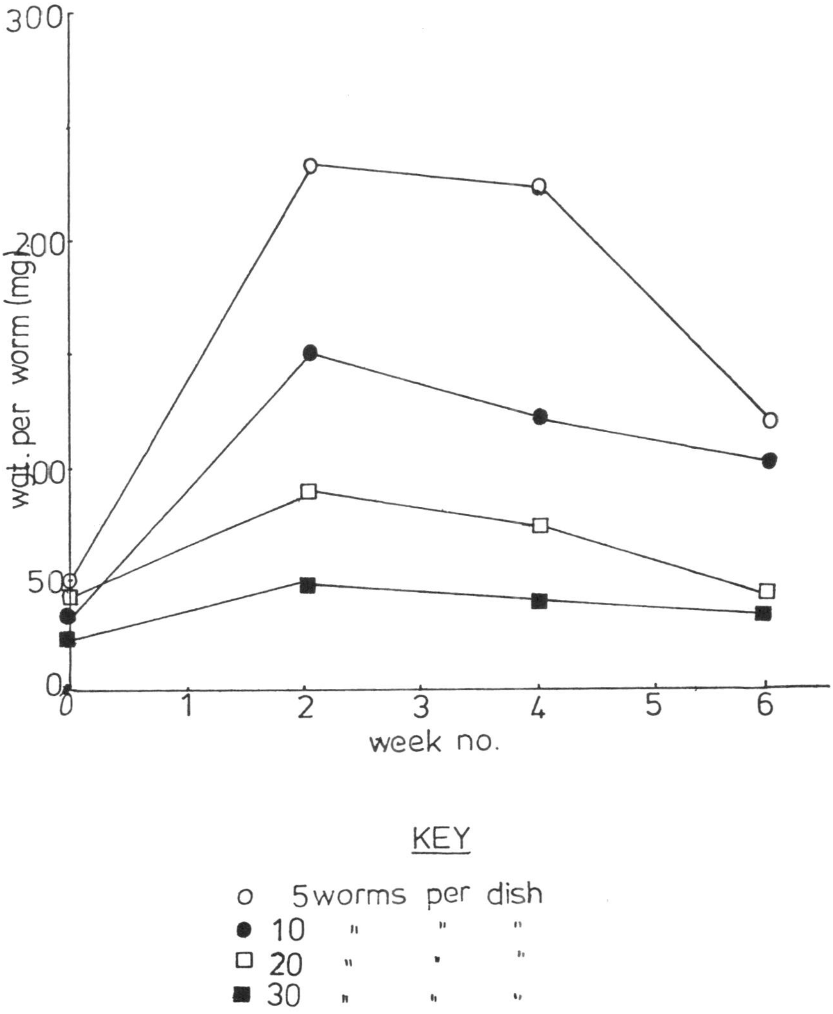

Fig. 5. Growth of *Eisenia foetida* in 30 gms. of cattle waste

From these observations there appears to be a relationship between micro-organisms and worms, rapid worm growth relying on a high number of micro-organisms. This relationship was obvious only when the worms were overcrowded and in small dishes that were undrained, and unfed. In normal conditions, in worm beds, worm numbers per gramme of waste are much lower so that there are no detectable differences in the microflora between waste with worms and waste without worms.

The addition of simple nutrients to cattle waste resulted in weight increases greater than in the unsupplemented control. The additions of low concentrations (< 0.1%

Table 1. E. foetida + cattle waste + simple nutrients, weight changes observed over four weeks 23°C ± 1°C

Nutrient added		Weight increase per worm ((mg) over four weeks)
Control (H_2O added)		214.6
Glucose	0.075% w/w	318.4
	0.3% w/w	298.9
	0.5% w/w	197.9
	1.0% w/w	206.6
	5.0% w/w	203.1
Sucrose	0.075% w/w	228.5
	0.3% w/w	279.3
Yeast extract	0.01% w/w	222.7
	0.05% w/w	226.0
Liver digest	0.01% w/w	224.7
	0.05% w/w	182.1
Autoclaved extract	3.0% v/w	190.0
	13.0% v/w	236.5
	20.0% v/w	190.0
Filtered extract	3.0% v/w	220.2
	13.0% v/w	201.3
	20.0% v/w	226.0

w/w)) of carbohydrates caused a greater increase in weight than in the control, whereas at carbohydrate concentrations over 0.3% (w/w), the worms attained lower weights than the control (Table 1). In low concentrations of protein sources and cattle waste extracts (*i.e.*, the liquid fraction from 1 kg cattle waste plus 1 litre of water) the worms increased in weight at values similar to or lower than the control.

Worms showed greater weight increases in the presence of carbohydrates which suggests that either the worms can utilize the carbohydrates or that the carbohydrates stimulate the micro-organisms on which the worms feed.

This was tested using microbially-free worms and autoclaved cellulose media. Cellulose was found to be the most suitable medium for studying changes in worm weights since, after an initial weight fluctuation, the worms lost weight very slowly over a period of eight weeks (Fig. 6). If the cellulose was substituted by sand or diatomaceous silica then a rapid weight loss was observed. The worms kept in sand died after only three weeks. The slow weight loss observed when the worms were kept in cellulose suggests that the worms were capable of utilizing cellulose as a food source and it has been reported that they may possess a cellulase (Tracey, 1951; Hartenstein, 1982). Worms ingested the cellulose avidly and deposited white castings around the edges of the dishes, whereas those in silica did not appear to ingest any of the particles over an eight-week-period.

When nutrients were added to the cellulose medium it was found that the concentrations of 0.15% (w/v) glucose and sucrose caused 100% mortality, whereas lower concentrations caused weight increases (Table 2). Nutrients were added every five days over fifteen days, and after this time the control showed the greatest weight loss and the worms fed on filtered extract the greatest weight gain. The filtered extract contained all the soluble fractions of the waste, but no micro-organisms, whereas the autoclaved extract contained dead micro-organisms and properties of some of the

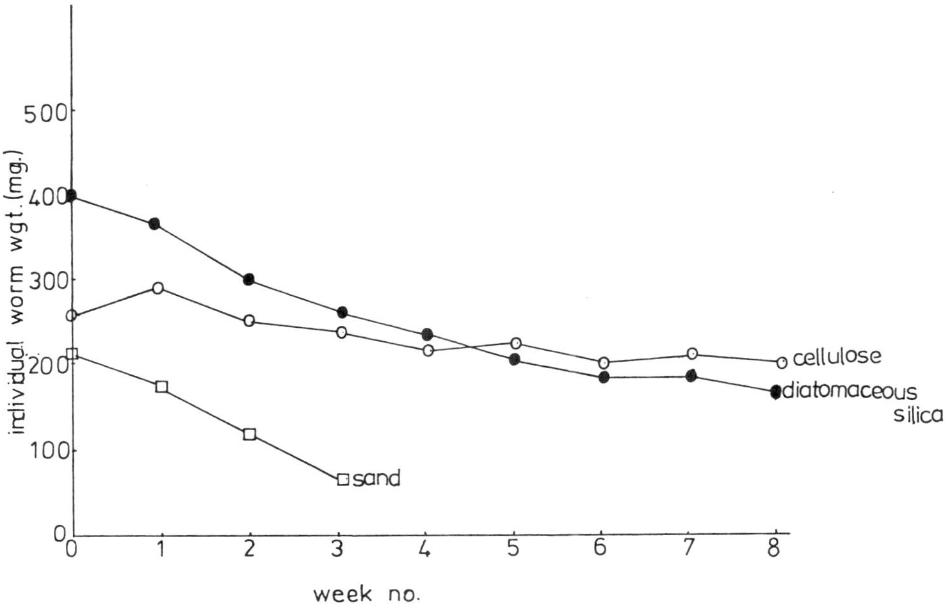

Fig. 6. Weight changes observed when *Eisenia foetida* is stored in inert media

Table 2. Weight changes observed when *E. foetida* was placed in autoclaved cellulose + simple nutrients for 15 days

Nutrient added		Weight change per worm (mg) over 15 days
Control (H_2O added)		− 6.0
Glucose	0.05% v/v	+ 7.3
	0.15% v/v	Dead
Sucrose	0.05% v/v	+10.5
	0.15% v/v	Dead
Yeast extract	0.02% v/v	− 4.3
Liver digest	0.02% v/v	− 2.7
Autoclaved extract	8.0% v/v	+ 7.0
Filtered extract	8.0% v/v	+20.7

constituents may have been altered. Thus, filtered extract contained the same spectrum of nutrients found in cattle waste in their unchanged form, so it would be expected to have the greatest effect on worm weights if, as the results suggest, worms are capable of utilizing nutrients in the absence of micro-organisms.

The worm weight increases observed (Table 2) were smaller than those seen in cattle waste with added nutrients (Table 1), but in all cases the worms grew better in the presence of low concentrations of nutrients than in their absence. The presence of live micro-organisms increased the growth rate of the worms.

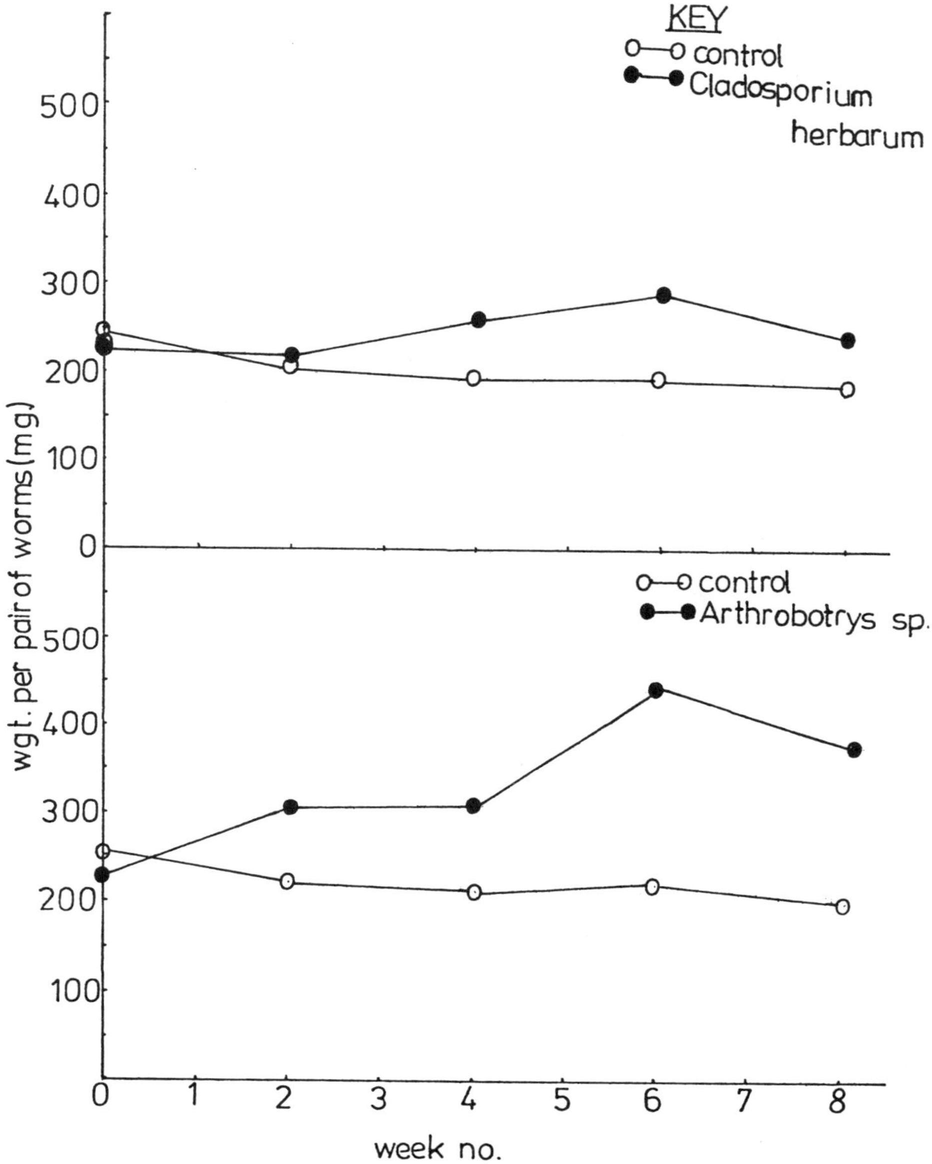

Fig. 7. Growth of *Eisenia foetida* on pure fungal cultures

Cultures of *E. foetida* fed on pure cultures of live micro-organisms showed weight increases in the presence of four species of fungi, but weight losses or death with all other organisms tested (Table 3). Death may have been due to the rigorous antibiotic treatment reducing their natural immunity to the potentially pathogenic organisms in the environment. Also the worms were confined with a single food source which

Table 3. Growth of *E. foetida* on pure cultures of micro-organisms

Name of organism		Weight of worms as a % of starting weight
Control (H_2O only)		97.2%
Bacteria	*Aeromonas hydrophila*	66.9%
	CDC gp. IIf	61.1%
	Serratia marcescens	65.2%
	Alcaligenes sp.	54.9%
	Flavobacterium sp.	47.7%
	CDC gp. V	72.2%
	Flav. meningosepticum	Dead within five days
	Pseudomonas aeruginosa	Dead within five days
	Acinetobacter lwoffi	92.7%
	Bacillus cereus var. mycoides	Dead within seven days
	Serratia rubidaea	98.6%
	Citrobacter freundii	Dead within seven days
Fungi	*Aspergillus terreus*	138.5%
	Asp. fumigatus (sp. 1)	Dead within 24 hours
	Asp. fumigatus (sp. 2)	122.7%
	Cladosporium herbarum	96.2%
	Arthrobotrys sp.	135.9%
	Penicillium sp.	105.1%
	F18	94.7%
	Asp. flavus	Dead within 24 hours
Protozoa	*Tetrahymena pyriformis*	Dead within three days
	Proterioochromonas minuta	Dead within three days

may not have been palatable, resulting in worm starvation, or the build up of toxic metabolites produced by the micro-organisms may have caused worm death. In cattle waste, a build-up of toxic metabolities from microbial activity presumably does not cause rapid death because the worms are able to move away from the more contaminated areas, and also the waste has a good buffering capacity and any toxins produced are probably destroyed or rendered harmless by the other organisms and components of the waste. All the organisms used in the feeding trials were isolated from the wastes and cultured in the laboratory, except for the protozoa which were obtained from the Institute of Terrestrial Ecology, Cambridge.

Worms grew very well on pure fungal cultures. The fungal species *Cladosporium herbarum* and *Arthrobotrys* sp. produced the greatest growth and in *Arthrobotrys* culture the worms doubled in weight (from 210 mg per pair of worms to 430 mg) in six weeks (Fig. 7).

The results obtained in these experiments indicate that the earthworm *E. foetida* is capable of utilizing both micro-organisms and simple nutrients for growth. This work needs to be expanded and more detailed experiments are required if earthworm nutrition is to be understood. Studies on nutrition could lead to better worm production and in increase in waste turnover benefitting all those involved in vermiculture. It is essential to perform detailed enzyme studies to establish if worms have the ability to utilize nutrients (as indicated in this work), or whether they rely solely on the activities of micro-organisms resident in their guts; also examination of the gut as a site for both digestion and absorption would be valuable. Some work on earthworm digestive systems is currently underway but much work is still required on the basic digestive abilities of the worm and the biochemical potential of the micro-organisms present in worm guts, before their precise nutritional requirements can be established.

References

Bassalik, K. 1913. On silicate decomposition by soil bacteria. Z. Gärungs-Physiol. 2: 1–32.

Cooke, A. and Luxton, M. 1980. Effect of microbes on food selection by *Lumbricus terrestris*. Rev. Ecol. Biol. Sol. 17(3): 365–370.

Day, G.M. 1950. Influence of earthworms on soil micro-organisms. Soil Science 69: 175–184.

Hartenstein, R. 1982. Soil macro-invertebrates, aldehyde oxidase, catalase, cellulase and peroxidase. Soil Biol. Biochem. 4: 387–391.

Khambata, S.R. and Bhatt, J.V. 1957. A contribution to the study of the intestinal flora of Indian earthworms. Arch. Mikrobiol 28: 69–80.

Neuhauser, E.F., Kaplan, D.L., Malecki, M.R. and Hartenstein, R. 1980. Materials supporting weight gain by the earthworm *Eisenia foetida* in waste conversion systems. Agric. Wastes 2: 43–60.

Parle, J.N. 1963. Micro-organisms in the intestines of earthworms. J. Gen. Microbiol. 31: 1–13.

Piearce, T.G. 1978. Gut contents of some lumbricid earthworms. Pedobiologia 18: 153–157.

Tracey, M.V. 1951. Cellulase and chitinase in worms. Nature (London) 167: 776–777.

SPECIFIC DIFFERENCES IN COCOON AND HATCHLING PRODUCTION IN *EISENIA FETIDA* AND *E. ANDREI*

P.S. SHEPPARD
Department of Zoology, University of Manchester

Introduction

The specific status of the worms most commonly used in vermiculture (Harman, 1955; Tomlin, 1983) remains unclear in much of the current literature. *E. fetida* (Savigny) comprises not one but two species which can be distinguished by pigmentation as described by André (1963). The first of the two species, *E. fetida*, is banded, the area around the intersegmental groove having no pigment and appearing pale or yellow. This gives the worm a brown and yellow striped appearance, hence its common names of 'brandling' or 'tiger' worm. The other species, *E. andrei*, is uniformly pigmented on its dorsal surface showing no such banding pattern and the common name "red" worm has been applied to it.

On biological criteria, two taxa are distinct species if they do not interbreed when syntopic. On this basis André (1963) was the first to demonstrate the specific status of these two worms. He created chimeras, by means of surgery, such that the male and female gonads in an individual chimera came from different species. By crossing such chimeras he was able to show that male *E. fetida* and female *E. andrei* gametes could not produce viable offspring. The reciprocal cross produced hybrid offspring. These hybrids had a banding pattern intermediate to that of their parents, resembling *E. andrei* in the midline but narrowing to resemble *E. fetida* laterally. They laid cocoons but these were all infertile.

Roch *et al.* (1980) and Valembois *et al.* (1982) showed that the two species were biochemically very different and suggested *E. andrei* could have been derived from *E. fetida* by the loss of certain alleles. Jaenike (1982) added further confirmation of their specific status using electrophoretic techniques. Five loci were examined and three of these had no alleles in common. Therefore, even within the mixed colony from which the specimens were collected, there was reproductive isolation.

Unfortunately, André (1963) did not give specific status to the worms, but referred to them as 'forms' *Eisenia foetida typica* and *Eisenia foetida unicolour*. Bouché (1972) amended this and gave them sub-specific status, naming them *Eisenia foetida foetida* and *Eisenia foetida andrei*. The two species are syntopic in England, mixed colonies in dung and compost heaps being common (personal observation), and from the evidence of the authors above, reproductively isolated. Sub-specific status is, therefore, inappropriate, and I propose to refer to the two species as *E. fetida* and *E. andrei* in this paper. *E. foetida* was an illegal emendation of the original *E. fetida* (Sims, 1983; Easton, 1983). The name *E. foetida* has been used by many authors, but which of the two species is being referred to has not always been clear. The name *E. foetida* should be restricted to the 'brandling' worm.

The reproductive performance of '*E. foetida*' has been well-documented (Hartenstein, *et al.* 1979; Herlant-Meewis, 1956; Mori and Okada, 1978; Neuhauser *et al.*, 1980; Tsukamoto and Watanabe, 1977; Vail, 1972) but, as explained above it is not

Earthworms in waste and environmental management
edited by C.A. Edwards and E.F. Neuhauser, pp. 83–92

clear which species is involved. This paper sets out to compare the cocoon and hatchling production of the two species. The differences found confirm their specific status and show that ecological, as well as biochemical differences, exist.

Materials and methods

Single pair matings were set up between *E. fetida* and *E. andrei* from two sources (Table 1). The breeding pairs of worms were kept in containers made from 100 mm lengths of 50 mm outside diameter P.V.C. drain pipe. The ends were sealed using nylon gauze and secured by locking rings made from a similar P.V.C. piping with a 50 mm internal diameter. When hatchlings were raised in these tubes the nylon gauze was replaced by cloth (65% Tetron, 35% cotton), the finer mesh preventing escape. The tubes were placed upright upon a moist gravel bed. Temperature was 24 ± 1°C., moisture 85 ± 1% using a substrate of reconstituted cow dung. Farm yard manure was dried at 40°C. When it was required it was rewet and allowed to stand for one week. If used before this time the substrate was toxic to the worms, possibly due to high ammonia concentrations initially liberated upon rewetting (Satchell personal communication). This substrate had the advantage of being easy to store.

Cocoons were kept at 24 ± 2°C. in stacked 50 mm crystal polystyrene petri dishes with a piece of moist filter paper to maintain a high humidity. Preclitellate specimens were selected at the start of the experiment to help ensure that the worms used were not storing spermatozoa from previous matings. It is possible for clitellate specimens to lose their sexual characteristics due to food shortage and thus appear preclitellate while still storing viable spermatozoa from previous matings (Herlant-Meewis, 1962). However, at the sites sampled there was an abundance of food, so that the preclitellate worms were almost certainly virgins.

The substrate was searched carefully twice weekly by means of a flat spatula and the worms weighed. Cocoons were removed, weighed and kept in the petri dishes described above until hatched. The hatchlings were in turn weighed upon emergence. Since the newly hatched worms were very delicate, the following procedure was devised to weigh them. The cocoon was recorded as having hatched once one worm had emerged, and any worms still in it were removed either by gently squeezing the cocoon or by cutting it open using fine scissors. Using a pasteur pipette, the worms were transferred to a piece of paper towelling which was blotted from underneath to remove as much water as possible.

The towelling, along with the worms, was placed on the balance which, by means of an automatic tare facility, was set to zero. Worms were removed one at a time and their weights recorded before again resetting the balance to zero. Evaporation from the towelling and worms while the balance stabilized, increased the apparent weight of the young worms so a correction factor was calculated daily. A piece of towelling was left on the balance for 12 seconds, the average time taken for the balance to stabilize and its weight loss due to evaporation recorded. This correction factor was subtracted from each of the young worm's weight on that day. The correction factor on occasion was as much as 1 mg, but normally was about half this amount. Although this represents a large potential error when considering the weight of a newly hatched worm, it is negligible in relation to adult worm weight, so that no correction factor was calculated for adult worms.

Table 1. The six inter- and intra-specific crosses set up between two species. *E. fetida* was obtained from the Lake District feeding upon sewage sludge, *E. andrei*, from a dung heap consisting principally of cow bedding in Cheshire and *E. andrei*$_2$ from a dung heap of horse bedding in Kent. Five replicates of each cross were used

	E. fetida	*E. andrei*$_1$	*E. andrei*$_2$
E. fetida	×	×	×
E. andrei$_1$		×	×
E. andrei$_2$			×

Table 2. Mean cocoon fertilities for the six crosses. Arrows indicate significantly different pairs of means. Q tables from Table 15 Snedeor and Cochrane 'Statistical Methods' 6[ed] 1967 Iowa State University Press. As errors attached to percentages are not equal an angular transformation was performed before analysis. (N.B. mean arcsin $\sqrt{\text{freq.}} \neq$ arcsin $\sqrt{\text{mean freq.}}$)

Cross	No. of pairs	Total no. of cocoons	Mean of the pairs cocoon fertility %	Mean of the pairs cocoon fertility arcs in freq.
E. fetida × *E. fetida*	5	147	82.2	65.3
E. fetida × *E. andrei*$_1$	5	173	44.3	39.1
E. fetida × *E. andrei*$_2$	5	100	15.7	17.9
E. andrei × *E. andrei*$_1$	5	133	71.4	58.5
E. andrei × *E. andrei*$_2$	4	71	75.7	60.5
E. andrei × *E. andrei*	5	118	73.4	60.8
		← $Q_{6,24}$ $p < 0.05$		
		← $Q_{2,24}$ $p < 0.05$		

Some adult mortality was experienced amongst the breeding pairs; when this occurred the dead worm was replaced with a preclitellate individual from the appropriate source. In 5 of the 30 crosses, time considerations after a death meant the pairs could not be kept for the full 10 weeks. (Two of the *E. fetida* × *E. fetida* crosses being kept for 52 and 53 days and three of the *E. andrei*$_2$ × *E. andrei*$_2$ crosses for 38, 56 and 59 days). In additional one of the five *E. andrei*$_1$ × *E. andrei*$_2$ replicates was lost due to mortality.

Results

Cocoon fertility

The worm pairs were kept for 10 weeks. Cocoons were removed from the substrate twice a week during that time and kept in petri dishes until sufficient time had elapsed to allow all fertile cocoons to hatch. There was no significant heterogeneity between the six crosses with respect to the numbers of cocoons produced per unit time per pair

Table 3. Mean hatchling weights and mean number of offspring per cocoon for the six crosses given with the associated mean and parental weights

Cross	No. of cocoons	Mean cocoon wt. gms. × 10^{-3}	Std. error	Mean parental wt. gms.	Std. error	Mean no. hatchlings per cocoon	Std. error	Mean hatchling wt. gms. × 10^{-3}	Std. error	No. of hatchlings
E. fetida × *E. fetida*	111	21.58	4.23	0.7654	0.0082	4.549	0.190	2.859	0.081	505
E. fetida × *E. andrei*$_1$	83	18.62	3.73	0.7595	0.0090	3.337	0.186	3.118	0.124	277
E. fetida × *E. andrei*$_2$	14	19.36	10.51	0.7706	0.0539	2.071	0.339	5.510	0.673	29
E. andrei$_1$ ×	90	17.60	2.82	0.7146	0.0102	3.067	0.237	3.288	0.420	276
E. andrei × *E. andrei*$_2$	59	16.37	4.73	0.6789	0.0138	2.559	0.075	3.321	0.191	151
E. andrei$_2$ × *E. andrei*$_2$	84	20.50	4.58	0.8199	0.0141	2.809	0.186	3.063	0.178	236

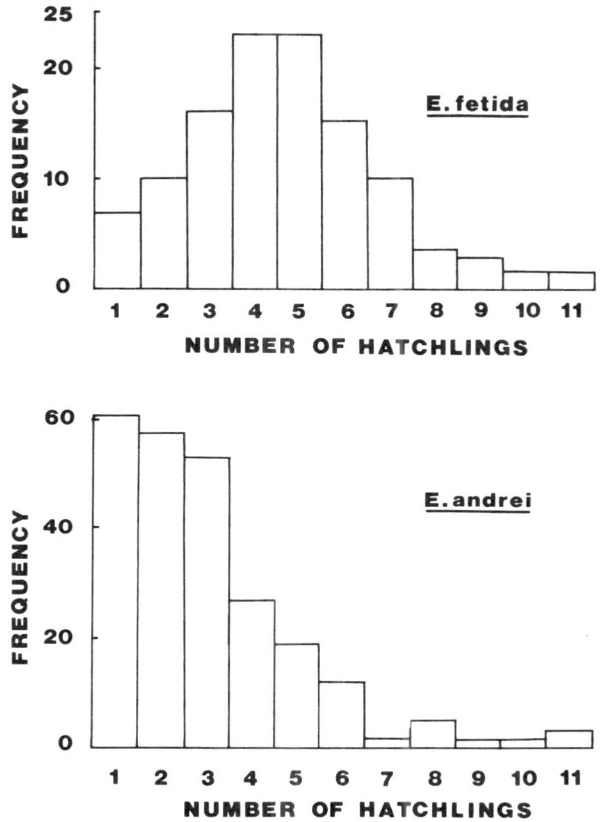

Fig. 1. Histograms showing the number of hatchlings per cocoon for *E. fetida* and *E. andrei*. The *E. andrei* histogram is a summation of the three intra-specific *E. andrei* crosses.

of worms ($F_{2,23}$ = 2.16 N.S.; Table 2). There was significant heterogeneity in the percentage fertility of the crosses ($F_{5,23}$ = 6.88 $p < 0.001$; Table 2). The intra-specific crosses did not differ, but the inter-specific *E. fetida* × *E. andrei*$_2$ cocoons were significantly less fertile than cocoons from all the other sources.

A cocoon's weight did not affect its chances of hatching. The mean weights from fertile and infertile cocoons were 19.23 mgs. (n = 441) and 19.70 mgs. (n = 335) respectively (T_{774} = 1.49 N.S.). There was no evidence for stabilizing selection as the variance ratios were not significant. $F_{334,441}$ = 1.09 N.S. The variances for fertile and infertile cocoons were 17.45 and 19.10 mgs. respectively.

Number of hatchlings per cocoon

E. fetida and *E. andrei* were highly fecund lumbricids, regularly producing more than one offspring per cocoon (Evans and Guild, 1948; Hartenstein *et al.*, 1979; Mori and Okada, 1978; Nakamura, 1973; Neuhauser *et al.*, 1980; Tsukamoto and Watanabe, 1977; Vail, 1972; Watanabe and Tsukamoto, 1976). There was a significant difference between *E. fetida* and *E. andrei* with respect to numbers of hatchlings per cocoon ($F_{1,413}$ = 62.11 $p < 0.001$). *E. fetida* produced a mean of 4.55 worms per cocoon (n = 111), compared with 2.86 worms per cocoon per *E. andrei* (Fig. 1).

This can be explained partly by the larger cocoon weight of *E. fetida* in this experiment (Table 3). In '*E. foetida*' the number of offspring per cocoon has been shown to increase with cocoon weight. Cocoon weight in turn increases with parental weight (Hartenstein *et al.*, 1979). Similar relationships were found when my data were analyzed, but the r values are much smaller (Table 4), thus, although cocoon weight influences the number of progeny per cocoon a large proportion of the variation remains unexplained.

The larger number of hatchlings per cocoon of *E. fetida* does not necessarily lead to a higher net reproductive output. There was no significant heterogeneity between the crosses, when numbers of offspring surviving to maturity per cocoon was considered ($F_{6,49} < 1$ N.S.) Worms from cocoons producing large numbers of progeny tend to be lighter at birth (Table 3) and this affects their chances of survival. The mean birth weight of worms surviving to sexual maturity was 4.4 mg (n = 84) compared with 2.8 mg (n = 104) for those that died ($F_{2,185}$ = 16.22 $p < 0.001$). Almost all mortality occurred within the first three weeks after hatching.

Pigmentation in F_1 generation

Both *E. fetida* and *E. andrei* in inter-specific crosses bred true for pigmentation pattern, their offspring resembling the parental type. Twenty worms from *E. fetida* crosses and 47 from *E. andrei* crosses were raised to sexual maturity.

From the inter-specific *E. fetida* × *E. andrei*$_1$ cross 17 worms were raised to sexual maturity. Ten of them resembled their *E. fetida* parent and seven of them their *E. andrei* parent. At no time did a single cocoon produce both *E. fetida* and *E. andrei* offspring.

Discussion

Interspecific comparisons

The two species of worm *E. fetida* and *E. andrei* had similar reproductive performances with respect to numbers of cocoons produced per unit time and cocoon viabilities. The values for cocoon viability were 82.2% (n = 147) for *E. fetida*, and 71.4% (n = 133), 75.7% (n = 71) and 73.4% (n = 118) respectively for the three *E. andrei* crosses. These are similar to those reported by Hartenstein *et al.* (1979) of 81.5% (n = 300) and Vail (1974) of 78.5% (n = 252). Reinecke and Kriel (1981), using incubators kept at either a constant or mean value of 20°C, obtained an overall cocoon viability of 94.9% (n = 98). This higher value could be explained by the work of Tsukamoto and Watanabe (1977), who found higher cocoon fertilities to be associated with lower temperatures. Temperature differences, along with the correlation between per cent cocoon hatching (r = 0.99 $p < 0.001$) found by Madhab and Bikram (1980) in tropical earthworms, may help to explain the seasonal variation in cocoon viability found by Watanabe and Tsukamoto (1976). The viability was 81%

Table 4. Regression equations for cocoon weight (CW) against parental weight (PW) and number of progeny against cocoon weight. (All measurements are in mg)

Cross	Regression equations	Value		Significance
E. fetida ×	CW = 0.01424 PW + 10.87	0.277	147	$p < 0.001$
E. fetida	No. of progeny = 0.0971 CW + 2.454	0.216	111	$p < 0.05$
E. fetida ×	CW = −0.00109 PW + 20.59	−0.0406	172	N.S.
E. andrei$_1$	No. of progeny = 0.0052 CW + 3.240	0.102	83	N.S.
E. fetida ×	CW = −0.00057 PW + 19.54	−0.0016	104	N.S.
E. andrei$_2$	No. of progeny = −0.0217 CW + 2.496	−0.0673	14	N.S.
E. andrei$_1$	CW = 0.00905 PW + 11.07	0.323	132	$p < 0.001$
E. andrei$_1$	No. of progeny = 0.1914 CW − 0.302	0.227	90	$p < 0.05$
E. andrei$_1$	CW = 0.01579 PW + 6.02	0.457	84	$p < 0.001$
E. andrei$_2$	No. of progeny = 0.0158 CW + 2.301	0.038	59	N.S.
E. andrei$_2$	CW = 0.03790 PW + 17.15	0.126	137	N.S.
E. andrei$_2$	No. of progeny = 0.0547 CW + 1.688	0.138	84	N.S.
Hartenstein	CW = 0.0106 PW + 6.48	0.97		
et al. (1979)	No. of progeny = 0.270 − 0.576	0.61		$p < 0.05$

$n = 18$) in the winter, dropping to a minimum of 33.3% ($n = 6$) in the summer.

There was a significant difference in the numbers of progeny per cocoon produced by the two species. Although part of this variation can be explained by a significant regression with cocoon weight, a large proportion of the variation remains unexplained. The number of progeny per cocoon varied with temperature. The higher the temperature the shorter the development time and the lower the number of hatchlings (Tsukamoto and Watanabe, 1977). For this reason, it is important to consider temperature when viewing figures for numbers of progeny per cocoon. Reinecke and Kriel (1981) have shown that a constant temperature of 20°C, as compared to a mean temperature of 20°C. (range 12–28°C) can cause significant changes in numbers of progeny per cocoon.

From Table 5 it can be seen that values for *E. andrei* are within the range reported by other authors (though which of the two species they were working with is unclear). The value for *E. fetida*, however, falls outside this range. It is important to note that in this study all progeny were recorded and weighed, and even those described by Vail (1974) as resembling 'an amorphous mass of protoplasm rather than worms'. She reported that such worms did not survive more than 24 hours. They are, however, sometimes capable of developing into adult worms. A worm weighing only 0.6 mg and another 0.8 mg at birth were both raised successfully to sexual maturity in this experiment. The other ten worms weighing 1 mg. or less failed to survive.

There appears, therefore, to be a decreasing probability of survival with decreasing weight down to a very low weight rather than a minimum birth weight that must be achieved if further development is to be successful. Mortality was almost totally confined to the first three weeks of life.

Interspecific matings

In the first experiment performed by André (1963), interspecific matings produced eight hybrid worms. These differed from the two parental types in that they were sterile and the pigmentation pattern was intermediate, resembling *E. andrei* in the

Table 5. *Overall mean from cocoons collected throughout the year, but hatched in an incubator at 20°C

Incubation temperature	Number of cocoons	Average number of progeny per cocoon	Author(s)
20°C	281	2.53*	Watanabe and Tsukamo (1976)
	198	2.6	Vail (1974)
	139	2.68	Evans and Guild (1948)
		2.9	Mori and Okada (1978)
18°C approx.	103	3.835	Nakamura (1973)
20°C constant	23	3.1	Reinecke and Kriel (1987)
20°C mean (range 12–28)	25	2.74	
24°C	111	4.55	*E. fetida*
24°C	232	2.86	*E. andrei*

mid-line but narrowing to an *E. fetida* like band laterally. The cocoon fertility was very low for this experiment, only five fertile cocoons being produced (n = 137), from eleven pairs of worms, one pair producing all the fertile cocoons. In a later experiment André (1963) showed such hybrids to be the result of male *E. fetida* gametes and female *E. andrei* gametes, the reverse cross being infertile.

In André's second experiment, however, 16 inter-specific crosses produced 61 fertile cocoons (n = 377, eight crosses being fertile). All these offspring resembled one or other of the parental types and were fertile. This was attributed to facilitated self-fertilization. My results are in agreement with André's (1963) second experiment in that the offspring consisted of fertile worms resembling the parental types. I have used the term 'facilitated self-fertilization' to indicate self-fertilization arising from spermatozoa mixing during normal mating. This distinguishes it from self-fertilization achieved by isolated worms, presumably as a consequence of self-insemination such as that shown in *Dendrobaena rubida f. subrubicunda* by André and Davent (1972). In the former a partner is needed, even a sterile partner sufficing; in the latter isolated worms are capable of producing fertile cocoons. Herlant-Meewis (1956) and André (1963) failed to find any evidence of non-facilitated self-fertilization. On the other hand, Hartenstein *et al.*, (1980) have shown that self-fertilization can occur and have raised worms in isolation to the F_2 generation.

There are, therefore, two methods by which the interspecific crosses in my experiment could have produced offspring in addition to hybridization. Which of the two methods was used is immaterial when considering the genetics of the two species as they are effectively reproductively isolated. The rare hybrids which could have been produced are sterile (André, 1963).

Summary

E. fetida and *E. andrei* were crossed both inter- and intra-specifically by means of single pair matings. Cocoons were collected over a 70-day-period and kept a 24 ± 2°C. The two species produce similar numbers of cocoons. *E. fetida* produced a higher number of progeny per cocoon of 4.55 (n = 111), compared with 2.86 (n = 232) for *E. andrei*. This was associated with a lower mean birth weight of 2.859 mg

(n = 505), compared with 3.576 (n = 663) in *E. andrei*. Weight at birth influenced a young worm's chance of surviving to maturity, the birth weights of worms failing to survive to maturity (mean = 2.8 mg n = 104) being significantly lower than the birth weights of those succeeding (mean = 4.4 mgs. n = 84). For this comparison, $F_{2,185} = 16.22$, $p < 0.001$.

The number of progeny surviving to maturity per cocoon was not significantly different between the crosses ($F_{5.50} < 1$ N.S.) The net result is that the differences tend to cancel each other, so that the number of offspring per worm surviving from egg to adult is approximately similar in the two species under these experimental conditions. Inter-specific matings showed reduced cocoon fertilities. These differences support the evidence for the specific status of the two species by demonstrating ecological distinctions between them.

Acknowledgments

I would like to thank my Doctoral supervisor, Dr. L.M. Cook, for his advice and support throughout the work. Mr. G. Blower and Dr. J.E. Satchell for their helpful comments. Saad A. Said assisted me in the daily collection of results and this was invaluable. The work was funded by N.E.R.C.

References

André, F. 1963. Contribution à l'analyse experimentale de la réproduction des lombriciens. Bull. Biol. Fr. Belg. 97: 1–101.

André, F., Davant, N. 1972. L'autofécondation chez les lombriciens. Observation d'un cas d'autoinsémination chez *Dendrobaena rubida f. subrubicunda* Eisen. Bull. Soc. Zool. France 97: 725–728.

Bouche, M.B. 1972. Lombriciens de France. Écologie et systématique. Ann. Zool. Ecol. Anim. (Num. spec.) 72-2 671 pp.

Easton, E.G. 1983. A guide to the valid names of Lumbricidae (Oligochaeta). *In* Earthworm Ecology from Darwin to Vermiculture. pp. 475–485. Edited by J.E. Satchell. Chapman and Hall, London.

Evans, A.C. and McL. Guild, W.J. 1948. Studies on the relationships between earthworms and soil fertility. IV. On the life cycles of some British Lumbricidae. Ann. Appl. Biol. 35(4): 471–484.

Harman, W.J. 1955. Earthworms of commercial importance and their effect on distribution. *In* Proceedings of the Louisiana Academic of Science. pp. 54–57.

Hartenstein, H., Neuhauser, E.F. and Easton, E.G. 1980. Growth and fecundity of F_2 *Eisenia foetida* derived from F_1's both reared in isolation from birth. Megadrilogica 3(11): 185–187.

Hartenstein, R., Neuhauser, E.F. and Kaplan, D.L. 1979. Reproductive potential of the earthworm *Eisenia foetida. Oecologia* (Berl.) 43: 329–340.

Herlant-Meewis, H. 1956. Croissance of réproduction du lombricien *Eisenia foetida* (Sav.). Ann. Sci. Nat. Zool. Biol. Anim. 18: 185–198.

Herlant-Meewis, H. 1962. Influence de la nutrition sur la réproduction chez *Eisenia foetida*. C.R. Acad. Sci. Paris 255: 2187–2188.

Jaenike, J. 1982. *Eisenia foetida* is two biological species *Megadrilogia* 4: 6–7.

Madhab, P.C. and Bikram, S.C. 1980. Cocoon morphology hatching and emergence pattern in tropical earthworms. Pedobiologia 20: 316–324.

Mori, T. and Okada, M. 1978. Studies on the feasibility of treatment of sewage sludge by earthworms. (1). Study of procedures for rearing earthworms in sewage sludge compost and of various factors related to productivity. Water and Sewage Research 1(2): 127–132 (Japanese).

Nakamura, Y. 1973. On the number of worms emerging from individual cocoons of several lumbricid species in Japan. Appl. Ent. and Zool. 8(3): 199–200.

Neuhauser, E.F., Hartenstein, R. and Kaplan, D.L. 1980. Growth of the earthworm *Eisenia fetida* in relation to population density and food rationing. Oikos 35: 93–98.

Reinecke, A.J. and Kriel, J.R. 1981. The influence of constant and diurnally fluctuating temperatures on the cocoon production, hatching time and number of hatchlings of *Eisenia foetida* (Lumbricidae, Oligocaeta). In Proc. 1st. Int. Workshop on Role of Earthworms in stabilizing organic residues. Western Michigan University, Kalamazoo. pp. 167–177.

Roch, Ph., Valembois, P. and Lassegues, M. 1980. Biochemical particularities of the antibacterial factor of the two subspecies *Eisenia fetida fetida* and *Eisenia fetida andrei*. Am. Zool. 20: 794.

Sims, R.W. 1983. The scientific names of earthworms. *In* Earthworm Ecology from Darwin to Vermiculture. pp. 467–474. Edited by J.E. Satchell. Chapman and Hall, London.

Tomlin, A.D. 1983. The earthworm bait market in North America. *In* Earthworm Ecology from Darwin to Vermiculture. pp. 331–338. Edited by J.E. Satchell. Chapman and Hall, London.

Tsukamoto, J. and Watanabe, H. 1977. Influence of temperature on hatching and growth of *E. foetida* (Oligochaeta, Lumbricidae). Pedobiologia 17: 338–342.

Vail, V.A. 1972. Contributions on North American Earthworms (Annelida). Observations on the Hatchlings of *Eisenia foetida* and *Bimastos tumidus* (Oligochaeta lumbricidae). Bull. Tall Timbers Res. Stn. No. 11.

Valembois, P., Roch, P., Lassegues, M., Davant, N. 1982. Bacteriostatic activity of a chloragogen cell secretion. Pedobiologia 24: 191–197.

Watanabe, H. and Tsukamoto, T. 1976. Seasonal change in size, class and stage structure of lumbricid *E. foetida* population in field compost and its practical application as the decomposer of organic waste matter *Rev. Ecol. Biol. Sci.* 13(1): 141–146.

II. Engineering waste management by earthworms

CONTENTS

STABILIZATION OF LIQUID MUNICIPAL SLUDGE USING EARTHWORMS*

RAYMOND C. LOEHR*, JOHN H. MARTIN, Jr. and EDWARD F. NEUHAUSER**
Cornell University, Ithaca, NY 14853

1. Introduction

Biodegradable organic matter, such as animal and vegetable wastes and municipal sludge, can be stabilized by earthworms (vermistabilization). The organic matter is inoculated with the appropriate species of earthworms and the worms derive their nourishment from the microorganisms that participate in the waste degradation process and from the organic matter that is being decomposed.

The worms maintain aerobic conditions in the mixture, ingest solids, convert a portion of the organics into worm biomass and to respiration products, and expel the remaining partially stabilized matter as discrete material (castings). The worms and the microorganisms act symbiotically to accelerate and enhance the decomposition of the organic matter. The degree to which the organic matter is degraded is a function of (a) the portion of the waste that is biodegradable, (b) maintenance of aerobic conditions, and (c) avoidance of toxic conditions.

Vermistabilization represents a technology that is environmentally sound, need not be energy, capital or equipment intensive, and should not require extensive management. It is a relatively new technology that can be classified as an innovative and alternative technology.

Earthworms have been used for waste stabilization for many years, especially in the Philippines and other countries in southeast Asia. The process also is being used in Italy, England, and the Netherlands not only to stabilize wastes but also to produce castings for horticultural purposes. In the United States, a feasibility study (Camp *et al.*, 1981; Pincince *et al.*, 1980) evaluated the process for sludge management and indicated that the operating costs of a practical system may be competitive with other sludge management options for certain communities.

Scientific studies have helped establish the technical basis for vermistabilization. Research (Hartenstein *et al.*, 1979; Neuhauser *et al.*, 1980; Kaplan *et al.*, 1980) has shown that aerobic sewage sludge can supply the nutrients necessary for the growth and reproduction of earthworms of the genera *Eisenia, Eudrilus* and *Pheretima*. In addition, studies have identified the effect of aged sludge on earthworm growth and the impact and bioaccumulation of metals on earthworm growth and reproduction (Hartenstein *et al.*, 1980; Malecki *et al.*, 1982). Earthworms also have accelerated the decomposition of sewage sludge in drying beds (Mitchell *et al.*, 1980).

The role of earthworms in vermistabilization is linked to the aerobic condition of the sludge and the moisture content. In sludges which are aerobic, the worms have a strong and positive effect. Earthworms can not exist in anaerobic sludges and will

* *Present address:* University of Texas, Austin, TX 78712.
** *Present address:* Niagara Mohawk Power Corporation, Syracuse, N.Y., 13202.

Earthworms in waste and environmental management
edited by C.A. Edwards and E.F. Neuhauser, pp. 95–110

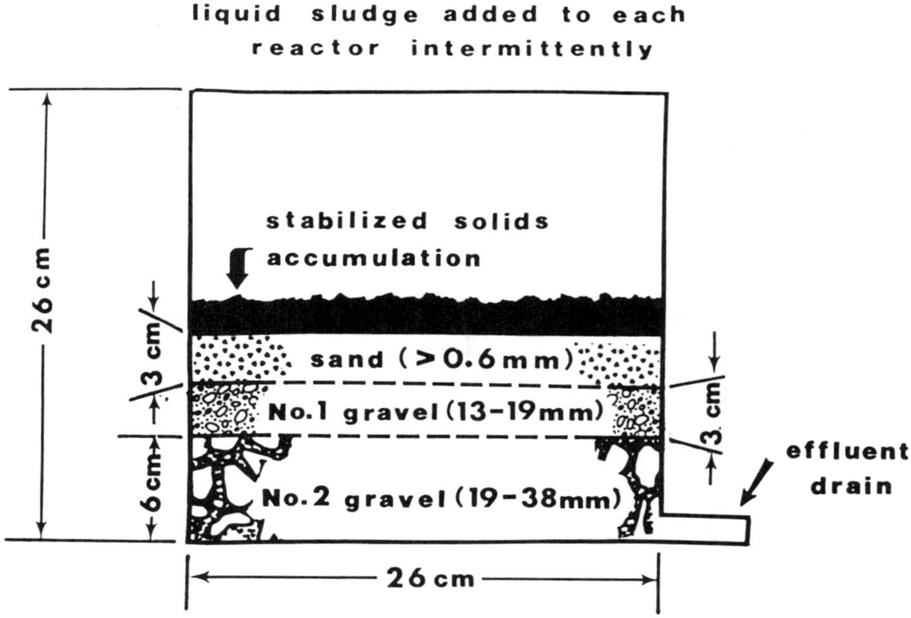

Fig. 1. Schematic of reactors used in the liquid vermistabilization studies.

not have a positive effect until aerobic conditions occur. Very high and very low moisture concentrations also will inhibit the earthworms.

2. Objectives

The research in this paper was part of a project to assess the feasibility of using earthworms to stabilize (vermistabilization) waste-water treatment sludges (Loehr *et al.*, 1984). The project (a) investigated the performance of the vermistabilization process using stabilized and unstabilized sludges and (b) developed a basis for the design and operation of vermistabilization processes. A major project goal was to develop a rational basis that would permit the performance and economics of this process to be compared to more conventional sludge stabilization processes.

Research was conducted along two complementary lines: (a) basic studies to identify fundamental factors that affect the performance of the vermistabilization process and (b) applied studies to determine design and management relationships. This paper emphasizes the results from the applied studies that used liquid municipal sludges.

3. Materials and methods

Liquid primary, waste-activated and aerobically digested sludges were evaluated in liquid sludge vermistabilization (LSVS) experiments. The sludges were obtained weekly from a well-operated publicly owned treatment works (POTW) in a municipality that contained virtually no industry. While being held in the laboratory, the liquid sludges were aerated mildly to prevent septic conditions.

Table 1. Operational details of the liquid sludge vermistabilization studies

Reactors	– Three with primary sludge, three with waste-activated sludge, three with aerobically-digested sludge and three controls (one with each of the sludges but with no worms added)
Substrate	– Primary, waste-activated and aerobically-digested sludges obtained weekly from a small municipal treatment plant. Little industry contributed wastes to this plant. The liquid sludge was added to the reactors at intervals varying from three to five times a week. No dewatering occurred before the liquid sludges were added to the reactors. While being held at the Cornell laboratories, the sludges were aerated mildly to prevent septic conditions.
Worms	– Between 80 and 200 grams (liveweight) of *Eisenia fetida* were added to each reactor except the controls. The worms were of mixed ages and sizes. No additional worms were added after the reactors were started.
Loading rate	– Each reactor received a different loading rate. After an equilibrium period lasting many weeks, the loading rate of some reactors was increased. This procedure continued until a reactor failed due to an excessive loading rate or until the project ceased.
Analyses	– The input liquid sludge and the liquid draining from the reactors was analyzed once per week. The accumulated sludge in the reactors that did not fail was analyzed at the end of the experiment.
Temperature	– Controlled at 20°C ± 1°C.

Eisenia fetida was the earthworm species used in these experiments. Other studies with mixed species (Loehr *et al.*, 1984) indicated that a mixture of earthworm species did not have any advantage over a single species in terms of the degree of waste stabilization. Previous research also had indicated that *E. fetida* was a species that performed well in the vermistabilization process.

Reactors about 10 inches (26 cm) in diameter and 10 inches (26 cm) high were used. The reactors had a solid bottom and a drain (Fig. 1). Layers of sand and gravel were placed in the reactors to retain the solids and to allow drainage of the water in the liquid sludge. The earthworms came from stock cultures maintained in the authors' laboratory. Worms of mixed ages and sizes were added to each reactor. No worms were added thereafter. The worms grew and reproduced in response to the substrate and conditions in each reactor.

Three reactors were set up for each type of sludge. In addition, three control reactors, one for each type of sludge but without worms being added, were set up. The details of the study are summarized in Table 1.

The sludges added to the reactors and the reactor contents were analyzed for a number of parameters to assess the performance of the LSVS process. The parameters and the analytical methods that were used are noted in Table 2.

4. Results and discussion

General

Current sludge management methods include dewatering, aerobic or anaerobic digestion and ultimate disposal. Many of these processes (dewatering, aerobic digestion) are energy intensive. Less energy and capital intensive sludge management processes could reduce sludge management costs.

Earthworms can exist under high moisture conditions and with many forms of

Table 2. Summary of analytical methods

Total solids	Total residue dried at 103–105°C – *Standard methods* (1980)
Volatile and fixed solids	Volatile loss and fixed residue at 550°C – *Standard methods* (1980)
Total Kjeldahl nitrogen	Semi-micro Kjeldahl technique – McKenzie and Wallace (1954) and *Standard methods* (1980)
Ammonia nitrogen	Steam distillation with phosphate buffer – *Standard methods* (1980) and Prakasam *et al.* (1975)
Nitrite nitrogen	Diazotization method – Montgomery and Dymock (1961)
Nitrate nitrogen	Salicylic acid method – Carone and Raquet (1936)
Chemical oxygen demand	Dichromate reflux method – *Standard methods* (1980)
pH	Glass electrode – *Standard methods* (1980)
Dissolved oxygen	Membrane electrode method – *Standard methods* (1980)

Table 3. Characteristics of the wastewater treatment sludges used for the liquid sludge vermistabilization studies – (Mean ± Standard deviation)

Parameter	Primary sludge*	Waste-activated sludge**	Aerobically-digested sludge**
Total solids, mg/l	12,600 ± 2,800	6,350 ± 1,230	8,530 ± 820
Volatile solids, mg/l	10,040 ± 2,350	4,780 ± 980	5,990 ± 770
Volatile solids, % of total solids	79.0 ± 3.8	75.3 ± 2.5	65.4 ± 3.8
Chemical oxygen demand, mg/l	14,760 ± 3,860	6,770 ± 1,440	7,760 ± 1,390
Total Kjeldahl nitrogen (mg/l)	540 ± 150	460 ± 90	470 ± 85
Ammonia nitrogen, (mg/l)	70 ± 30	40 ± 15	9 ± 7
Nitrite & nitrate (mg/l)	0	2 ± 6	77 ± 55
pH	6.7 ± 0.5	7.4 ± 0.2	6.6 ± 1.4

* $n = 28$
** $n = 29$

organic matter. These LSVS experiments were undertaken to determine the extent to which earthworms could stabilize liquid sludge, directly from the treatment plant. The objective was to determine whether dewatering is necessary for vermistabilization and to determine the performance of the process with both primary and biological sludges. Primary, waste-activated and aerobically-digested sludges were used as obtained. The characteristics of the sludges added to the reactors were determined throughout the study. The average characteristics of the specific sludges are summarized in Table 3.

The loading rate of the various reactors varied during the study. The liquid sludge loading rates used and the frequency of sludge application are indicated in Table 4. The period of time the reactors operated and the changes that occurred are identified in Table 5.

The liquid sludge was applied at frequencies of three and five times a week to determine if the frequency of application was an important operational consideration (Table 4). Sludge was applied three times per week to the noted reactors for the first

Table 4. Application of sludge to the liquid sludge vermistabilization units

Type of sludge	Reactor no.	Frequency of sludge application (No. of times/week)	Average loading rates to reactors			
			Liquid (l/m²/wk)	VS (gm/m²/wk)	COD (gm/m²/wk)	TKN (gm/m²/wk)
Primary	30 (Control)	5	37	367	620	23
	31	3	37	381	553	21
	32	3	58	583	838	34
		5	69	577	781	37
	33	3	77	775	1115	45
		5	84	780	1184	50
	33A	5	98	984	1598	61
	34	5	126	1132	1672	69
Waste activated	20 (Control)	5	93	347	641	38
	21	3	79	337	473	39
	22	3	114	526	734	58
		5	100	550	765	54
	23	3	152	702	979	77
		5	158	744	1084	76
	23A	5	214	887	1359	102
	24	3	222	1087	1511	115
		5	187	1110	1524	108
Aerobically digested	10 (Control	5	70	376	516	38
	11	3	74	431	616	43
	12	3	103	607	868	61
		5	119	591	770	60
	13	3	138	809	1157	81
		5	146	767	1035	77
	14	3	162	956	1368	95
		5	183	961	1296	97
	15	5	205	1134	1574	114

(See text for explanation of numbering of reactors and frequency of application)

eight to ten weeks of the period of operation and was applied five times per week for the remainder of the period.

The LSVS reactors were operated to identify the maximum loading rate that would permit successful performance. Low loading rates were used initially. If, after a suitable equilibrium period, the reactors seemed to function satisfactorily at a specific loading rate, the loading rate was increased. For example, reactors 31, 32, and 33 were able to treat (stabilize) liquid primary sludge at the initial loading rates. After a 76-day-period of operation, the loading rate of reactor 31 was increased almost four times. To avoid confusion, reactor 31 was renumbered and became reactor 34. Ultimately that loading rate caused failure; (Table 5) *i.e.*, the worms died. However, failure did not occur until 71 days after the loading rate was increased.

In a similar manner, after 124 days of successful operation, the loading rate to reactor 32 was increased and reactor 32 was renumbered reactor 33A to reflect the fact that the loading rate was between that of reactors 33 and 34. The comparable changes that were made with the reactors stabilizing waste-activated sludge and liquid aerobically-digested sludge are identified in Table 5.

Table 5. Period of operation of the liquid sludge vermistabilization reactors

Reactor number	Period of operations (days)	Operational changes
Primary sludge		
30 (Control)	20	Failed
31	76	No failure, became reactor 34
32	124	No failure, became reactor 33A
33	182	No failure, project ceased
33A	58	No failure, project ceased
34	71	Failed
Waste-activated sludge		
20 (Control)	20	Failed
21	48	No failure, became reactor 24
22	140	No failure, became reactor 23A
23	198	No failure, project ceased
23A	51	Failed
24	30	Failed
Aerobically-digested sludge		
10 (Control)	57	Failed
11	48	No failure, became reactor 14
12	140	No failure, became reactor 15
13	189	No failure, project ceased
14	140	No failure, project ceased
15	49	No failure, project ceased

The control reactors, *i.e.*, those containing no worms, failed in a short period of time (Table 5). However, many of the LSVS reactors functioned satisfactorily for long periods of time (140–198 days) and several operated successfully until the project ceased. For example, a reactor receiving liquid primary sludge (reactor 33) operated successfully for 182 days; reactor 23 operated successfully for 198 days, and reactor 13 operated successfully for 189 days.

The studies did identify loading rates at which failure occurred and therefore identified the upper limit of loading rates at which the LSVS process should function successfully. Failure occurred in the reactors when the liquid no longer would flow through the accumulated sludge solids, ponding resulted and anaerobic conditions developed. When the LSVS reactors functioned well, most of the water added with the liquid sludge drained quickly (within 60 minutes) and aerobic conditions were maintained. The total quantity of drainage and its characteristics were determined routinely.

Several symptoms of incipient failure occurred before the reactors actually failed. These symptoms were: (a) drainage of the applied water slowed and (b) the worms exhibited signs of stress, *i.e.*, some escaped through the drain and some died. These symptoms, and the change in characteristics of the drainage liquid, served as clues to the onset of LSVS process failure.

Because of the changes in loading rates that were made and the fact that failure occurred with several reactors, only six reactors were operating at the conclusion of the project, reactors 13, 14, 15, 23, 33, and 33A. The quantity and characteristics of the residual solids in five of these reactors (13, 14, 15, 33 and 33A) were analyzed

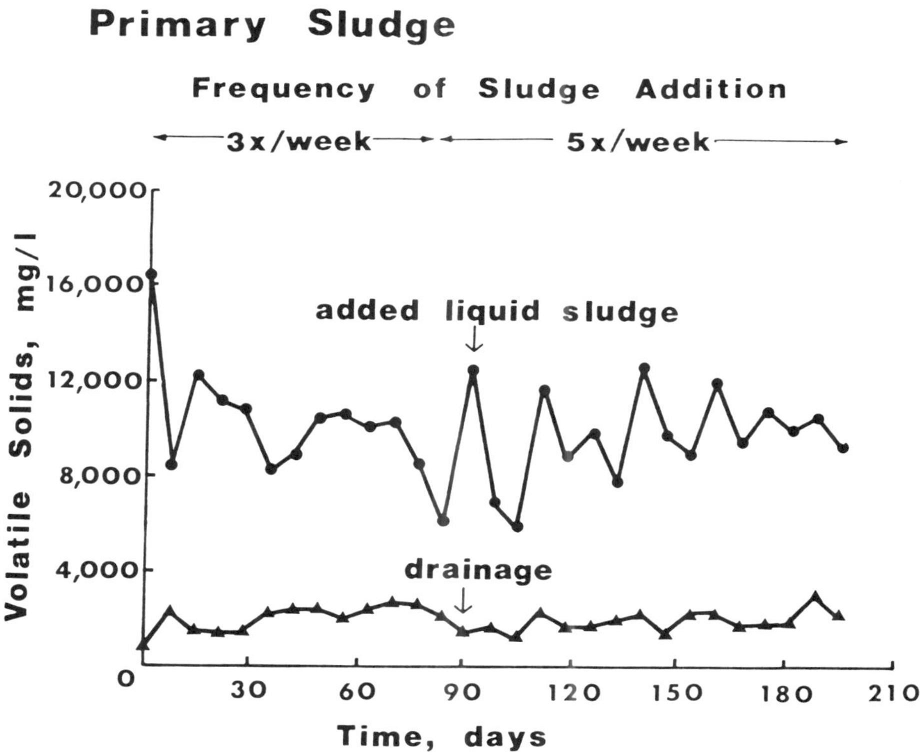

Fig. 2. Volatile solids removal pattern in LSVS units stabilizing primary sludge.

to determine the degree of sludge solids stabilization that had occurred. Care was taken to separate the residual solids from the sand and gravel and to remove the worms. The characteristics of the residual solids in reactor 23 could not be analyzed due to the small amount that had accumulated.

The performance of the LSVS reactors clearly indicated that the earthworms are a key component of the process since the controls failed more quickly than the reactors with the worms. Properly loaded LSVS reactors can operate for over six months at appropriate loading rates.

Of interest are the characteristics of the liquid that drained from the reactors and the characteristics of the vermistabilized sludge. The drainage characteristics are of interest since with a large scale vermistabilization system, the drainage will have to be returned to the wastewater treatment plant for additional treatment. The characteristics of the vermistabilized sludge are of interest because ultimately the sludge must be disposed of.

Drainage

The liquid that drained from the reactors following the liquid sludge applications contained the by-products of the stabilization that occurred in the reactors. Figures

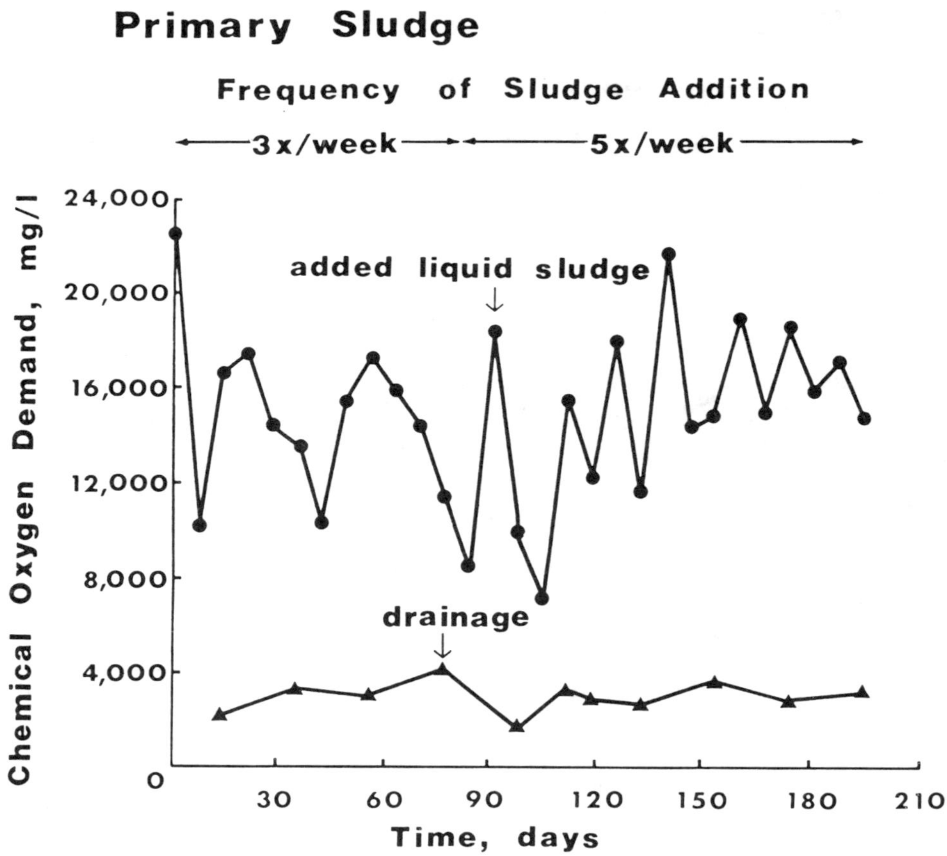

Fig. 3. COD removal pattern in LSVS units stabilizing primary sludge.

2 and 3 illustrate the pattern of characteristics of the applied liquid primary sludge and the resultant drainage.

With both the liquid waste-activated sludge and the liquid aerobically-digested sludge reactors, there was a short lag period as the reactors were started before equilibrium conditions became established. No lag period occurred when liquid primary sludge was used.

The average characteristics of the drainage from each reactor were determined. The averages represented steady state characteristics, *i.e.*, after any lag or start-up period. Except for ammonia and nitrite and nitrate nitrogen, no obvious differences were apparent as a function of frequency of sludge application. The drainage characteristics seemed to increase as the loading rate increased. Figure 4 illustrates the average results that were obtained. The LSVS reactors retained the applied solids satisfactorily and the drainage consisted of soluble constituents in the applied sludge and of by-products of the earthworm and microbial stabilization in the reactors.

The average characteristics of the drainage from the LSVS reactors are summarized in Table 6. The drainage from the activated sludge reactors had a greater con-

Table 6. Average characteristics of the drainage from the LSVS reactors

Parameter (mg/l)	Average drainage characteristics		
	Primary sludge	Waste-activated sludge	Aerobically-digested sludge
Total solids	3210	4100	3550
Volatile solids	1750	2750	2040
Chemical oxygen demand	3120	3730	1840
Organic nitrogen	160	230	115
Ammonia nitrogen	38	40	11
pH	7.6	7.3	6.9

centration of the constituents that were examined than did the drainage from the primary or aerobically-digested sludge reactors. Although the liquid primary sludge was the least stable and the most concentrated of the sludges added to the reactors, the drainage from the primary sludge reactors was not much different than that from the aerobically-digested sludge reactors. In addition, although there was a large difference in loading rates, there was not a large difference in drainage characteristics.

Table 7. Percent* of the applied parameters in the drainage from the LSVS reactors

Parameter (mg/l)	Primary sludge (%)	Waste-activated sludge (%)	Aerobically-digested sludge (%)
Total solids	25	65	42
Volatile solids	17	57	34
Chemical oxygen demand	21	55	24
Total Kjeldahl nitrogen	37	59	27
Organic nitrogen	34	53	25

* The parameter in the drainage (mg/l) as a percent of the concentration in the liquid sludge applied to the LSVS reactors.

The characteristics of the drainage can be put into perspective by comparing them to the characteristics of the liquid sludges applied to the LSVS reactors. This comparison is noted in Table 7 in terms of the average concentration of the parameter in the drainage as a percent of the average concentration in the liquid sludge applied to the reactors. With each of the liquid sludges, a sizable fraction of the applied material was retained in the reactors and stabilized by the worm and microorganisms.

There were, however, differences in the percentage of the applied material that appeared in the drainage. These differences may be due to the characteristics of the applied liquid sludges, such as the soluble fraction of the noted parameters. Except for nitrogen (Table 7), the amount of material retained in the primary sludge LSVS reactors was greater than that retained in the aerobically-digested and waste-activated sludges reactors. The waste-activated sludge reactors retained the least amount (35–50% retention) of the constituents in the applied sludges.

The drainage from the LSVS reactors is not of the quality for direct discharge

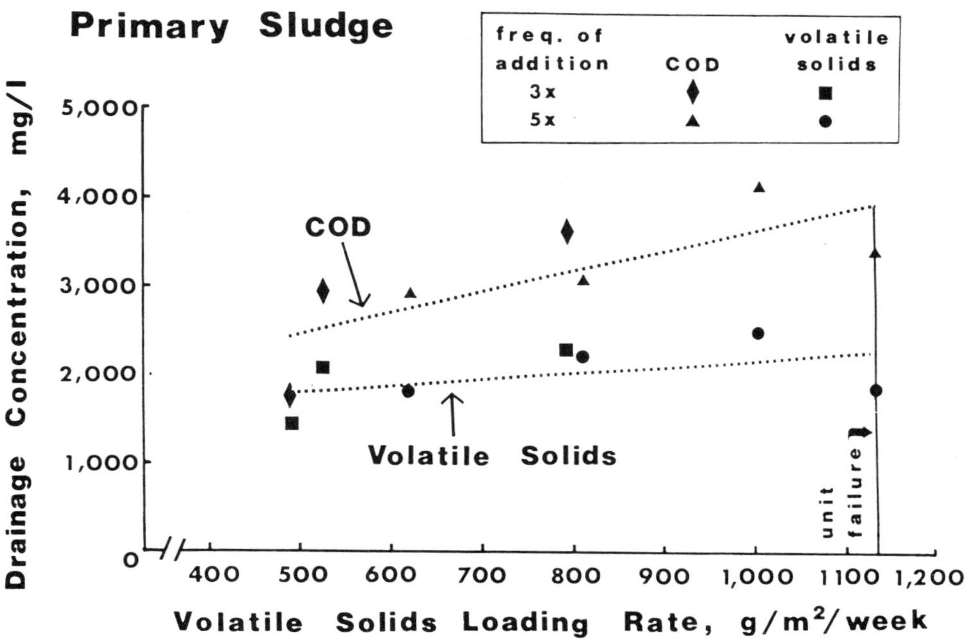

Fig. 4. Drainage characteristics as a function of loading rate for LSVS units stabilizing primary sludge.

(Table 6) and should be returned to the waste treatment plant for further treatment. Although the concentrations in the drainage are high, the solids and COD probably consist of stabilized constituents that will not add an extraordinary load to the waste treatment facilities. The organic and ammonia nitrogen will add an oxygen demand to the treatment facilities.

The actual load that the drainage will place on the treatment facilities is related to both concentration and flow. Water loss did occur in the LSVS reactors as a result of evaporation and the respiration of the worms and microorganisms. The average water losses ranged from 7–20% for primary sludge, 4–9% for waste-activated sludge, and 7–16% for aerobically-digested sludge. Generally, higher water losses occurred when the reactors received the liquid sludge less frequently (three times per week). The lowest water losses (3–4%) occurred in the reactors that failed.

Several of the reactors were loaded at rates that caused process failure (Table 5). However, drainage characteristics such as total solids, volatile solids, COD and organic nitrogen did not indicate any obvious difference when failure occurred. Thus, these characteristics are not good indicators of potential process failure.

The ammonia, nitrite and nitrate nitrogen concentrations as a function of loading rate provided an interesting pattern (Figs 5 and 6). Again the frequency of waste application did not cause obvious differences. However, variations in loading rate did. As the loading rate increased, the nitrite and nitrate nitrogen concentrations in the drainage decreased. In the primary sludge reactors (Fig. 5), the ammonia concentration in the drainage increased at the higher loading rates and was the highest from the reactor that failed.

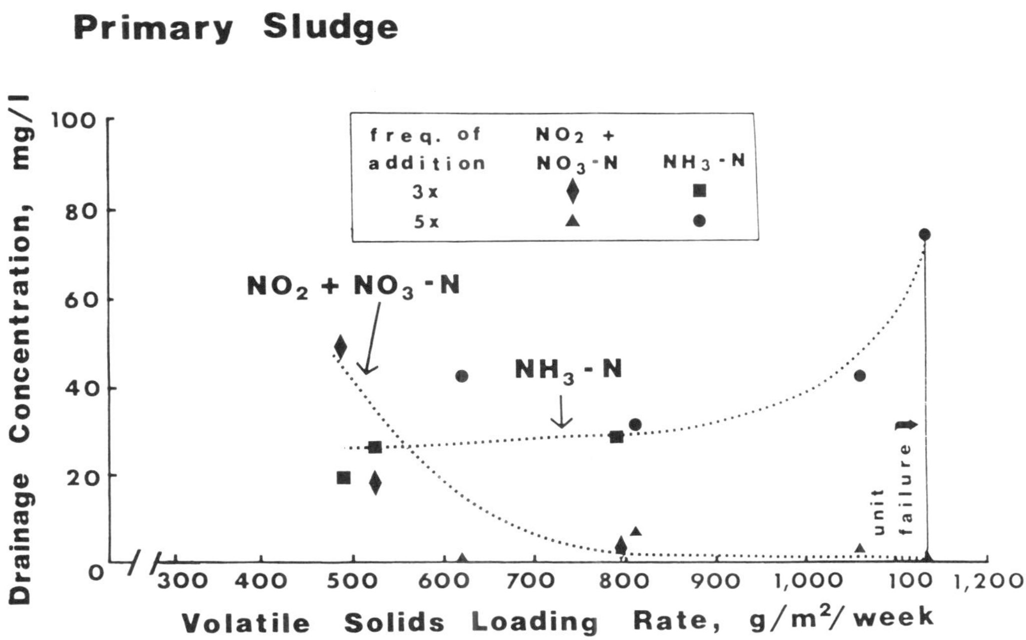

Fig. 5. Nitrogen concentrations in drainage from LSVS units stabilizing primary sludge.

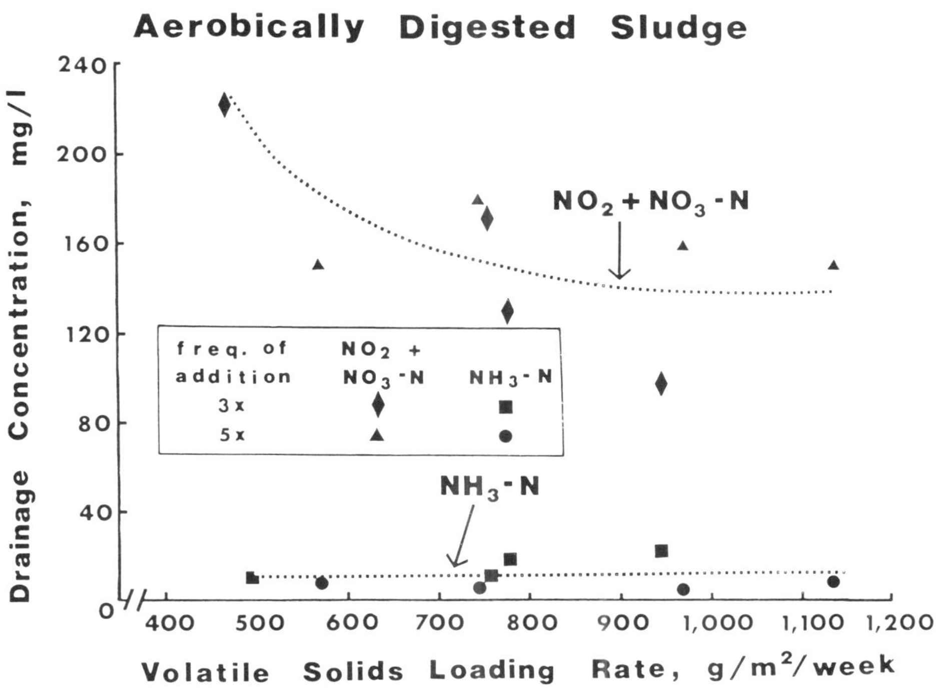

Fig. 6. Nitrogen concentrations in drainage from LSVS units stabilizing aerobically-digested sludge.

This pattern is consistent with the basic concepts of a vermistabilization reactor. In order for these reactors to function successfully, they must be kept aerobic. Under aerobic conditions, nitrification will occur. If anoxic conditions occur, the earthworms will die and the process will fail. The nitrite and nitrate concentrations of the drainage can be an indicator of general process performance as well as of possible process failure of the LSVS process.

If there is at least a trace of oxidized nitrogen in the drainage, aerobic conditions can be assumed in the LSVS reactors and the process should be functioning satisfactorily. All of the LSVS reactors that had high concentrations of oxidized nitrogen in the drainage did not fail and some that had low but measurable concentrations were able to function for extended periods of time.

The data indicated that:

a. loading rates as high as 1000 grams of volatile solids/m^2/week may result in satisfactory operation of LSVS reactors stabilizing liquid primary sludge (Fig. 5);
b. loading rates about 1000 grams of volatile solids/m^2/week or less will result in satisfactory operation of LSVS reactors stabilizing waste-activated sludge, and
c. loading rates considerably greater than 1200 grams of volatile solid/m^2/week will result in satisfactory operation of LSVS reactors stabilizing aerobically-digested sludge (Fig. 6).

Vermistabilized sludge

To estimate the biodegradability of the sludges used in the LSVS studies, batch aerobic digestion studies (without worms) were conducted. Samples of the primary, waste-activated and aerobically-digested liquid sludges that were applied to the LSVS reactors were aerobically-digested in separate reactors for thirty days. Two separate batch studies were run with each type of sludge and the contents of the batch digesters were analyzed periodically throughout the thirty days. Distilled water was added daily to compensate for evaporative losses. Dissolved oxygen measurements also were made daily to insure that a residual concentration always was present.

The biodegradable fraction of the sludges that were evaluated are presented in Table 8. The sludge that had the least amount of stabilization prior to these studies, *i.e.*, primary sludge, had the greatest biodegradable solids fraction.

The stabilization that occurred due to vermistabilization was determined by analyzing the residual solids in the primary and aerobically-digested sludge reactors that were operating satisfactorily at the end of the project (reactors 13, 14, 15, 33, and 33A), (Table 5). The amount of a constituent that was retained in each reactor

Table 8. Biodegradable fractions of the sludges used in the liquid sludge vermistabilization studies

Sludge	Biodegradable fraction			
	Total solids	Volatile solids	COD	Organic nitrogen
Primary	0.43	0.55	0.60	0.44
Waste-activated	0.27	0.35	0.69	0.63
Aerobically-digested	0.19	0.28	0.47	0.50

Table 9. Sludge stabilization[a] that occurred in the LSVS reactors

Parameter	Liquid aerobically-digested sludge[b]	Liquid primary sludge[c]
Total solids	0.24–0.42 (0.19)[c]	0.60–0.66 (0.43)
Volatile solids	0.36–0.60 (0.28)	0.76–0.81 (0.55)
COD	0.58–0.70 (0.47)	0.80–0.84 (0.60)
Organic nitrogen	0.60–0.75 (0.50)	0.71–0.77 (0.44)

[a] Stabilization of the material retained in the reactors. Numbers identify the biodegradable fraction as determined from the LSVS study.
[b] Range of data from reactors 13, 14, and 15
[c] Range of data from reactors 33 and 33A
[d] Data in parentheses indicates the biodegradable fraction as determined in batch studies (Table 8)

was determined by subtracting the average amount of that parameter in the drainage from the average amount that was added to each reactor. The amount that was degraded was obtained by subtracting the amount remaining at the end of the study from the amount retained, *i.e.*

amount retained = amount added – amount in the drainage

amount degraded = amount retained – amount recovered in the residue

The degradation (stabilization) that occurred in the vermistabilization reactors is summarized in Table 9. A significant amount of sludge stabilization occurred in the LSVS reactors. For comparison purposes, the biodegradable fraction of the sludge, as determined from the aerobic digestion studies (Table 8), also is presented in Table 9.

A greater degree of stabilization occurred in the LSVS reactors than in the batch aerobic digestion reactors. A possible reason for this difference is the long solids retention time that occurs in a vermistabilization unit. Several of the vermistabilization units functioned for long periods of time (reactor 33 – 182 days, reactor 23 – 198 days, reactors 13 and 14 – 189 and 140 days (Table 5)). Because liquid sludge continually was added to the reactors, the average solids retention time was about half of the time of operation. In any event, the solids retention times in some of the vermistabilization reactors were considerably greater than the retention times of the batch aerobic digesters (30 days).

The solids in vermistabilization reactors do not have to be removed and disposed of as often as those in conventional aerobic or anaerobic digesters. The rate of solids accumulation in the vermistabilization reactors depend on the liquid sludge application rate and the amount of inert and nonbiodegradable solids that are retained. Experience indicates that the rate of accumulation will be slow and accumulated solids removal can be as infrequent as 12 months depending upon the rate of accumulation and the available storage space for the residual solids in the LSVS reactors.

Because the LSVS process is a stabilization system, the characteristics of the material that drained from the reactor also should be considered in determining the overall amount of stabilization that occurred. When the degree of stabilization in the LSVS reactors includes the material in the drainage and is compared to that achieved by aerobic digestion (Table 10), the overall stabilization achieved by the LSVS reactors is about the same as that achieved by the batch aerobic digestion studies. Additional studies are needed to determine if there are any real differences.

Table 10. Overall sludge stabilization[a] that occurred in the LSVS reactors (%)

Parameter	Liquid aerobically-digested sludge[b]	Liquid primary sludge[c]
Total solids	15–22 (19)[d]	45–50 (43)
Volatile solids	24–33 (28)	59–66 (55)
COD	33–48 (47)	60–69 (60)
Organic nitrogen	46–52 (50)	50–60 (44)

[a] Stabilization that occurred in the LSVS reactors in terms of both the material retained and the material in the drainage as a percent of the amount added.
[b] Range of data from reactors 13, 14, and 5
[c] Range of data from reactors 33 and 33A
[d] Data in parentheses indicates the biodegradable fraction as determined in the batch studies (Table 8)

The average characteristics of the residual stabilized solids are compared in Table 11 and identify the degree of dewatering that occurred. The liquid primary sludge that was applied had a total solids content of about 1.3% and the vermistabilized solids at the end of the study had a total solids content of about 21–24%. Although there were not enough residual stabilized solids in the liquid waste-activated sludge reactors to obtain meaningful data, the stabilized solids content of such a reactor also should be in the range of the data reported in Table 11. On this basis, the dewatering that would occur would be from about 0.6% total solids to about 14–24% total solids.

Table 11. Physical and chemical characteristics of residual stabilized solids resulting from vermistabilization of liquid sludge

Parameters	Aerobically-digested sludge reactor			Primary sludge reactor	
	13	14	15	33	33A
Total solids, %wb[a]	16.1	14.4	16.2	23.6	21.4
Volatile solids, g/kg TS	478	549	488	457	490
Chemical oxygen demand, g/kg TS	622	729	622	606	626
Organic nitrogen, g/kg TS	30	35	27	31	34
NH_3-N, g/kg TS	0	0	0	0	0
NO_2 + NO_3-N, g/kg TS	Tr[b]	Tr	Tr	Tr	Tr
pH	6.6	6.6	7.1	6.9	7.0

[a] wb = wet basis
[b] Tr = trace

Except for the total solids content, the other parameters (expressed as grams per kilogram TS) are about the same suggesting that: (a) regardless of the type of the sludge, the vermistabilized solids will have similar characteristics and (b) the applied solids are stabilized to about the same extent with the LSVS process.

5. Summary

The results of these liquid sludge vermistabilization studies have provided useful information on the performance of the LSVS process and related design and operational characteristics.

- Earthworms are a key component of the LSVS process. Control reactors that did not contain worms failed in a short period of time, 20 days for liquid primary and waste-activated sludge and 57 days for liquid aerobically-digested sludge.
- The LSVS reactors that were not overloaded functioned successfully for long periods of time (140–198 days) and were stopped only because the end of the project occurred.
- Oxidized nitrogen (nitrates) in the drainage from the LSVS reactors indicates that aerobic conditions are being maintained in the reactors. A decrease or lack of nitrate in the drainage can be an indicator of possible process failure.
- Liquid primary sludge and liquid waste-activated sludge can be stabilized by the LSVS process.
- Liquid primary sludge was stabilized to about the same degree as was liquid aerobically-digested sludge in the LSVS process.
- Moisture balances indicated an overall moisture loss in the range of 4–20%.
- In terms of drainage characteristics (with the exception of nitrogen), the frequency of sludge application had no obvious effect.
- In terms of residual solids characteristics, the frequency of sludge application had no obvious effect.
- Loading rates of about 1000 grams of volatile solids/m^2/week or less resulted in satisfactory operation of LSVS reactors stabilizing liquid primary and liquid waste-activated sludge.
- Loading rates greater than 1200 grams of volatile solids/m^2/week can be used for LSVS reactors stabilizing liquid aerobically-digested sludge.
- The solids stabilization that occurs in LSVS reactors appears to be equal to or better than the stabilization that occurred in a laboratory batch aerobic sludge digestion evaluation.
- With LSVS reactors, the disposal of residual stabilized solids occurs at long intervals.
- The total solids content of the stabilized residual solids in the LSVS reactors was in the range of 14–24%, a considerable increase from the 0.6–1.3% that was added. The LSVS process is a successful dewatering as well as stabilization process.
- The stabilized residual solids had approximately the same characteristics regardless of the type of liquid sludge added to the reactors.

Acknowledgements

This report is based on research supported by a grant from the Program in Appropriate Technology, National Science Foundation, NSF Grant No. ISP-8016764. Any opinions, findings, summaries or conclusions expressed in this report are those of the project investigators and do not necessarily reflect the views of the National Science Foundation.

The support of the National Science Foundation, the cooperation of Mr. Gary Gleason, Gleason Environmental Processes, Trumansburg, New York, and Mr. Richard Bowslby, Department of Public Works, Ithaca, New York, and the assistance of Mr. Michael R. Malecki and of the students who participated in the project are gratefully appreciated and acknowledged.

References

Camp, Dresser and McKee, Inc. 1981. Engineering Assessment of Vermicomposting Municipal Wastewater Sludges, prepared for the Municipal Environmental Research Laboratory, Office of Research and Development, Environmental Protection Agency, Cincinnati, Ohio, EPA 600/2-81-075.

Carone, H. and Raquet, D. 1936. Colormetric determination of nitrates in water in presence of chlorides. (in French). J. Pharm. Chem., Paris 22: 446–447.

Hartenstein, R., Neuhauser, E.F. and Kaplan, D.L. 1979. The reproductive potential of the earthworm *Eisenia foetida*. Oecologia 43: 329–340.

Hartenstein, R., Neuhauser, E.F. and Collier, J. 1980. Accumulation of heavy metals in the earthworm *Eisenia foetida*. J. Env. Qual. 9: 23–26.

Kaplan, D.L., Neuhauser, E.F., Hartenstein, R. and Malecki, M.R. 1980. Physicochemical requirements in the environment of the earthworm *Eisenia foetida*. Soil Biol. Biochem. 12: 347–352, 1980.

Malecki, M.R, Neuhauser, E.F. and Loehr, R.C. 1982. The effect of metals on the growth and reproduction of *Eisenia foetida*. Pedobiologia 24: 129–139.

Loehr, R.C., Martin, J.H., Neuhauser, E.F. and Malecki, M.R. 1984. Waste Management Using Earthworms – Engineering and Scientific Relationships. Final Project Report, Project ISP-8016764, National Science Foundation, 76 pp.

McKenzie, H.A. and Wallace, H.S. 1954. The Kjeldahl determination of nitrogen: a critical study of digestion conditions-temperature catalyst, and oxidizing agent. Austral. J. Chem. 7: 55–71.

Mitchell, M.J., Hornor, S.G. and Abrams, B.I. 1980. Decomposition of sewage sludge in drying beds and the potential role of the earthworm, *Eisenia foetida*. J. Env. Qual. 9: 373–378.

Montgomery, H.A.C. and Dymock, J.F. 1961. The determination of nitrites in water. Analyst 86: 414–416.

Neuhauser, E.F., Kaplan,D.L., Malecki, M.R. and Hartenstein, R. 1980. Materials supportive of weight gain by the earthworm *Eisenia foetida* in waste conversion systems. Agricult. Wastes 2: 43–60, 1980.

Pincince, A.B., Donovan, J.F. and Bates, J.E. 1980. Vermicomposting municipal sludge: an economical stabilization alternative. Sludge 23: 30.

Prakasam, T.B.S., Srinath, E.G., Yang, P.Y. and Loehr, R.C. 1975. Analyzing physical and chemical properties of liquid wastes. *In* Standardizing Properties and Analytical Methods Related to Animal Waste Research. pp. 114–166. American Society for Agricultural Engineers, St. Joseph, Michigan.

Standard Methods for the Examination of Water and Wastewater, 15th Ed. 1980. American Public Health Association, Washington, D.C.

ENGINEERING PROBLEMS IN THE BREAKDOWN OF ANIMAL WASTES BY EARTHWORMS

V.R. PHILLIPS
Farm Buildings Division, National Institute of Agricultural Engineering, Wrest Park, Silsoe, Beds, U.K.

1. Introduction

The different contributions of the main collaborators, in the work referred to below, are outlined in Fig. 1. Because of commercial agreements between these collaborators, it will not be possible in this paper to describe all details of the equipment which has been developed for conducting vermiculture on a commercial scale.

2. The relative profitabilities of the two types of product arising from vermiculture

Vermiculture gives rise to various forms of worm protein (from processing of the worms themselves) and also to various forms of high grade horticultural compost (Fig. 2) (from processing the worm-worked waste). According to calculations made at NIAE (Fieldson, 1988) the major source of profit under U.K. conditions are the horticultural composts.

3. On-farm stages and depot stages of vermiculture

It is planned to operate vermiculture commercially in the U.K. (Fig. 3). Preparation of wastes and worm-working of them will occur on the livestock (or other) enterprises on which the wastes arise. In fact worm-working is already under way on a number of U.K. farms. The worm-rich waste will be sold at a guaranteed price to British Earthworm Technology Ltd., who will transport it to regional depots for worm/waste separation and further processing of each. This division between on-farm stages and depot stages, with depots so separated that each serves an area accessible within one day's return travel by truck, is estimated to be the most efficient operating strategy under U.K. conditions.

The on-farm stages have proven a more difficult design problem than have the depot stages, because of the great variations in throughput and in type of waste between different farms. It was originally hoped to offer a standard design for the on-farm stage, applicable to all farms, but it has been necessary to offer a range of designs to suit different farm situations. In contrast, the throughputs of the depot stages are more defined, and the design of these stages is an easier problem.

4. Design of the on-farm stages

As indicated above, the organizations wishing to take up vermiculture vary greatly

Earthworms in waste and environmental management
edited by C.A. Edwards and E.F. Neuhauser, pp. 111–118

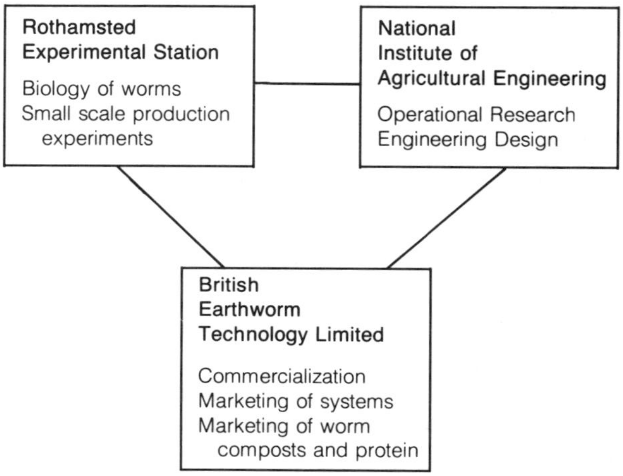

Fig. 1. The organization of the project.

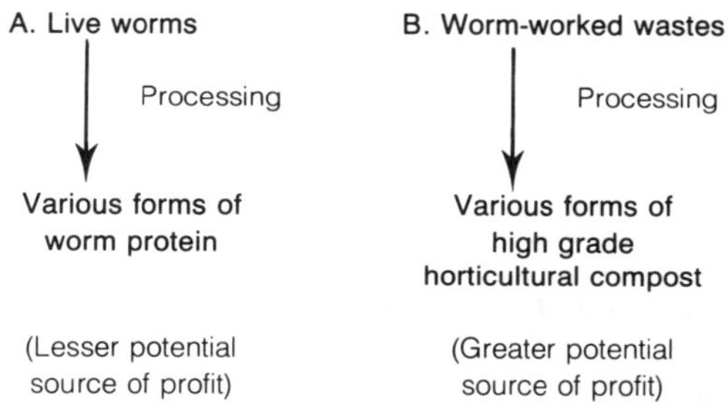

Fig. 2. Products arising from vermiculture

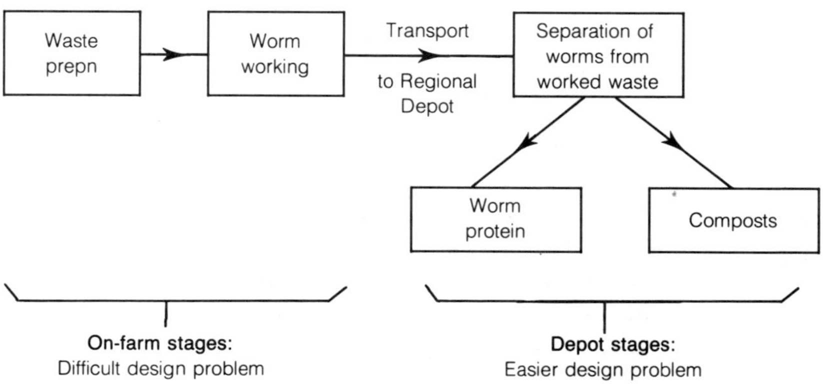

Fig. 3. On-farm stages and depot stages.

1. Aerobicity	
2. Ammonia content:	< 0.5 mg/g waste wet basis
3. Salts content:	< 0.5%
4. Temperature:	15–25°C preferred
	12–30°C tolerable
5. Moisture content:	80–90% w.b. preferred
	60–90% w.b. tolerable
pH:	5–9

Fig. 4. Necessary conditions for breeding *E. foetida* in waste.

in the type of their wastes, and their rate of waste production. Examples of types of waste in which *E. foetida* has been grown successfully are: separated cattle solids, separated pig solids, straw-based duck waste and waste sludge from paper-making. Examples of rates of production are 100 t/a separated cattle solids from a 100 cow dairy unit and at the other end of the scale, 40,000 t/a of straw-based wastes from a large intensive duck unit.

The necessary conditions for successful growth of *E. foetida* in wastes, and the problem of designing the on-farm stages, is of course the problem of providing the worms with these conditions in the most cost-effective way (Fig. 4).

Some examples of 'waste preparation' stages which may be necessary are given below:

a. For animal slurries, mechanical separation of the coarse fraction (over 1 mm). This coarse fraction is then used for vermiculture. Whole slurries are unsuitable for vermiculture – unless bulking agents are added – because of the problems of maintaining aerobicity throughout a volume of slurry.
b. For straw-based wastes, mechanical homogenization and comminution to a mean particle size of a few millimeters. This has been achieved using a side discharge muck-spreader or, on a smaller scale, a horticultural shredder.
c. If the ammonia content in the waste is initially too high, a period of storage outdoors, until the ammonia content has fallen to 0.5 mg/g waste wet basis.

Turning to the worm-working stage itself, shallow beds of waste are estimated to be the best present technique under U.K. conditions. (A system based on trays of waste was considered, but rejected as being too labour-intensive on all but the smallest scales.)

The main factors in bed design are outlined below:

a. Ease of access. Good access is essential for the operations of initial bed laying, bed feeding and bed harvesting. These needs are best accommodated by having linear beds of width not over four meters.

A gantry equipped with a hopper, costing typically a total of £1,000, is then a worth-while labour-saving investment for all but the smallest operations.

b. Aerobicity. The requirement to maintain aerobicity throughout the volume of waste limits the maximum bed depth to 0.5–1.0 meter, depending on the type of waste.

The beds must be laid on free-draining bases, again to maintain aerobicity.

c. Temperature control. For year-round operation in Britain beds laid under cover are preferable. To guarantee that worm-working continues smoothly throughout the winter, the beds should be equipped on all sides with sheets of insulating materials and their bases should be equipped with electric soil-warming cable.

Conversely, in the summer, precautions are necessary to prevent excessive solar radiation reaching the beds, *e.g.*, if the beds are laid inside a polythene-covered tunnel house, then an opaque coating on the polythene is desirable.

d. Moisture content. Holding this within the desired range is not normally a problem in the U.K., provided the beds are equipped with some form of simple loose-fitting but impervious cover. Under the hottest summer conditions, it may be necessary to spray the waste with water.

e. Materials of construction. A wide range of materials is acceptable for worm beds, *e.g.*, wood, concrete, plastic, steel mesh, the general rule being to make best use of what is available to each farmer.

Lists of unit costs have been compiled at NIAE for each component of a worm bed, *e.g.*, base, walls, insulation sheet, weather protection. From these lists the most cost-effective bed design for a given situation can be identified. An average overall unit capital cost for a worm bed is £20/m^2.

Although the kind of beds described above are considered to be the best present technology, they are already known to be relatively inefficient, because worm-working proceeds only near to the surface of the waste in a bed as fresh waste is added.

Studies have been made at NIAE to design and prove a continuous 'reactor' for worm-working of waste. Waste would be contained in a vessel, with fresh waste being steadily added at the top and worked waste being steadily removed at the bottom. Thus the waste would be steadily moving downwards. The worms in the waste would be working upwards steadily, and the rate of descent of the waste would be adjusted to keep the worms within the reactor. This work has proved to be very promising and problems being addressed include:

a. How to ensure aerobicity throughout the depth of waste in the vessel.
b. The finite (but still not reliably quantified) lifetime of *E. foetida*. Will a recycle of cocoons be necessary?
c. The inherently difficult flow characteristics of the wastes.
d. The need to keep the equipment simple, especially as on-farm operation is envisaged.

5. Design of the depot stages

The first depot stage is separation of worms from worm-worked waste. A number of worm growers have developed independently various trommel-type separators based on an inclined rotating cylinder made of woven wire mesh. Indeed such a separator has been built at NIAE and used at Rothamsted Experimental Station on a pilot scale (Fig. 5).

Fig. 5. Trommel-type worm/waste separator at Rothamsted Experimental Station.

However, this type of separator was considered too labour-intensive and of too low a throughput for commercial operations, and so other separation techniques, both mechanical and physiological, were sought. Gentle heating of the waste is effective in driving the worms from it without damaging them, but this technique appears expensive, having an estimated heating cost of £2/m^3 of waste. (Nevertheless this technique has proven very useful on a lab scale.)

For commercial scale operations, a new mechanical separator has been developed. The principle is again rotary, but no screens are employed. Figure 6 shows the results of a performance test on this separator. Since this test was run, a modification has been made to the separator, which has certainly further increased its efficiency, but this increase has not yet been quantified. The performance of the separator of course depends on the properties of the waste treated, particularly the moisture content, which for good performance should not be above 80% wet basis.

Many of the processing stages downstream of worm/waste separation (Fig. 7) can be achieved using conventional horticultural equipment. A typical throughput of a first generation processing plant would be 22 tonnes waste (wet basis)/day + 2 tonnes worms (wet basis)/day. In what follows, the figures in brackets are the estimated

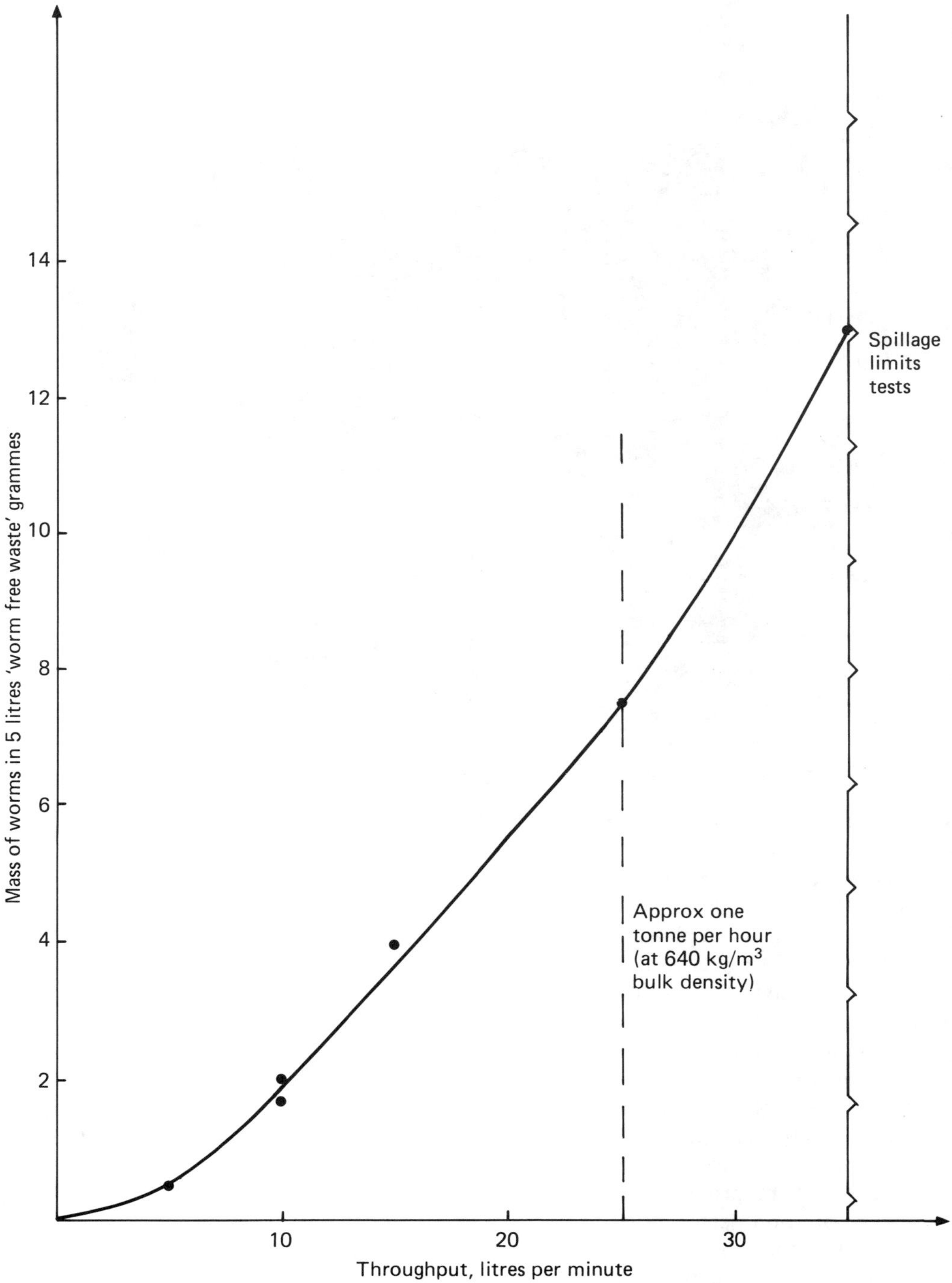

Fig. 6. Performance of new separator separating *E. foetida* from cattle solids at 80% moisture content (wet basis).

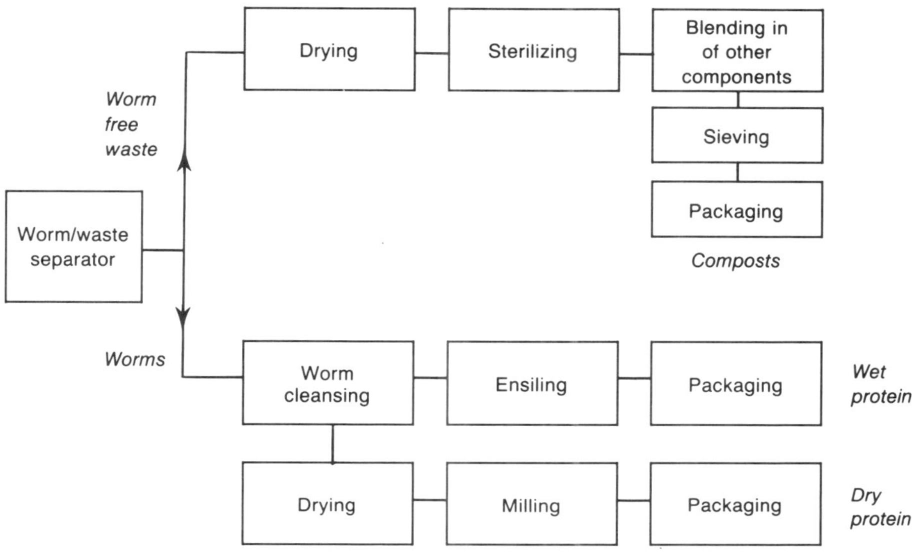

Fig. 7. The depot stages of the process.

capital costs of the process equipment for each stage at the above throughput.

To prepare the composts, the worm-free waste is first dried. A mechanical dewaterer (£10K) can reduce the moisture content to 60% wet basis: thereafter passive drying (no capital cost) is used as far as possible and final forced drying, if it is necessary, is by gas-fired rotary tunnel dryer (£18K). Sterilization would be achieved in this same device. This latter stage kills worm cocoons (which are not removed by the worm separator), kills any few worms which have escaped the separator and kills weed seeds.

Other components are next blended in with the worm worked waste in a paddle blender (£4K). The nature (and quantity) of these other components depends on which grade of compost is being prepared, but peat and shredded tree bark are two examples. The next stages use coarse sieving equipment (£1K) to remove any lumps and improve product appearance, and, finally, a packaging unit (£4K).

To prepare worm protein, the worms are first rinsed externally (£2.5K) with water to remove adhering waste, then cleansed internally by a period of immersion in cold water (£1.5K), during which they void the contents of their gut. If wet protein is required, the worms are then ensiled by maceration with 3 wt% of pure formic acid (£7K) to give a paste which is finally packaged. If dry protein is required, then worms are, after cleansing, dried down to a low moisture content in a batch tray dryer at 80°C (£10K).

For certain applications, *e.g.*, preparation of fish feed from *E. foetida*, a blanching stage (£4.5K) in boiling water is necessary before this drying, to extract the unpalatable coelomic fluid from the worms. After drying, the worms are milled in a hammer mill (£2.5K) and the resulting powder can if necessary be pelletized using an extrusion pelleting machine (£7.5K), and as above, finally, packaged (£4.0K).

6. Summary

a. The design of a process has been outlined for growing *E. foetida* in wastes and subsequently producing therefrom both horticultural composts and high protein animal feed supplements.

b. Of these two products, the composts are expected to yield the major profits (Fieldson, 1988).

c. It is estimated that, under U.K. conditions, it is most economic to operate the worm growing stage on farms and the worm/waste separation and subsequent processing stages at regional depots.

d. Because of the variability of type and quantity of wastes on different farms, it has been necessary to offer a range of bed designs and waste-handling techniques.

e. In contrast, the depot stages are better defined. The major pieces of equipment necessary there are identified in this paper and their estimated capital costs are quoted.

Acknowledgement

I thank my colleagues R.S. Billington, J.S. Price and A.L. Wilkin for their help in the preparation of this paper.

Reference

Fieldson, R.S. 1988. The economic viability of earthworm culture on animal wastes. This volume (p. 145)

A POPULATION MODEL FOR THE EARTHWORM *EISENIA FOETIDA*

I.R. JEFFERIES and E. AUDSLEY
National Institute for Agricultural Engineering, Silsoe, Beds, U.K.

1. Introduction

The lifecycle of the earthworm *Eisenia foetida* may be divided into three distinct phases: the cocoon phase, the young (immature) phase and the adult (mature) phase. These phases are summarized in Table 1.

Table 1. Characteristics of phases

Phase	Characteristics
Cocoon	Each cocoon contains N_c infant worms When the cocoon hatches N_c worms enter the young phase
Young	Hatched cocoons provide young worms Young worms are growing physically Young worms cannot produce cocoons
Adult	Young worms become adult when they mature, *i.e.*, become capable of producing cocoons Adult worms may be growing physically

The model divides the phases further into a number of stages and defines probabilities of a worm moving from stage i to stage j, l_{ij}, in the basic time interval (one iteration of the model). These probabilities are then used to build up a probability matrix, L, which may be used to calculate the population distribution after one basic time interval, given the initial population. By successive iterations the population distribution at any time may be calculated from an initial population distribution. If the initial population vector is S having elements s_i and the population vector after one iteration is F having elements f_j then

$$F = L^T S \tag{1}$$

Alternatively, equation (1) may be written as:

$$f_j = \sum_{i=1}^{m} l_{ij} s_i \tag{2}$$

where m is the order of L, *i.e.*, the total number of stages.

A worm in a particular stage is limited to a number of stages into which it may develop in one iteration, which means that L contains many zero elements. In mathematical terms the three phases discussed above are not necessary to the model structure as the individual stages describe the lifecycle of the worm fully. However, they

Earthworms in waste and environmental management
edited by C.A. Edwards and E.F. Neuhauser, pp. 119–134

Table 2. Possible transitions between stages

Phase	Possible transitions	
Cocoon	$c_i \rightarrow c_{i+1}$	Worm in cocoon remains unhatched, ages by 1 stage
	$c_i \rightarrow y_1$	Cocoon hatches, worm enters first young stage y_1
Young	$y_i \rightarrow y_{i+1}$	Remains immature, worm ages by 1 stage
	$y_i \rightarrow a_1$	Matures, worm enters first adult stage (a_1)
		Dies, worm removed from model
Adult	$a_i \rightarrow a_{i+1}$	*and* N_i in c_1. Worm ages by 1 stage, producing N_i worms in cocoons
		Dies, worm removed from model

do provide convenient labels for groups of stages with similar characteristics and so the names will be retained. The possible transitions for a worm in each phase are summarized in Table 2. The methods used to actually calculate the probabilities for each of these transitions are described in detail below.

The model also relates numbers of worms to the weight of the population, which is the parameter easily measured in practice. This leads to a slight complication as the weight of an individual worm is governed by its age measured from hatching, whereas its reproductive ability is controlled by its age measured from maturation and the relationship between these two ages varies from worm to worm. Throughout the model an adequate supply of food and non-restrictive moisture levels are assumed.

2. Notation

a_i	$1 < i < P$	i^{th} adult stage.
b_1		growth curve parameter.
C		average number of cocoons produced/adult/week.
c_i	$1 < i < M$	i^{th} cocoon stage.
C_1		growth curve parameter.
d_i	$1 < i < N$	probability of worm in i^{th} young stage dying.
E_g		energy available to worm for conversion into growth.
e_i	$1 < i < P$	probability of worm in i^{th} adult stage dying.
F		initial population vector.
f_i	$1 < i < m$	elements of F
h_i	$1 < i < M$	probability of cocoon in i^{th} stage hatching.
I		iteration interval in degree days.
L		probability matrix.
l_{ij}	$1 < i, j < m$	elements of L.
M		number of stages in cocoon phase.
m		total number of stages in model (M+N+P) = order of L
m_i	$1 < i < N$	probability of worm in i^{th} young stage maturing.

m_1	growth curve parameter.
N	number of stages in young phase.
N_i	number of cocoons produced by adults in stage in one iteration.
N_c	number of infant worms in cocoon.
N_h	number of infant worms hatching from cocoon.
N_{iT}	number of cocoons produced per week by adults in i^{th} phase at temperature T°C.
n_i	number of cocoons per iteration produced by cocoons in stage i.
P	number of adult stages.
P_{NB}	proportion of non-barren cocoons.
$p_i \quad 1 < i < M$	probability of a cocoon hatching in i^{th} interval (from χ^2 tables).
$q_i \quad 1 < i < N$	probability of a young worm maturing in i^{th} interval (from χ^2 tables).
S	final population vector.
$S_1 \quad 1 < i < m$	elements of *S*.
T	temperature in °C.
t	time in degree days.
t′	time in days.
t_1	time variable in growth curve.
w	weight of worm.
w_1	weight variable in growth curve.
w_{max}	maximum weight attainable.
$y_i \quad 1 < i < N$	i^{th} young stage.
$\alpha, \beta, \gamma,$	parameters in growth model.
ν	degrees of freedom in χ^2 curve.

3. The model

3.1. *The cocoon phase*

New cocoons enter the model in stage c_1 when they are laid by adult worms. A number of these cocoons are barren, *i.e.*, contain no worms or worms that will never hatch, and the proportion of barren cocoons increases as temperature increases. A suitable proportion of cocoons in c_1 are removed from the model to take account of these barren cocoons. It was found that the proportion of non-barren cocoons was best described by:

$$P_{NB} = 0.39 \left[1 - \frac{1}{1 + \exp(-0.34(T-26.2))} \right] \tag{3}$$

This function is fitted to data obtained from E. Neale (unpubl.) (Entomology

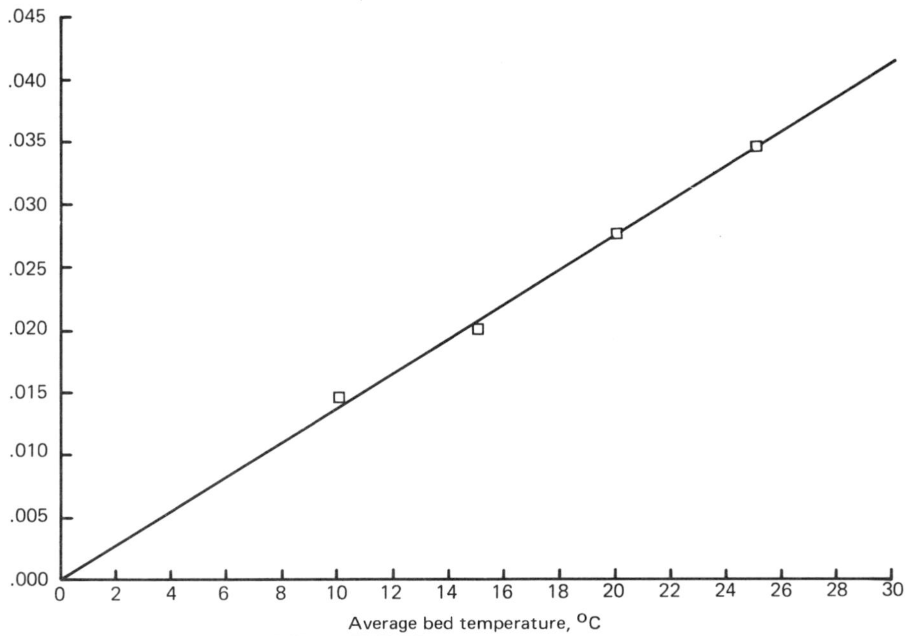

Fig. 1. 1/(Mean time to hatch) vs temperature.

Department, Rothamsted Experimental Station) as are all other equations in the model unless otherwise stated.

After these cocoons have been removed it is assumed that all cocoons will hatch. For a cocoon in stage c_i there are two possible stages into which it may progress in one iteration: it may hatch, to which we assign the probability h_i, or it may simply age by one stage into c_{i+1} with probability $1-h_i$. It is therefore necessary to calculate the values h_i $1 < i < M$, where M is the number of cocoon stages.

Data were available from experiments carried out in the laboratory on the time taken for cocoons to hatch at various temperatures in the range 10–25°C. A plot of 1/(mean time to hatch) vs temperature is given in Fig. 1. This linear relationship led to the adoption of a degree day timebase for the model, *i.e.*, the unit of 'time' became the product of time in days and the mean bed temperature in °C. This enabled data corresponding to different temperatures to be included in a single relationship and was found to hold for maturation times and growth curves as well as for hatching times.

The hatching times for four different temperatures are plotted against degree day time in Fig. 2, along with the fitted χ^2 curve defined by:

$$\chi^2 = \frac{t}{21.8} \qquad \nu = 34$$

It is now possible to calculate the values h_i, given the degree day time interval corresponding to one iteration. The end of the cocoon phase is fixed by making the as-

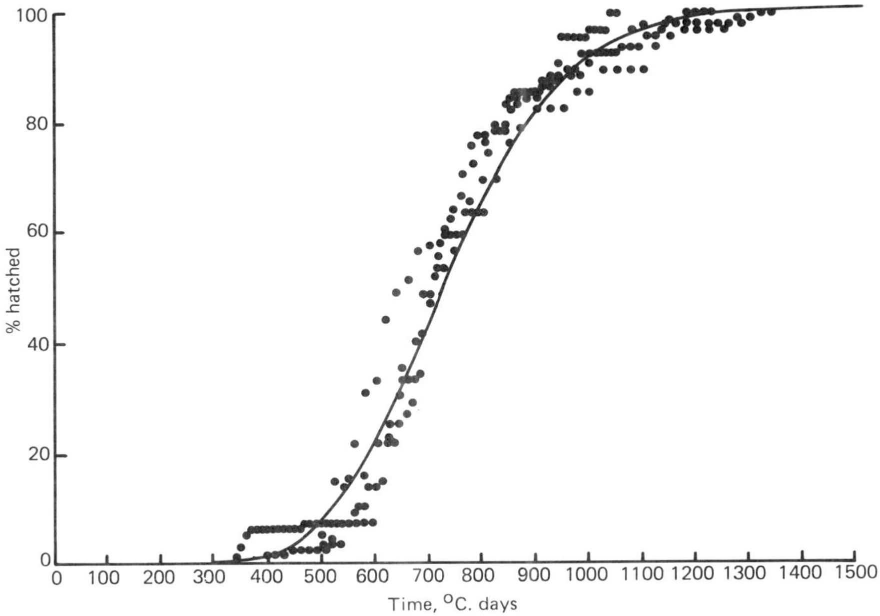

Fig. 2. Hatching times for *E. foetida*.

sumption that all cocoons will have hatched when the fitted curve reaches 99.9% of its maximum value.

Choosing the time interval for one iteration required some care because reducing the interval achieves greater accuracy in the simulation of the populations at the expense of increased amounts of calculation. 200 degree days was chosen as the interval to give a level of accuracy consistent with the data and the timescale to be simulated. Notice that the probabilities h_i are defined as the probability of a cocoon hatching in the ith stage, given that it does not hatch in a previous stage, whereas the χ^2 values are the probability of a cocoon hatching in the ith stage, hence some calculation is necessary in order to derive the correct values from the fitted χ^2 curve. If the probability of a cocoon hatching in the ith interval is p_i (found from χ^2 tables), then h_i is given by:

$$h_i = \frac{p_i}{\prod_{j=1}^{i-1} (1 - h_j)} \qquad 2 < i < M \tag{14}$$

$$h_1 = p_1$$

Starting from h_1, h_i $2 < i < M$ may be calculated iteratively. With an iteration interval of 200°C days for *E. foetida* the above procedure led to the model containing seven cocoon stages, *i.e.*, M = 7. The values calculated for h_i are given in Table 3, h_M is set to 1.0.

It is assumed that the number of worms in the cocoon has no effect on hatching time as no data is available. Rate of cocoon production is considered in the adult phase.

Table 3. Hatching probabilities for *E. foetida*

i	h_i
1	0
2	0.013
3	0.215
4	0.542
5	0.761
6	0.790
7	1.0

h_i is the probability of hatching in the ith stage, given that the cocoon does not hatch previously. Iteration interval 200°C days.

The mean number of infant worms hatching from individual cocoons was found to vary with temperature. An inverse sigmoid curve was fitted to data, giving the mean number of hatchlings per cocoon as:

$$N_h = \frac{4}{1 + \exp(0.19(T-28.1))} \tag{5}$$

All cocoons are assumed to yield this mean number of young worms on hatching as in practice a large number of cocoons will hatch in each time interval.

3.2. *The young phase*

On hatching, infant worms enter the first young stage, y_1. For a young worm in stage y_i there are three possible outcomes of an iteration: the worm may die, to which we assign the probability d_i, it may mature, to which we assign the probability m_i, or it may simply age by one iteration to stage y_{i+1} with the probability $1-d_i-m_i$.

Very little information is available on mortality rates in earthworm populations. No data were available from the RES experiments but Hartenstein *et al.* (1979) quote a mortality rate of 1% per week for *E. foetida* and this value is used in the model. Hence if there are N young stages:

$$d_i = \frac{0.01}{7T} I \qquad 1 < i < N \tag{6}$$

where I is the iteration interval in °C days.

The maturation time vs degree day time curve was found to be of the same form as for hatching and a similar χ^2 curve was fitted, defined by:

$$\chi^2 = \frac{t}{54.8}, \qquad \nu = 20$$

this curve is compared with experimental data in Fig. 3.

Following the same procedure as for h_i, the probability of a young worm maturing in the ith stage, given it has not yet matured may be calculated as:

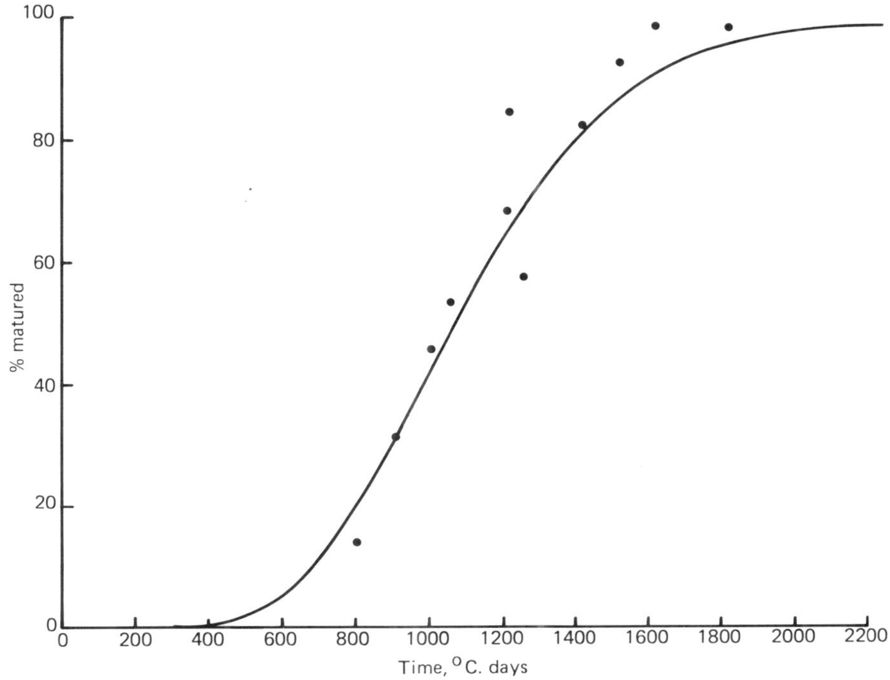

Fig. 3. Maturation times for *E. foetida*.

$$m_i = \frac{q_i}{\prod_{j=1}^{i-1} (1 - m_j)} \qquad 2 < i < N \tag{7}$$

$$m_1 = q_1$$

where q_i is the probability of maturing in the ith interval, from χ^2 tables. Again an iterative procedure is followed to calculate all m_i. In the case of *E. foetida* with a 200°C day interval, the values calculated for m_i are given in Table 4. As with M for cocoons, N is determined by the assumption that all young have matured by the time the fitted curve has reached 99.9% of its maximum level, and the value of m_N is set to 1.0.

3.3. *The adult phase*

When they mature, young worms enter the first adult stage a_1. An adult worm in stage a_i may either die, to which we assign the probability e_i, or it may just age by one stage into a_{i+1} with probability $1 - e_i$. In the latter case the adult worm will also produce a number of cocoons.

As with young worms, little information is available on mortality rates. Again the rate of 1% per week as quoted by Hartenstein *et al.* (1979) is used, *i.e.*,

Table 4. Maturing probabilities for *E. foetida*

i	m_i
1	0
2	0.004
3	0.049
4	0.156
5	0.285
6	0.394
7	0.476
8	0.516
9	0.558
10	1.0

m_i is the probability of maturing in the ith stage, given that the worm does not mature previously. Iteration interval 200°C days.

$$e_i = \frac{0.01}{7T} I \qquad 1 < i < P \tag{8}$$

There are P adult stages and in the current model adults leaving the P^{th}stage are assumed to have died and take no further part in the model. For *E. foetida* P was chosen to be 23 which at 25°C is equivalent to about 40 weeks – approximately the time quoted by Hartenstein *et al.* (1979) for the termination of reproduction. Experiments at RES indicate that the rate of cocoon production is best modelled by an inverse sigmoid curve. The number of cocoons produced by an adult in stage a_i in one iteration is given by:

$$n_i = \frac{I}{7T} C(1 - \exp(-0.5i)) \qquad 1 < i < P \tag{9}$$

The value of C was found to be 4 for *E. foetida*. The number of cocoons produced per adult per iteration is further modified by a temperature dependent function to make the final equation for cocoons per adult per iteration:

$$N_{iT} = \frac{4}{7T} I \frac{(1 - \exp(-0.5i))}{(1 + \exp(-0.72(T - 15)))} \tag{10}$$

Equations (1)–(10) together describe the reproductive cycle of *E. foetida*.

3.4. *Relating worm numbers to weights*

It is assumed that the amount of food that an individual worm can consume, and hence its energy intake, is proportional to its bodyweight w. In order to find food the worm must use up some of this energy in moving – this amount will also be proportional to w. In addition, energy is used to maintain the normal static bodily functions of the worm and again the amount will be proportional to w. The remaining energy is available for growth of the worm.

Let the consumed energy be αw, the energy used in moving per unit energy found be βw and that used to maintain static bodily processes be γw. If the temperature is lowered, it is assumed that all the bodily functions of the worm are slowed by a proportion μ therefore the energy available for conversion to growth is:

$$E_g = \mu\,(\alpha w(1-\beta w)-\gamma w) \tag{11}$$

This energy E_g is converted to bodyweight in some ratio λ which may be set to unity by suitable choice of units for α, β and γ, hence

$$\frac{dw}{dt'} = \mu(\alpha w(1-\beta w)-\gamma w) \tag{12}$$

The factor μ may be incorporated into t', consistent with the use of a degree day timescale for growth.

$$\frac{dw}{dt} = (\alpha-\gamma)w-\alpha\beta w^2 \tag{13}$$

If food is scarce then for the same amount of energy consumed, β is increased, controlling the growth rate at different population densities. From equation (13) the maximum weight attainable is:

$$w_{max} = \frac{\alpha-\gamma}{\alpha\beta}, \tag{14}$$

growth ceases if $\alpha-\gamma < \alpha\beta w$, and weight is given by:

$$w_1 = \frac{C_1}{1+\exp(-b_1(t_1-m_1))} \tag{15}$$

$$\text{where } b_1 = \alpha-\gamma \text{ and } C_1 = \frac{\alpha-\gamma}{\alpha\beta} \tag{16}$$

A curve of the type described by equation (15) was fitted to data on the growth of *E. foetida* and the result is shown in Fig. 4.

The current model assumes an ample food supply so that α remains constant throughout. β, which is associated with the type of food available is also assumed to remain constant, as is γ which should be constant for worms of the same species. Hence the parameters for equation (15) were fixed as:

$$b_1 = 0.00247 \qquad C_1 = 1.422 \qquad m_1 = 1564.3$$

The data for this growth curve was taken from experiments at RES growing *E. foetida* in the laboratory on excess pig manure. Neuhauser *et al.* (1980) show growth curves obtained when growing *E. foetida* in cow manure, horse manure and activated sludge in the laboratory. These curves have the same form as that used in the model

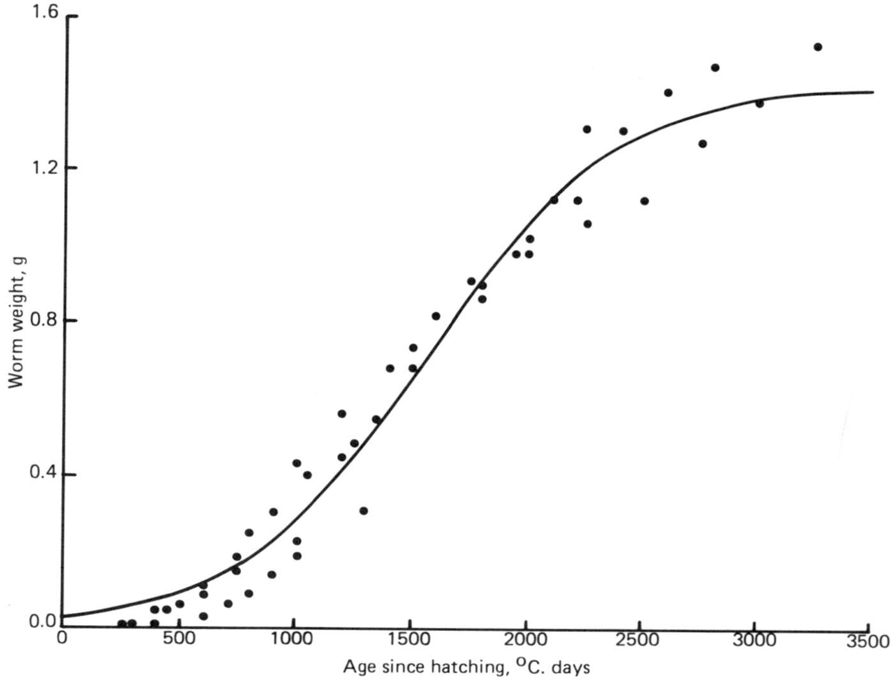

Fig. 4. Growth curve for *E. foetida.*

although Neuhauser's worms seem to grow to a slightly larger maximum weight on all three types of food. However, the current model equation is based on local data and is adequate for the present.

4. Use of the model

The model described in sections 1–6 forms the model for a computer simulation of an earthworm 'farm'. The system simulated is a series of worm beds, processed in batches, into which a known population distribution of worms is introduced and allowed to grow. The worked waste and the worms have then to be separated, the worms being sold as protein and the worked waste as a soil amendment or compost. Some portion of the worm population is retained to be returned to the worm beds to initiate the next batch of waste treatment and the system should require no further worms from outside.

The amount of substrate consumed (*i.e.* waste processed) by the worms in unit time is proportional to the total mass of the population at that time. The mass of worms available for use as a protein source is the harvested mass. The choice of which portion of the population to harvest must be taken with care as it is not only desirable to maximize protein output but also to ensure that there are enough worms left to return to the bed for the next growth period. The way in which the model will be used to obtain information on a real system like that just described may be illustrated by

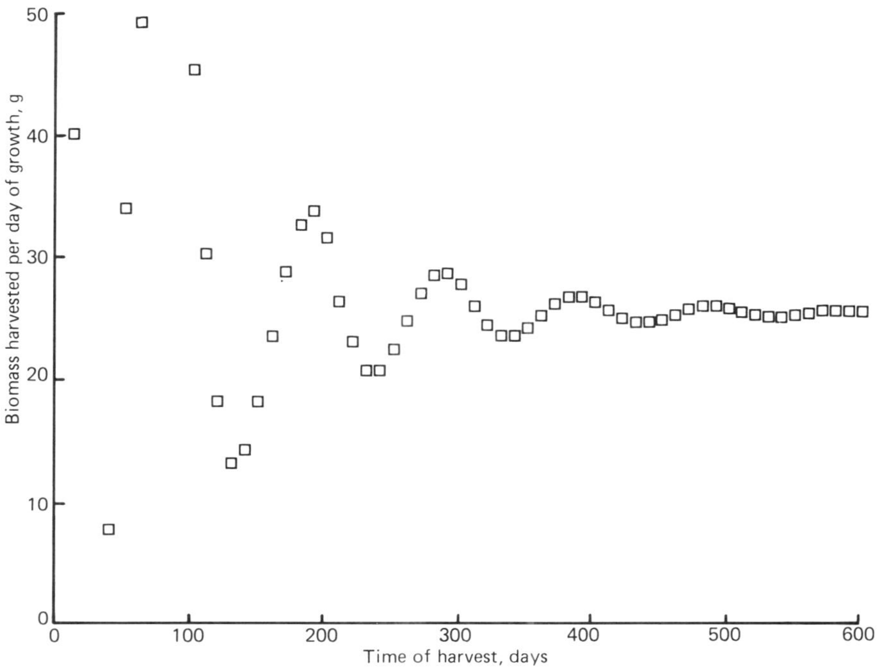

Fig. 5. Weight of worms collected at each harvest.

two examples. In the first, the model is used to find the best average bed temperature for population growth with a set harvesting strategy. The second determines the minimum size of worm that may be harvested whilst still leaving a large enough population to ensure the survival of the worm colony through successive batches of waste. This time the temperature as well as the harvesting strategy is fixed.

4.1. *Example 1*

The model was run with an average simulated bed temperature of 10, 15, 20, 25 and 30 degrees centigrade. The population at the start of each run was 1000 g of fully grown adult worms and the harvesting strategy chosen was to return 1000 g in total of worms after each harvest, keeping the proportion of worms in each stage of the model constant before and after harvest. If there was not 1000 g of worms available for harvest at the required time, then no harvest took place and the population was left untouched. For each temperature, harvest was attempted at a range of regular intervals and the weight of worms harvested recorded. This harvesting strategy corresponds to leaving a fixed proportion of the worm bed behind (*i.e.*, processed waste + worms) to initiate the next batch of waste to be processed.

A typical result of this simulation, a plot of weight of worms harvested per day of growth versus time of harvest, is given in Fig. 5, where the interval between successive harvests is 10 days. This typical result exhibits several interesting characteristics and is worthy of some discussion.

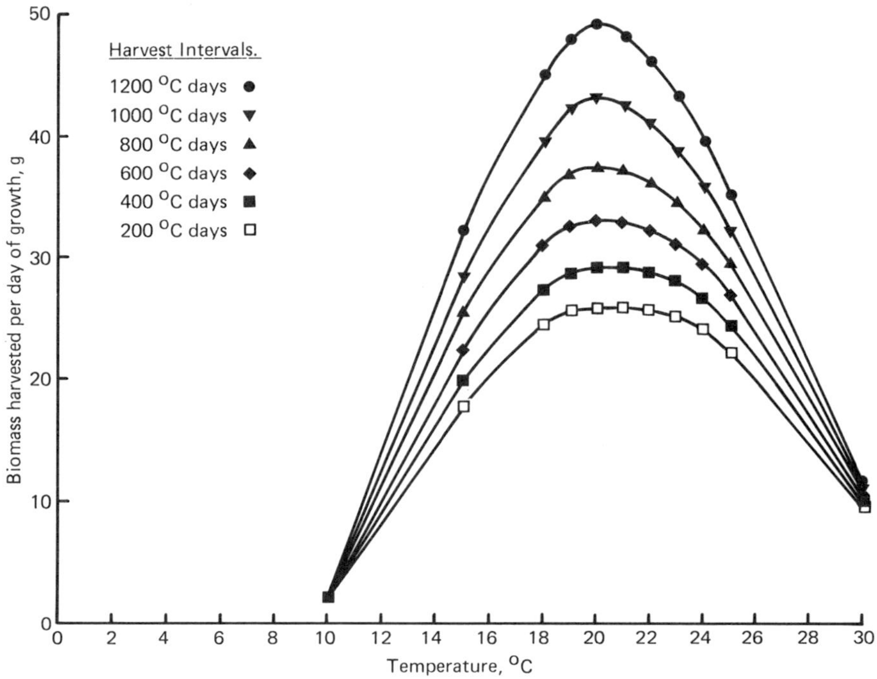

Fig. 6. Steady state biomass harvested.

The weight of worms harvested per day of growth approaches a steady state value as time progresses. The initial large fluctuations are due to the starting population differing from the steady state population which is approached as time passes. An operator could think that after 130 days his worm bed had failed. He would be rewarded for perseverence by day 200, and would then think his worm bed was now a success. He would however be wrong. These initial variations could be reduced by a suitable choice of starting population, ideally the steady state population, which depends on the harvesting strategy and harvest interval, as well as the characteristics of the worm themselves. An approximation to the weight of worms harvested per day of growth in the steady state was found by taking an average of the values found over the last third of the simulation. As these averages are measured in weight of worms harvested per day of growth they are directly comparable across harvest intervals, temperatures and harvesting strategies. A comparison of six different harvesting intervals is given, over the range of temperature simulated, in Fig. 6. This shows clearly that the optimum temperature for harvested weight is between 20 and 21°C for all intervals.

Another feature of Fig. 6 is that the amount of worms harvested per day of growth increases with the interval between harvests. This is a result of the assumption within the model that the worms always have an excess food supply which implies a larger worm bed for the longer harvest intervals. In practice, the size of the worm bed, along with the rate of supply of fresh waste would restrict the growth of the worm population, establishing an optimum harvesting interval.

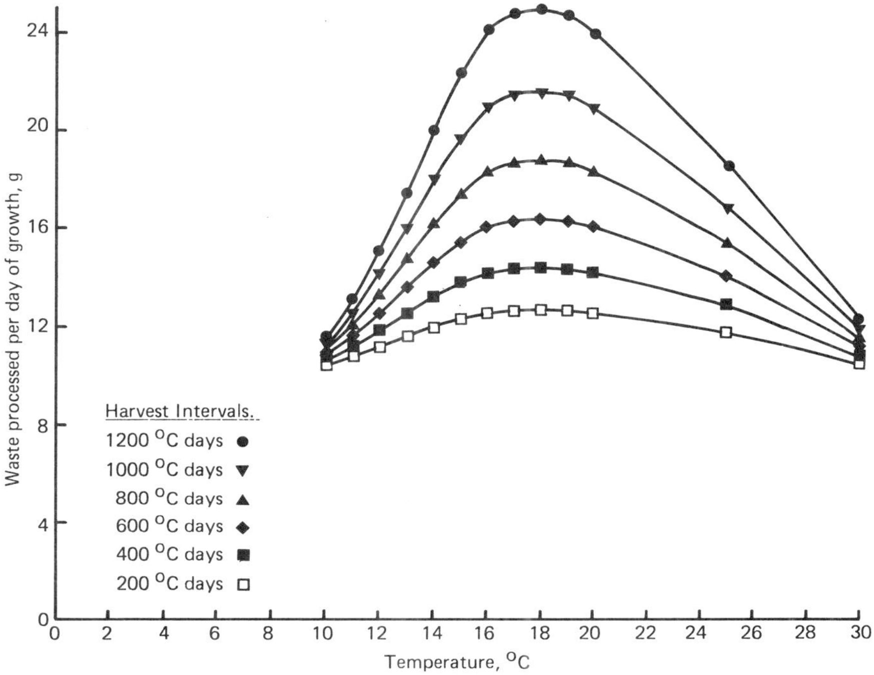

Fig. 7. Steady state waste processing rate.

Foote and Fieldson (1982) concluded that the most important contribution to the economic viability of an earthworm farm would be made not by the sale of earthworms as a protein source, but by the conversion of animal wastes into soil amendments. Consideration should therefore be given here to the rate of conversion of waste as well as to the rate of harvest of worms. Figure 7 shows the average units of waste converted per day for the simulation described above, assuming the rate of waste converted is proportional to worm population weight. This is of very similar form to Fig. 6, the harvested weight per day, even the proportional increase with interval is almost identical. The only major difference is the optimum temperature for processing which is about 18°C for all intervals, as opposed to 21°C for protein production. It is worth noting that it is unlikely that, in practice, the worms environment will be controlled to within even an average bed temperature fluctuation of ± 3°C. In any case, only a small loss in protein production is incurred by operating at the waste processing optimum.

4.2. *Example 2*

The model was run at the near optimum temperature of 20°C over a range of harvesting intervals. The harvesting strategy was to leave 1000 g of all worms less than a particular size, which could be varied. The proportion of worms and cocoons in each stage was kept in proportion to the preharvest numbers. This corresponds to separating the worms from the waste mechanically in a machine that can only remove worms

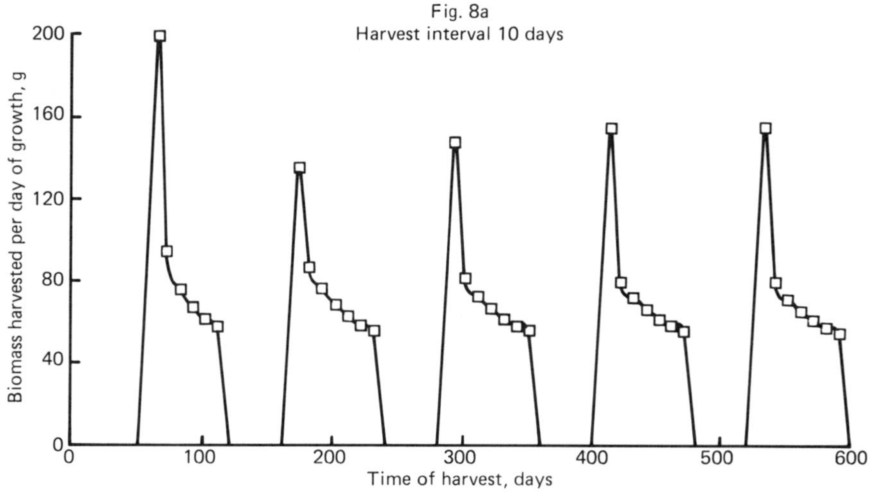

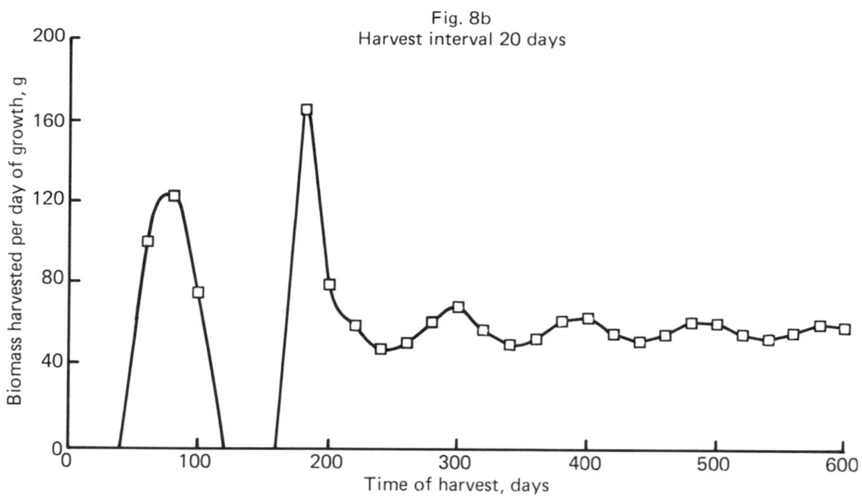

Fig. 8. Weight of worms collected at each harvest.

above a certain minimum size and returning a fixed portion of the remaining worked waste with small worms to the beds to initiate the next batch. The intention behind this simulation was to determine any limit to the size of worm which could be harvested whilst still leaving behind enough worms to ensure continuous processing of waste and to find any relationship between this limit and harvesting interval. The results quoted are for a steady state system; the large fluctuations in the harvested weight at the start of the simulation are ignored. Sample plots of biomass harvested per

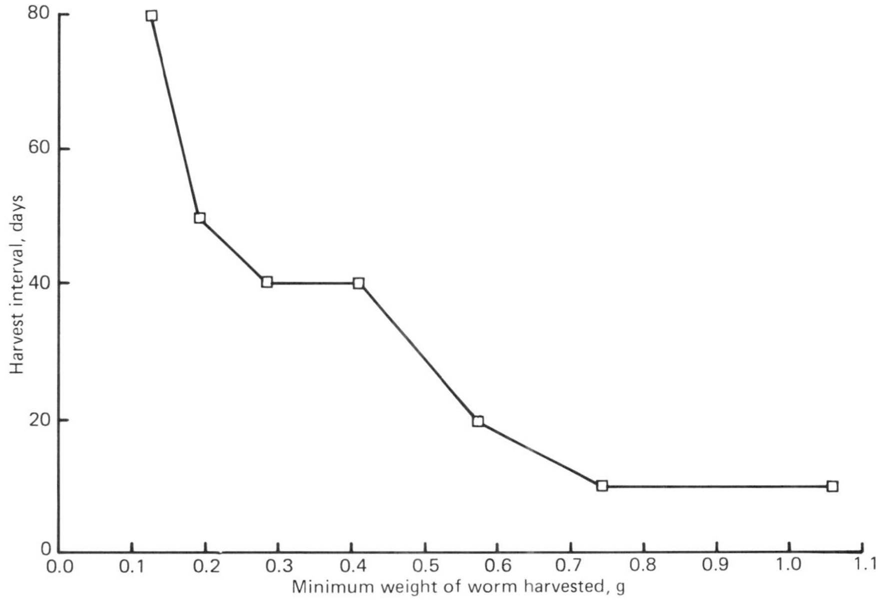

Fig. 9. Last feasible harvest interval.

day of growth versus time of harvest are given in Fig. 8.

Figure 8 shows that at a 10 day harvest interval insufficient numbers of worms are left for the following batch of waste to ensure continuous processing in the long term, for this particular harvesting interval. However, at a 20 day harvesting interval processing is continuous and would be acceptable in a practical situation. Both of these plots are for a harvest of worms greater than 0.57 g at 20°C.

Simulations were carried out with the minimum weight of individual worm to be harvested set to coincide with each of the weight classes included in the model. The results, showing the first continuous harvesting interval for each minimum weight are shown in Fig. 9. The strategies above the line are feasible, and those below it are not.

The relationship implied by Fig. 9 is one of two regimes, the 'hump' in the middle of the plot is the transition between the two. At the shorter harvesting intervals it is not possible for enough worms to evolve from the cocoon phase to being sufficiently large to be harvested in a single interval. However, when the harvest interval is much longer, this becomes possible and there is a shift in the steady state population so there is the 'hump' in Fig. 9 where the change between these two occurs.

It is clearly not possible to demonstrate all the uses of the model here, but it is hoped that the examples quoted above illustrate the sort of information that the model can provide.

5. Conclusion

The model described forms a basis on which to build a more complete model of an earthworm farm operating on a commercial basis. Several areas have been found in which little is known about *E. foetida*, especially on the effects of food quality and population density on growth and reproductive ability. The omission of any population density effects from the model is acceptable if, in the real case, worm beds are regularly supplied with fresh layers of unworked waste satisfying the 'excess food' assumption. If this conditon is not met then growth may be restricted towards the end of the harvesting interval and worms may even lose weight if they are left too long in the substrate causing results to differ from those predicted by the model.

The model of the reproductive cycle fits well to published data (*e.g.*, Hartenstein *et al.* (1979), Tsukamoto and Watabe (1977)) as well as to recent experimental data from RES. The only area where little data is available is on mortality rates. Those used in the model are taken from laboratory experiments and probably provide a lower limit for mortality rates in worm beds. All the data used in the formulation of the model relate to laboratory scale experiments and as yet little information is available on how well the model simulates an earthworm population in a large scale worm bed.

In general the model probably provides an upper limit to the potential production of a real system, as it is based on data from small scale experiments in almost ideal, well-controlled environments. It does, however, provide a convenient and quick way of examining the effects of changing any of the conditions in which the worms live and under which they are farmed.

References

Foote, N. and Fieldson, R. 1982. A Study of the Economic Feasibility of Earthworm Culture for Feed Production and Animal Waste Conversion on Livestock Units in England and Wales. Div. Note DN/1141, National Institute for Agricultural Engineering, Silsoe (unpubl.).

Hartenstein, R., Neuhauser, E. and Kaplan, D. 1979. Reproductive potential of the earthworm *Eisenia foetida*. Oecologia (Berl.) 43: 329–340.

Neale, R.E. 1989. The Production of Earthworms and Plant Growth Media from Pig Manure. Ph.D. Thesis (in preparation).

Neuhauser, E., Hartenstein, R. and Kaplan, D. 1980. Growth of the earthworm *Eisenia foetida* in relation to population density and food rationing. Oikos 35: 93–98.

Tsukamotomo, J. and Watanabe, H. 1977. Influence of temperature on hatching and growth of *Eisenia foetida*. Pedobiologia 17: 338–342.

THE CASE FOR TEMPERATURE CONTROL IN VERMICULTURE

R.J. ASTON
CERL Freshwater Biology Unit, Ratcliffe on Soar, Nottingham NG11 OEE

1. Introduction

The rate of population growth of oligochaetes like other poikilotherms is directly dependent on temperature, the rates of the various biological processes involved increasing with temperature up to various optima above which they are likely to decline. It is important therefore that vermiculturists should attempt, when it is economically possible, to use temperatures at which population growth is most rapid and/or at which the efficiency of conversion of substrate to worm tissue is highest. The temperatures best suited for culture are likely to depend on the temperature requirements of four biological processes in particular. These are (1) somatic growth (2) reproduction (3) mortality and (4) efficiency of food conversion, all of which are likely to vary from species to species.

Here some attempt is made to summarize information on the temperature relations of worms recorded by a number of authors and to point out where there are significant gaps in our knowledge. It will be seen that the optimal temperatures for culture of a number of species are frequently above the seasonal ambient in temperate regions, particularly in autumn, winter and spring and the possible use of reject heat from power station cooling growth to achieve required temperatures is considered here.

Temperature relationships in oligochaetes

Table 1 summarizes studies by a number of authors showing the probable optimum temperatures for growth and reproduction for various spaces of terrestrial and aquatic oligochaetes. Also included are temperature preferenda for worms allowed to move freely through temperature gradients. In cases where the highest temperature used gave the highest growth or reproductive rate an asterisk has been placed after the data. In studies where optimum temperature data for culture are based on apparently reliable long term experience, but not necessarily on controlled experiments, the data are given in parenthesis in between the columns for somatic growth and reproduction.

Somatic growth and reproduction

It can be seen (Table 1) that the temperatures at which the highest rates of somatic growth have been recorded range from 12–30°C while highest rates of cocoon production have been recorded over the range 15–25°C. When interpreting results of this type, consideration should be given to interactions between somatic growth and cocoon production, which may change with temperature as well as the state of maturity of a worm. In experiments on the temperature relationships of the tubificid *Branchiura sowerbyi* (Aston 1968), it was found that the optimum temperature for

Earthworms in waste and environmental management
edited by C.A. Edwards and E.F. Neuhauser, pp. 135–143

Table 1. Effect of temperature on somatic growth and cocoon production of earthworms

Species	Temperature °C			
	Somatic	Cocoon	Preferenda	Author
Allolobophora chlorotica	20° (hatching only)			Gerard (1967)
Allolobophora chlorotica		(15)		Graff (1953)
Allolobophora caliginosa		(12)		Graff (1953)
Allolobophora rosea		(12)		Graff (1953)
Branchiura sowerbyi (imm)		(25–30)		Aston *et al.* (1982)
(mat)	15	25		Aston (1968)
Dendrobaena attemsi		(18–20)		Graff (1953)
Dendrobaena rubida		(18–20)		Graff (1953)
Eisenia foetida	20–29			Kaplan *et al.* (1980)
Eisenia foetida		(25)		Graff (1953)
Eisenia foetida		(16–21)		Gaddie and Douglas (1978)
Eisenia foetida			15.7–23.2	Grant (1955)
Eisenia foetida	25*			Tsukamoto and Watanabe (1977)
Eisenia rosea			24.1–25.6	Reinecke (1974)
Enchytraeus albidus		(17–18)		Ivleva (1973)
Enchytraeus albidus		(15–20)		Learner (1972)
Enchytraeus coronatus	20*	15–20		Learner (1972)
Enchytraeus bucholzi	15–20	20*		Learner (1972)
Limnodrilus hoffmeisteri	≃18–22			Appleby and Brinkhurst (1970)
Limnodrilus hoffmeisteri		25		Aston (1973)
Lumbricillus rivalis	20*	20*		Learner (1972)
Lumbricus rubellus		(15–18)		Graff (1953)
Lumbricus rubellus	12	15*		Evans and Guild (1944)
Lumbricus rubellus	12–15	15–18		Wei-Chun Ma (1984)
Octolasion cyaneum		(15)		Graff (1953)
Peloscolex multisetosus	≃14			Appleby and Brinkhurst (1970)
Tubifex tubifex	≃18			Appleby and Brinkhurst (1970)
Tubifex tubifex		15–22		Aston (1973)

Data in parentheses are estimates based on general experience with cultures. Asterisks are used in cases where the temperature given is the highest used in an experiment.

somatic growth in sexually mature worms was about 15°C whereas in immature worms this was 25–30°C (Fig. 1). This difference appears to be related to the rate of cocoon production at different temperatures, the rate increasing with temperature up to 25°C. It seems likely that in mature worms maintained at this temperature a smaller proportion of energy is used for somatic growth than for cocoon production than at 15°C. When the growth-suppressing effect of cocoon production is not present as in immature worms, the growth rate increases with temperature up to 25°C–30°C. It seems likely, on the basis of these observations that the optimum temperature for somatic growth will fall as a worm becomes sexually mature. This principle may well apply to other species of oligochaete, since it has been recorded in invertebrate groups even less closely related such as crustaceans (Aston and Milner 1982) and triclads (Abeloos 1929; Reynoldsen *et al.* 1965).

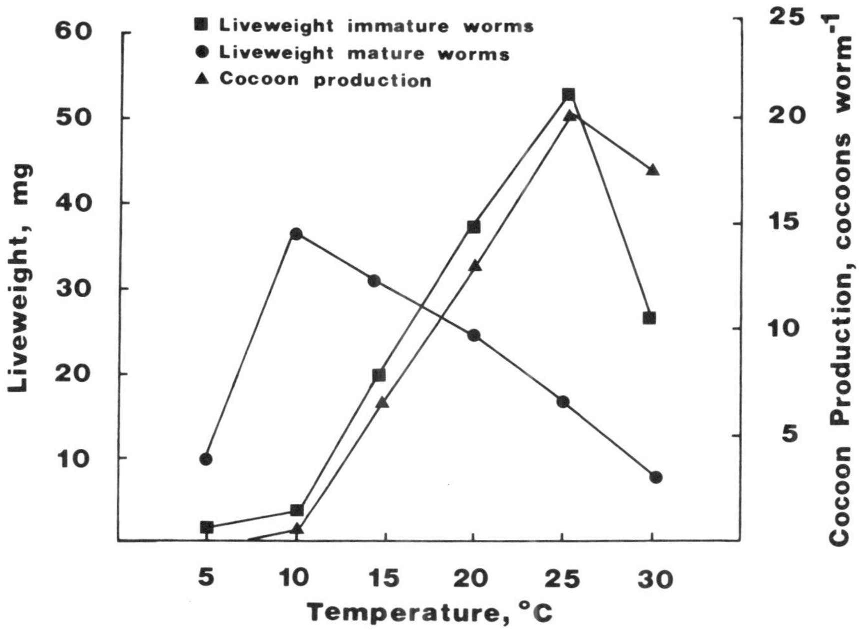

Fig. 1. Average increase in the live weight of immature worms (--------) sexually mature worms (———) and cocoon production (· · · · · ·) in cultures of *Branchiura sowerbyi* maintained over a period of 28 days.

Population growth

From a practical standpoint a vermiculturist is likely to be more interested in the effects of temperature on population growth than on somatic growth or reproduction. The optimum temperature for population growth will be an integration of the temperature effects on not only somatic growth and reproduction but also on mortality and all the sub-processes involved in these major body processes.

A useful index of population growth is the Intrinsic Rate of Natural Increase, 'r', sometimes known as the Malthusian parameter (Pianka, 1978) from which can be derived the time taken for a population to double. The Intrinsic Rate of Natural Increase applies to populations with a stable age structure and is a constant in the equation

$$N_t = N_o e^{rt}$$

where: t = time (in weeks in the following examples).
N_o = number in initial population.
N_o^t = number in final population at time t.

From this equation we can predict the size of a population after a period of time, t.

The value of r can be calculated by iteration from Euler's implicit equation:

$$\Sigma e^{-rx} \ell_x m_x = 1$$

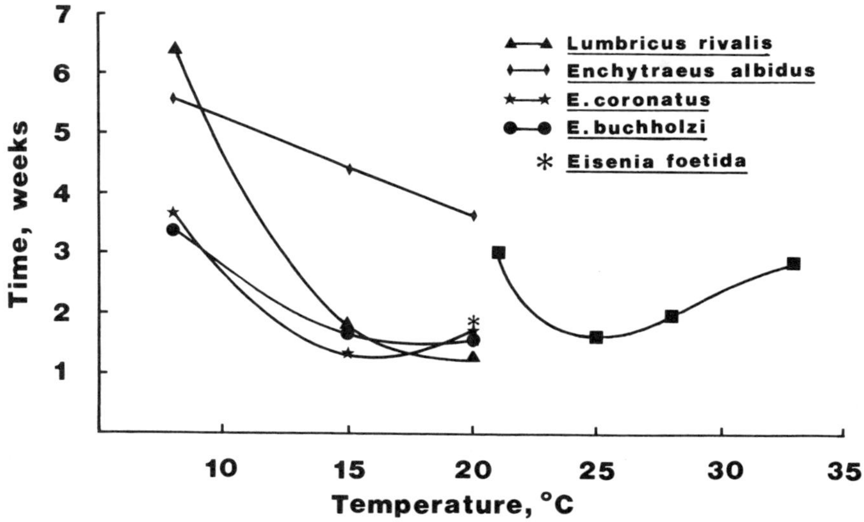

Fig. 2. Time taken for populations to double in relation to temperature in various species of oligochaetes (◇ = *Lumbricillus rivalis*; ◆ = *Enchytraeus albidus*, ○ = *E. coronatus*, ● = *E. buchholzi* in Learner (1972)) (* = *Eisenia foetida* in Hartenstein *et al.* (1979)) (■ = *Branchiura sowerbyi* in Aston *et al.* (1982)).

where: x = subscript representing age
ℓ_x = probability of survival to age x
m_x = number of offspring produced by an average organism of age x per unit time.

Methods for finding ℓ_x and m_x are described in Aston *et al.* (1982) and Pianka (1978).

The time taken for a population to double, T, can be derived for r using the equation

$$T = \frac{\log_{e^2}}{r}$$

A comparison of population doubling times at different temperatures for various oligochaete species is shown in Fig. 2. The data shown are for a limited number of species most of which are not used commercially. The doubling times are very small and probably represent ideal conditions unrestricted by population density and related pressures. Clearly there is a need for accurate information of this type covering a range of temperatures for commercially viable species.

Lethal temperatures

As in other poikilotherms the lethal temperatures of oligochaetes are likely to be affected by a number of factors, but three of these may be of particular importance.

Firstly lethal temperatures are likely to be affected by the temperature at which the worms were acclimatized. In experiments on *Pheretima hupeiensis* for example, Grant (1955) found an average gain in heat tolerance of about 0.3°C for every 1°C rise in conditioning temperature. The minimum time required for conditioning was about 12 days, but a month was allowed in most of the experiments carried out in this study. Secondly, lethal temperatures are likely to vary according to the duration of exposure. For example Grant (1955) in the same set of experiments recorded 100% mortality at 30°C after 10 hours exposure at 27.5°C after 29 hours, and at 25.6°C after 48 hours exposure. Thirdly, in the light of the observation by Hogben and Kirk (1944) that the body temperature of a worm may depend on the relative humidity of the surrounding air, it seems likely that lethal temperature will also be affected by this parameter. Hogben and Kirk (1944) recorded that in air with a relative humidity above 37%, the body temperatures were not greater than 2°C above that of a wet bulb thermometer but at a relative humidity of 18% or lower, the worms were less efficient at maintaining low body temperatures because they became dessicated.

In view of considerable differences in the experimental conditions and difference in the criteria used by different authors, lethal temperatures are not summarized in Table 1. In general, however, it appears that oligochaetes common in temperate regions have rather low lethal temperatures in the range 25–33° (Grant, 1955; Miles, 1963; Reinecke, 1974; Wolf, 1938) while the tropical or sub-tropical species examined have higher lethal temperatures in the range 34–38°C (El Duweini and Ghabbour, 1965; Madge, 1969). It seems likely, that more stringent tests could shed some light on apparent anomalies which now exist. For example, Grant (1955) records that under certain conditions *Eisenia foetida* will die at 24.7°C while Tsukamoto and Watanabe (1977) record best somatic growth rates for this species at 25°C.

Temperature preferenda

The temperature preferenda of some poikilotherms, particularly fishes, have been shown to correspond closely to the optimum temperature for growth (Jobling, 1981), but Edwards and Lofty (1977) draw attention to the fact that the temperatures preferred by earthworms are often lower than the temperatures for rapid growth. This may be a response to greater evaporative losses which are likely to occur at higher temperatures rather than a direct response to temperature. We might expect therefore a closer coincidence of preferred temperatures and the temperature for most rapid somatic growth in aquatic worms, where there are no evaporative losses, than in terrestrial worms.

Temperature and food conversion

The effect of temperature on the efficiency of food conversion, *i.e.*, the proportion by weight of worm tissue produced by a given amount of substrate, has received little attention. There are, however, several studies on fishes (Brett *et al.* 1969; Elliot 1976) in which a close relationship between temperature and food conversion has been found and it seems likely that other poikilotherms including oligochaetes will show a similar relationship. The studies indicate that at the relatively high temperatures suitable for rapid somatic growth the proportion of energy lost in other metabolic

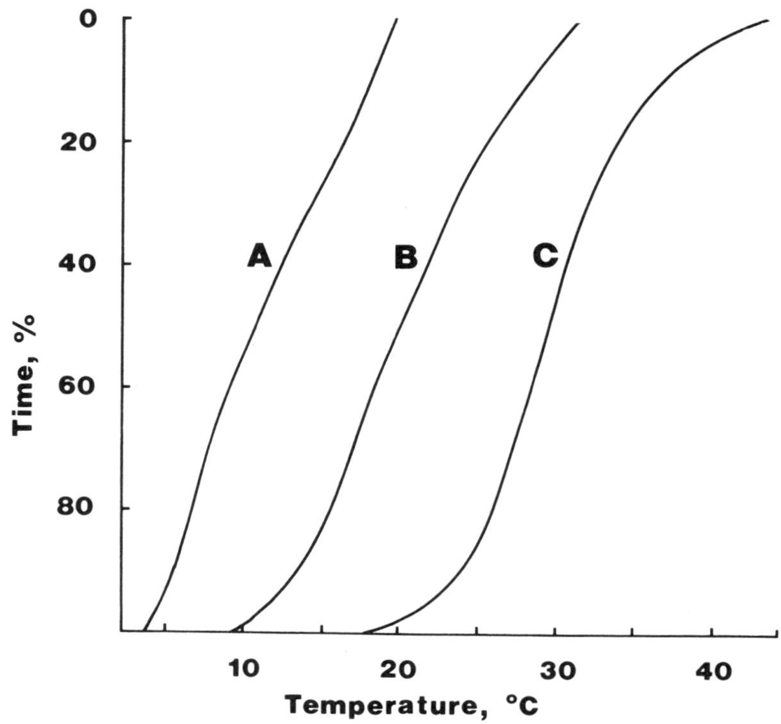

Fig. 3. The proportion of time when water temperatures were above specified levels over a five-year-period in (A) a British Midlands river at natural temperatures (B) the condenser outlets from Bradwell power station (C) the condenser outlets from Ratcliffe power station.

processes is higher than at lower temperatures, highest conversion efficiencies occurring several degrees below the optimum temperature for somatic growth. The implications could be considerable since the vermiculturist could be faced with a choice between a temperature for most rapid growth on the one hand or for best food conversion on the other. This is clearly an area which requires more research and could result in more economic use of worm food substrates.

Temperatures in power station cooling water

Heat is rejected in greater quantities during the generation of electricity than in any other industrial process. The major part of this heat is extracted from steam condensers by cooling water taken from and returned to a river, a lake or the sea. During its passage through the condensers the temperature of the cooling water normally increases by 8–10°C. The temperature of the warmed water leaving the steam condensers is dependent on a number of factors including seasonal temperature, the location, the type of cooling system used and the generating load of the power station. Ratcliffe, *e.g.*, an inland power station sited in the U.K. on the R. Trent, uses closed circuit cooling with cooling towers.

The temperature of the warm water discharged from the condensers ranges from 12°C in winter to about 40°C in summer with a mean of about 28°C. Bradwell power station, also in the U.K. but sited on the coast, uses direct cooling and has lower condenser discharge temperatures ranging from 10°C in winter to about 30°C in summer with a mean of around 20°C. Graphs comparing the proportion of time above various temperatures in a natural river and the condenser discharges from Ratcliffe and Bradwell power stations are shown in Fig. 3.

The temperature range of water discharged from the condensers amply covers the range for optimum growth and reproduction and population growth for most of the oligochaete species shown in Table 1 and Fig. 2. At times, however, the temperature of the cooling water is higher than required and cooling would be necessary. This could be achieved by adding water at natural temperatures from the local river, lake or the sea, from water which has passed through cooling towers or by allowing the culture to be cooled by the surrounding air.

Using reject heat from power stations for vermiculture

Experience on the use of reject heat for vermiculture is limited to small scale culture of aquatic species in the cooling water from Ratcliffe power station. This, together with experience gained from horticultural and fish farming projects at power stations, provides a reasonable basis from which to speculate on the likely structure and operation of a vermiculture project at a power station. It seems likely that the basic methods used to transfer heat from the cooling water to the worm culture will depend on whether terrestrial or aquatic species are used.

Terrestrial species

Heat could be applied from power station cooling water to terrestrial cultures using either water/air heat exchangers to heat the surrounding air or water/substrate heat exchangers in the form of pipes buried in or below the substrate. Efficient water/air heat exchangers have been developed by research workers from CEGB, NE Region (Statham, 1979) for heating a greenhouse complex at Drax Power Station. These are in the form of large thin-walled radiators which carry the warm water from the power station. The radiators are fitted with efficient electric fans which maintain the air at required temperatures with the aid of thermostat controllers. Methods of distributing power station cooling water for soil heating are reviewed in Kawaratani (1978).

Aquatic species

It is envisaged that aquatic species such as tubificids could be cultured in substrates submerged in slow flowing cooling water. A possible advantage of direct contact with the cooling water, apart from maintaining a required temperature is that it would supply dissolved minerals to the culture. In an experiment at Ratcliffe power station which takes cooling water high in dissolved minerals from the Trent, worms were provided only with cellulose as a basic food source (Aston, 1984). The cellulose was colonized and broken down by microorganisms which provided food for cultures of tubificid worms. In the process the microorganisms took up nitrate and other dissolved minerals from the cooling water. Clearly, this was a system capable of fixing an inorganic form of nitrogen into bacterial and eventually worm protein. It is conceivable therefore that vermiculture at an inland power station such as Ratcliffe could provide a process for combining three major waste products – (1) reject heat from a power station; (2) dissolved minerals, particularly nitrate, derived mainly from sewage effluent and land run-off into rivers and (3) waste organic materials providing a source of energy for the metabolism and growth of the culture. One possible disadvantage of holding the cultures under cooling water is that the worms are likely to be stressed or become contaminated by any chemical pollutants which may be present. Some aquatic forms are nevertheless able to tolerate and at the same time accumulate high levels of heavy metals. In such cases there would be a need for fairly regular monitoring of the worm tissue if it were to be incorporated into food chains ending in man.

Acknowledgements

This paper is published by permission of the Central Electricity Generating Board.

References

Abeloos, M. 1929. Influence de la temperature sur la croissance des planaires. Séance Acad. Sci. Paris 188: 881–883.

Appleby, A.G. and Brinkhurst, R.O. 1970. Defecation rate of three tubificid oligochaetes found in the sediment of Toronto Harbour, Ontario. J. Fish. Res. Bd. Canada 27: 1971–1972.

Aston, R.J. 1968. The effect of temperature on the life cycle, growth, and fecundity of *Branchiura sowerbyi* (Oligochaeta: Tubificidae). J. Zool. Lond. 154: 29–40.

Aston, R.J. 1973. Field and experimental studies on the effects of a power station effluent on Tubificidae (Oligochaeta-Annelida). Hydrobiologia 42: 225–242.

Aston, R.J. 1984. The use of cellulose substrate for the culture of the worm *Branchiura sowerbyi* Beddard (Oligochaeta, Tubifidae) Aquaculture (in press).

Aston, J.R. and Milner, A.C. 1980. A comparison of populations of the isopod *Asellus aquaticus* above and below power stations in organically polluted reaches of the R. Trent. Freshw. Biol. 10: 1–14.

Aston, R.J., Sadler, K. and Milner, A.G.P. 1982. The effect of temperature on the culture of *Branchiura sowerbyi* (Oligochaeta, Tubificidae in activated sludge. Aquaculture 29: 137–145.

Brett, J.R., Shelbourn, J.E. and Shoop, C.I. 1969. Growth rate and body composition of fingerling Sockeye salmon, *Onchorhynchus nerka*, in relation to temperature and ration size. J. Fish. Res. Bd. Canada 26: 2363–2394.

Edwards, C.A. and Lofty, J.R. 1977. Biology of Earthworms. Chapman and Hall, London. 333 pp.

El Duweini and Ghabbour 1965. Temperature relations of three Egyptian oligochaete species. Oikos 16: 9–15.

Elliot, J.M. 1976. The energetics of feeding metabolism and growth of brown trout (*Salmo trutta* L.) in relation to body weight, water temperature and ration size. J. Anim. Ecol. 45: 923–948.

Evans, A.C. and Guild, W.J.M. 1984. Studies on the relationships between earthworms and soil fertility IV. On the life cycles of some British Lumbriculidae. Ann. Appl. Biol. 35: 471–484.

Gaddie, P.E. and Douglas, D.E. 1978. Earthworms for Ecology and Profit. Vol. I. Scientific Earthworm Farming. Bookworm Publishing Co. Ontario, California. 50 pp.

Gerard, B.M. 1967. Factors affecting earthworms in pastures. J. Anim. Biol. 36: 235–252.

Graff, O. 1953. Die Regenwürmer Deutschlands. Schrift Forsch. Land Braunschweig-Volk 7: 81.

Grant, W.C. 1955. Temperature relationships in the megascolecid earthworm *Pheretima hupeiensis* Ecology 36: 412–417.

Hogben, L. and Kirk, R.L. 1944. Body temperatures of worms in moist and dry air. Proc. Roy. Soc. Londond 132: 239–252.

Ivleva, I.V. 1973. Mass cultivation of invertebrates. Biology and methods. Academy of Sciences of the

USSR. All-Union Hydrobiological Society Izdatel'stvo 'Nauka' Moskva 1969. Israel Programme for Scientific Translations Jerusalem. 34 pp.

Jobling, M. 1981. Temperature tolerance and the final preferendum – rapid methods for the assessment of optimum growth temperatures. J. of Fish Biol. 19: 439–456.

Kaplan, D.L., Hartenstein, R., Newhauser E.F. and Malecki, M.R. 1980. Physiochemical requirement in the environment of the earthworm *Eisenia foetida*. Soil Biol. Biochem. 12: 347–352.

Kawaratani, R.K. 1978. State-of-the Art. Wast Heat Utilization for Agriculture and Aquaculture. Technical Planning Study Contract Nos SPS 77–700. TV 45928A, TPS 77734. TV 47396A. Electrical Power Research Institute. 111 pp.

Learner, M.A. 1972. Laboratory studies on the life-histories of four enchytraeid worms (Oligochaeta) which inhibit sewage percolating filters. Ann. Appl. Biol. 70: 251–266.

Ma, Wei Chun. 1984. Sublethal effects of copper on growth, reproduction and litter breakdown activity in the earthworm *L. rubellus*, with observation on the influence of temperature and soil pH. Environ. Pollut. Ser. A. 33: 207–220.

Madge, D.S. 1969. Field and laboratory studies on the activities of two species of tropical earthworms. Pedobiologia 9: 188–214.

Miles, H.B. 1963. Heat death temperature in *Allololobophora terrestris* f. *longa* and *Eisenia foetida*. Nature (Lond.) 199: 826.

Pianka, E.R. 1978. Evolutionary Ecology. Harper and Row, New York, 2nd Ed. 397 pp.

Reinecke, A. 1974. The upper lethal temperature of *Eisenia rosea* (Oligochaeta) Wetenskaplike Bydraes van die PU vir CHO Reeks B: Natuurwetenskappe 62: 1–4.

Reynoldson, T.B., Young, J.O. and Taylor, M.C. 1965. The effect of temperature on the life cycle of four species of lake dwelling triclads. J. Anim. Ecol. 34: 24–45.

Roots, B.I. 1956. The water relations of earthworms. II Resistance to dessication and immersion, and behaviour when submerged, and when allowed choice of environment. J. Exp. Biol. 33: 29–44.

Statham, J. 1979. Existing and proposed projects with the CEGB to demonstrate and evaluate the potential horticultural and other uses of reject heat in cooling water. Paper presented to workshop conference Low Grade Heat, a Source in Cold Climates 6–10 October, Chalk River, Ontario, Canada.

Tsukamoto, J. and Watanabe, J. 1977. Influence of temperature on hatching and growth of *Eisenia foetida* (Oligochaeta, Lumbricidae) Pedobiologia 17: 338–342.

Wolf, A.V. 1938. Studies on the behaviour of *Lumbricus terrestris L.* to dehydration and evidence for a dehydration tropism. Ecology 19: 232–242.

THE ECONOMIC VIABILITY OF EARTHWORM CULTURE ON ANIMAL WASTES

R.S. FIELDSON
National Institute of Agricultural Engineering, Silsoe, Beds, U.K.

1. Introduction

Small-scale earthworm production on a variety of municipal, industrial and agricultural wastes has been demonstrated to be technically feasible, and the hardware and other requirements for expanding the scale of operations are becoming better understood. A variety of entrepreneurs in Britain, the rest of Europe and the USA, have set up profitable worm-breeding businesses, and some municipal authorities have tested vermiculture systems as a cost-effective method of waste treatment and disposal. The question now is whether the technical and engineering knowledge for the expansion of worm production can be matched by attractive economic returns to large-scale production. Of course, economic returns are not the only motivation for the introduction of vermiculture. Some installations are primarily for research purposes, and there may be valid political or environmental reasons for installing vermiculture units in areas with severe waste handling problems. Nevertheless, vermiculture must normally be expected, if not actually to show a financial profit, at least to be cost-effective when compared to alternative methods of treating or disposing of the wastes on which the worms are bred.

This paper is based on work which has been carried out at the NIAE into the economics of earthworm culture on animal wastes (Foote and Fieldson, 1982). However, many of the factors which influence the viability of the systems under review are also important in other vermiculture systems and in other countries, and so attempt is made to draw out the general implications of the study rather than go into details of the specific example. The paper should therefore be of relevance to all interested in assessing the economic viability of earthworm production.

2. The animal waste problem

Although research has taken place in the UK into earthworm production on a variety of different waste substances, animal manures and slurries were chosen at an early stage as suitable material for vermiculture because of their ready availability and their suitability (when properly prepared) as a breeding environment for worms.

A wide range of figures are quoted on the quantity of animal manures and slurries produced in the UK, but the values given in Table 1 are reasonable estimates of the amounts produced by housed animals (*i.e.*, excluding that which occurs in the field).

Although dairy farms are the largest source of such wastes, their disposal problems are not normally severe as most units have sufficient land on which to spread slurry or FYM at safe levels of application. If there is enough grassland to feed the herd there is enough to permit the disposal of the waste. However, the costs of storage and field-spreading commonly exceed the fertilizer value when applied to the crops.

Earthworms in waste and environmental management
edited by C.A. Edwards and E.F. Neuhauser, pp. 145–156

Table 1. Summary of main sources of farm animal wastes[a]

	Dairy cattle	Pigs	Poultry
Excreta/animal/day (kg)	41	4	0.125
Total from housed animals (million tonnes p.a.)	22	9	1.8
% as liquid or semi-liquid slurry	60	80	Low
Severity of disposal problems	Low	High	Medium

[a]Figures summarized from Larkin *et al.* (1982).

The pig industry contains many more large-scale intensive units with a small land base, so disposal problems are more serious. It was calculated that 45% of the pigs in England and Wales are on farms where the number of pigs per ha is too high for safe land-spreading of wastes within the farm. This was consistent with an earlier survey in Scotland which found that 49% of farms had stocking rates which represented a pollution risk (Brownlie, 1974). Informal contracts between pig producers and neighboring farmers for the disposal of slurry are common, especially in areas with high concentrations of pigs, and the receiving farmer usually meets the costs of transport and spreading. The pig industry has been the cause of pollution problems in some areas, notably as a result of odors released during agitation or emptying of slurry stores. It has been subject to by far the largest number of pollution complaints against agriculture, and many farmers have had to install expensive slurry treatment plant to overcome pollution problems.

The poultry industry is also characterized by very large units on a small land base. However, chicken manure has a high dry matter content so is easy to handle and stack, and has a high fertilizer value making it suitable for land spreading. Most poultry farmers have little difficulty disposing of their waste to neighboring farmers, some of whom pay for it. Nevertheless, for some large operators the organizational, storage, and transport problems are so acute that alternative waste disposal systems could be attractive.

From this brief overview it can be seen that, from the point of view of waste disposal problems and resultant pollution risk, large slurry-producing pig farms appear to offer the largest potential demand for vermiculture. There are already high costs attached to the disposal of slurry for some farmers, so the savings which could result from vermiculture are greatest. Even where land spreading of animal manure or slurry is not a problem, the costs of storing wastes from animals housed over winter and then spreading in the spring and summer normally exceed the manurial value calculated on the NPK content of the waste. When the investment in waste handling, lagoon or above ground storage and the time and equipment needed for spreading are taken into account, it was calculated that it typically costs over four times as much to get the slurry on to the land as it is worth as fertilizer (Foote and Fieldson, 1982).

This is an important point for vermiculture systems based on other waste products. Even if, as often seems likely, the process cannot be justified solely on the value of the products which it creates, it may be economically viable when compared with alternative costs of disposing of the wastes on which worms are bred. This is consistent with experience on sludge-based vermiculture systems in the USA. The products are sometimes not marketed at all, or if they are marketed then at relatively low prices, but the system may still be cost-effective compared to other techniques of sludge disposal (Alpert and Johnson, 1983).

3. Methods of vermiculture on animal wastes

Information on the technical and scientific aspects of worm production was provided by Dr. Edwards' unit based at Rothamsted Experimental Station (R.E.S.). Inevitably there are gaps in the data as so many aspects of worm production are still being researched. In addition, results from laboratory trials are commonly not representative of what occurs under farm-scale operations. In many respects, the whole system is still at the 'learning stage' and so results from trials carried out in the past should be improved upon as the understanding of environmental control and other influences on worm growth improve.

Most of the analysis is on separated slurry systems as these have received most attention at R.E.S. and so are more fully understood, and also slurry is the most common form of livestock waste. Whole slurry requires separation or the addition of a bulking agent such as straw in order to raise its dry matter to the level required for worm culture. Slurry separators are uncommon in the UK, with only about 300 in operation. Capital costs are high, and there have been reliability problems, especially with the ancillary equipment. Worm culture on slurry plus chopped straw requires that the mixture is left to compost before the addition of earthworms, and also increases the time the worms take to process the material. This increases the bed area needed for a given amount of waste, but the method may nevertheless be appropriate to smaller farms with slurry based systems.

Calculations have been based on the assumption that worm breeding occurs at the farm level, and that the worm-worked mixture (including worms) is transferred to centralized units for processing and packaging. Transport of unseparated slurry for centralized worm breeding would be uneconomic, as only about 20% of the volume is used for worm production. But if each farm had a slurry separator and only the separated solids were transported to a central breeding area, then the scope for economies of scale would be much reduced. An attractive possibility might be to have a mobile separator which moved from farm to farm and then to transport just the separated solids to the worm-breeding center. No such system has been developed yet, but it merits closer study.

It is assumed that worms would be bred in a bed-system in which the waste is gradually added in thin layers and the worms work their way up through it. The rate at which worms work the waste is difficult to estimate. This is most conveniently expressed in kg per m^2 per day which can be converted by the worms, and this allows the time needed to convert a given volume of waste to be calculated directly (App. 1). In previous trials at R.E.S. the rate of conversion has not exceeded about 3 $kg/m^2/day$ for separated pig slurry, and 4 $kg/m^2/day$ for separated cattle slurry. Trial beds have often been interrupted by the need to remove part of the worm population or retarded by variations in temperature or moisture. It is therefore assumed that an undisturbed worm population in an environment designed for maximum conversion could readily achieve a rate of 4 $kg/m^2/day$ (or about four months for the conversion of 0.5 m of separated slurry), and that eventually this might be raised to 8 $kg/m^2/day$ (two months to convert 0.5 m). These figures can be verified operationally, and the results of the analysis are dependent on such verification.

The optimum breeding environment for worms will be discussed in detail in other papers, so will only be touched on here. There are many alternative bed designs using different materials and of varying degrees of sophistication (these are considered in detail in Billington and Price, 1984). The design which was costed in the economic analysis consisted of a concrete base with underbed heating, wooden walls, and bed surface insulation, the whole bed being housed in a relatively cheap insulated polytunnel. A gantry system is provided to reduce labor inputs. Calculations of the bed area and costs are at Appendix 2. There are no doubt cheaper systems, but it is important to recognize that they only improve the overall economics if they do not bring about a proportionate reduction in productivity (which is assumed to be at quite a high level). Removal of the worms and worm-worked waste from the beds is assumed to be done by conventional materials handling equipment already available on the farm.

Processing of the worm-worked waste depends on the end use to which it is to be put. If it is applied directly to the land as a soil conditioner, then no further processing is required. However, it seems inevitable that for the vermiculture process to be economically viable the worms and worked waste will have to be marketed, in which case separation and processing of the worms, and drying, blending, and packaging of the compost will be required. Machines have been developed at NIAE specifically for separating worms from the compost.

The processing stage is not costed in detail: it is assumed that this would be done at central plants, and that the value to the farmer of the worms and compost can be reduced from the market price by an amount to cover the costs of transport, processing, and marketing. The processes and costs of such centralized plants have been examined in more detail by other NIAE researchers (Ducker, 1983; Billington, 1984).

4. The benefits of vermiculture on animal wastes

There are three areas of potential economic benefits of vermiculture on animal wastes:

1. the worms;
2. the worm compost;
3. reductions in waste disposal costs.

4.1. *The worms*

There have been a number of profitable outlets for small-scale worm producers such as fishing bait, feed for birds, fish, etc., and, not least, sales to newly established worm breeders! Such markets are easily saturated and do not represent a long-term demand for the output of larger-scale producers. Limited markets may develop for the application of worms to the land, *e.g.*, newly-reclaimed areas. But the main large-scale market for worms is likely to be as a source of protein in animal feed rations. The quality of earthworms as animal feed will be discussed in detail by other contributors. For the purposes of the economic analysis it was assumed that processed worm protein could be substituted for alternative protein sources, such as fishmeal, at values of around £300 per tonne dry weight. An allowance of £50 per tonne was made for the costs of processing the worms, leaving the value to the farmer at £250 per tonne dry weight. The dry weight of worms was assumed to be 20% of wet weight, and the proportion of worms (wet weight) in the worm-worked waste was assumed to be 5%.

4.2. *The worm compost*

The worm compost appears, from the trials which have been carried out so far, to be potentially valuable in compost mixes for the horticulture industry. Though varying slightly according to the type of animal waste from which it was produced, the worm compost has higher nutrient content than peat and good slow release characteristics. There is room for further studies of the characteristics and composition of the material, but there should be a ready market for it among existing compost purchasers if the price is right. Areas of potential use include nurseries, landscaping, tree-planting etc.

As with the worms, the smaller-scale producers in several markets have been able to sell worm compost at very high prices for specific purposes, *e.g.*, more than £1 per kg for blends including worm compost sold as organic pot-plant additive. Such markets will expand only if there is clear evidence of the advantages of worm compost over alternative blends, and it is unlikely that such benefits will be large enough to command the large price premia that have existed so far.

Examples of prices for growing media are given in Table 2. These are prices from the wholesaler exclusive of transport. Peat prices vary very much according to type and quality, and the figure given is the minimum level quoted. Levington composts are peat based, balanced with nutrients and trace elements, and the John Innes are soil-based composts. The organic composts include those based on composted FYM, horse manure, poultry manure, and household wastes.

With all types of growing media it is important for many applications to ensure that the material is reasonably uniform, and then the nutrient balance can be adjusted as required by the grower. If this can be achieved in the case of worm-worked waste, then mixes based on worm-compost should be able to compete with blended com-

Table 2. Wholesale prices of various growing media

	£ per 25 kg bag	£ per m^3
Peat		12.50 upwards
Levington compost		20–35
John Innes compost	1.45–1.70	58–68[b]
Organic composts	2.50–3.50	100–140[b]

[a] Source: Anon (1984).
[b] Derived from previous column.

posts such as Levington's and John Innes. Therefore, worm compost was assumed to be worth £25–£50 per tonne after processing, blending and drying. Allowance of £10 per tonne is made for transport and processing costs (including the material added during blending) although this is clearly an approximation making the price to the farmer £15–£40 per tonne. Higher prices, up to those of the organic composts now on sale, may be achievable for limited volumes, but cannot realistically be expected to continue if worm production accelerates quickly.

There is the danger that the expansion of worm production could lead to saturation of the compost market and depression of the existing price level. The present UK market for peat is about 250,000 tonnes per year, so if only 50 pig farmers with an average of 2,500 pigs, each converted all their slurry into compost, they would provide about 5% of market requirements. In the long-term, there may be shortages of peat, but there is no reason to expect them in the near future unless availability is limited for political reasons.

The effects of transport costs on competitiveness are significant: worm growers who have a ready local market for composts have a major advantage. It must be emphasized that thorough market research, product development, quality maintenance, and sales effort, are essential if worm compost is to compete effectively. Literature from the USA indicates that failure to develop adequate markets for composts is one of the biggest problems facing municipal waste treatment plants with composting or vermiculture units (Russell, 1982; Franz, 1978). This adds further weight to the argument for centralized processing plants whose management can develop the expertise in these areas that the individual farmer would not normally have.

4.3. *Reductions in waste disposal costs*

As costs of handling, storage, and disposal of animal wastes are substantially greater than their manurial value, the reduction in volume that occurs when worms are grown on it brings about a reduction in waste disposal costs. There may also be qualitative improvements in animal waste resulting from vermiculture and its associated processes. For example, worm-worked waste is odor free and readily stackable, and separated liquid is more easily pumped and less prone to crusting and stratification than raw slurry.

The extent of such savings depends very much on the individual farm situation. For example, on an established unit, slurry storage facilities may already exist which would become partially unnecessary, but on a new unit, a smaller capacity would be constructed from the start. It also depends on the details of the system used: if whole slurry could be used, perhaps with a bulking agent, then the savings would be much greater than if only the separated solids can be utilized.

On farms with severe waste disposal problems at present, the potential savings may provide a powerful economic incentive for the introduction of vermiculture. However, in the average case, they will be very small, and so no allowance for them has been included in the economic analysis.

5. The results of the analysis

Calculations of the economic costs and benefits of separated slurry based vermiculture systems are summarized in Appendix 3.1 and 3.2 for pig and cattle units respectively. Appendix 3.3 gives the net annual returns at the two assumed rates of conversion and the alternative values for compost. While these results are specific to the particular systems and assumptions on which they are based, a number of interesting conclusions emerge.

First, the viability of vermiculture depends heavily on the value of the compost. The proportion of total benefits attributable to the worms and the worm compost is determined by their proportions in the worm-work waste and the relative prices assumed, and so does not vary with farm size or type. The worms contribute only 13% of total benefits with compost at £40 per tonne, or 35% with compost at £15 per tonne. The success of the system is therefore more sensitive to changes in compost prices than prices obtained for the worms, and this makes effective marketing of the compost a key factor.

Capital costs make up a high proportion of the total for smaller units but for large units, operating costs become dominant. As a result of the high costs of beds and slurry separators, capital costs form about two-thirds of total costs on the cattle-based and smallest pig-based system, but for medium and large pig units operating costs represent 50–62%. Labor costs become more prominent on the larger units at about a third of total costs, and on the 5000 pig farm, one man would be required virtually full-time to operate the vermiculture unit. Actual labor costs depend on the individual farm's labor availability. In some cases, work on the vermiculture unit could be carried out during slack periods when the opportunity cost of labor is low, but in general, farmers are unlikely to want to take on additional workers, or to want the vermiculture operation to interfere with their existing enterprises. A low labor input system will therefore be most attractive, and there may be further scope for mechanization, *e.g.*, of the harvesting stage.

There are substantial economies of scale in the assumed system, particularly relating to slurry separation. Therefore, only the larger systems show profits (2500 and 5000 pigs and 200 cow herds). On smaller units slurry separators would clearly be underutilized and the introduction of a smaller scale cheaper slurry separator than those currently available on the UK market would make the economics of vermiculture much more attractive on such holdings. It was calculated that if the price of the separator is reduced by 50% *all* the systems are profitable with the price of the compost at £40 per tonne and the higher conversion rate. Alternatively, vermiculture systems for smaller farms will have to be based on FYM or slurry plus chopped straw. Preliminary calculations have been made of a FYM-based system, assuming that prechopping of the material would be needed and that the rate of conversion would be half that of separated slurry. For a 100 cow herd the system could just be viable with compost at £40 per tonne, though no allowance was made for the greater decline in waste volume during processing or the probably inferior quality of the compost.

Detailed calculations of the economics of centralized worm production have not yet been made, but a simple model of one possible system showed it to be economically attractive if the liquid fraction of separated slurry could be utilized within the system, or disposed of at little cost. This might be difficult in areas of high livestock densities. In addition, the model was highly sensitive to distances over which slurry was transported, and to variations in transport costs.

6. Conclusions

The results of the analysis rest on the assumptions which go into it, including assumptions on the techniques to be used, on the productivity achieved, and on the value of the products. All of these require much further research and will, no doubt, be subject to modification. However, incorporating existing information into an overall economic analysis of the system draws attention to gaps in present understanding and to variables which are crucial to the long-term expansion of vermiculture. For example, the rate at which worms can convert different types of waste under given environmental conditions is not yet reliably known, but this influences the area of beds and housing and the inputs of labor, so has a major effect on profitability.

The study also highlights the importance of the state of the waste to the viability of the process. If expensive pre-treatment is required before worms can be introduced, then the whole system may become economically unattractive. This is a challenge to adapt vermiculture to the more difficult breeding media. For example, the need to expand research into worm breeding on slurry (with different bulking agents) rather than on separated solids of FYM is indicated.

Finally, the study draws attention to the vital role of establishing markets for the products of vermiculture. A sign at the farm gate will clearly not do! The industry must produce well-blended, uniform, well-packaged products at appropriate levels of moisture content and consistency. This demands a good understanding of the commercial horticulture industry, and may be the point at which the researcher, like the farmer, has to hand over to the experienced commercial processing and marketing company.

References

Anon. 1984. Pots Fertilisers and Growing Media. Supplement to Gardeners' Chronicle/Horticultural Trade Journal, Jan 1984.

Alpert, J.E. and Johnson, P. 1983. Cost Analysis of Static Pile Composting in the United States. Paper presented at the Conference on Composting of Solid Wastes and Slurries. University of Leeds, 1983.

Billington, R.S. and Price, J.S. 1984. An Economic Assessment of the Worm Culture Stage for Farm Scale Vermiculture Systems in the United Kingdom. Div. Note DN/1233, NIAE, Silsoe, UK (unpubl.) 51 pp.

Billington, R.S. 1984. An economic assessment of the unit operations needed for a plant, processing earthworm compost and earthworm protein. Divisional Note DN/1252, NIAE, Silsoe, U.K. 57 pp.

Brownlie, T.G. 1974. Department of Agriculture for Scotland. Personal Communication.

Ducker, J.R. 1983. A Preliminary Process Flowsheet for an Earthworm Compost Processing Plant. Div. Note DN/1211, NIAE, Silsoe, UK (unpubl.) 29 pp.

Foote, N. and Fieldson, R.S. 1982. A Study of the Economic Feasibility of Earthworm Culture for Feed Production and Animal Waste Conversion on Livestock Units in England and Wales. Div. Note. DN/1141, NIAE, Silsoe, UK (unpubl.) 68 pp.

Franz, M.F. 1978. The Economic Future of the Earthworm in Recycling. *In* Compost Science/Land Utilisation. 19(6).

Larkin, S.B.C., Morris, R.M., Noble, D.M. and Radley, R.W. 1981. Resource mapping of agricultural wastes and residues. NCAE, Silsoe, UK. 693 pp.

Russell, S.H. 1982. Resources Recovery Economics. Marcel Dekker Inc., New York.

Appendix 1

Calculation of days required for conversion of waste

Assumptions: Final bed depth = 0.5 m
Loss of volume of waste in processing = 25%
Loss of weight of waste in processing = 10%

Depth of bed = 0.5 m

∴ Final volume of worked waste per m^2 = 0.5 m^3
Loss of volume in processing = 25%
∴ Total volume of waste added = 0.667 m^3
Initial density of waste = 0.7 t/m^3
∴ Initial weight of waste added = 0.467 t
= 467 kg

Rate of conversion kg/m^2/day:	2	4	6	8	10
Days required for conversion:	233.5	116.8	77.8	58.4	46.7

Appendix 2.1. Calculation of bed and housing costs: pigs

Animals per herd	500		2,500		5,000	
Slurry/animal/day (liters)	5		5		5	
Sep. fiber/animal/day (liters)	1		1		1	
Sep. fiber/herd/day (liters)	500		2,500		5,000	
Days housed per year	360		360		360	
Sep. fiber while housed m^3	180		900		1,800	
Density of sep. fiber t/m^3	0.7		0.7		0.7	
Sep. fiber while housed t	126		630		1,260	
Rate of conversion $kg/m^2/day$	4	8	4	8	4	8
Number of harvests per year	3	6	3	6	3	6
Area of beds required m^2	87.5	43.8	437.5	218.8	875	437.5
Assumed bed width m	3	3	3	3	3	3
Length of beds m	29.2	14.6	145.8	72.9	291.7	145.8
Number of beds	1	1	7	4	15	7
Area of housing m^2 (Bed area × 1.5)	131.3	65.7	656.3	328.2	1,312.5	656.3
Concrete base @ £11/m^2	962.5	481.8	4,812.5	2,406.8	9,625	4,812.5
Underbed heating @ £2/m^2	175.0	87.6	875.0	437.6	1,750	875.0
Bed walls @ £5/m	322.0	176.0	1,668.0	849.0	3,367	1,668.0
Surface insulation @ £1/m^2	87.5	43.8	437.5	218.8	875	437.5
Insulated polytunnels @ £8/m^2	1,050.4	525.6	5,250.4	2,625.6	10,500	5,250.4
Bed feeding hopper system	1,000.0	800.0	2,200.0	1,400.0	3,600	2,200
Total costs of beds and housing	3,597.4	2,114.8	15.243.4	7,937.8	29,717	15,243.4

Appendix 2.2. Calculation of bed and housing costs: dairy cattle

Animals per herd	100		200	
Slurry/animal/day (liters)	41		41	
Sep. fiber/animal/day (liters)	8.2		8.2	
Sep. fiber/herd/day (liters)	820		1,640	
Days housed per year	180		180	
Sep. fiber while housed m^3	148		295	
Density of sep. fiber t/m^3	0.7		0.7	
Sep. fiber while housed t	103.6		206.5	
Rate of conversion $kg/m^2/day$	4	8	4	8
Number of harvests per year	3	6	3	6
Area of beds required m^2	72	36	144	72
Assumed bed width m	3	3	3	3
Length of beds m	24	12	48	24
Number of beds	1	1	2	1
Area of housing m^2 (Bed area × 1.5)	108	54	216	108
Concrete base @ £11/m^2	792	396	1,584	792
Underbed heating @ £2/m^2	144	72	288	144
Bed walls @ £5/m	270	150	540	270
Surface insulation @ £1/m^2	72	36	144	72
Insulated polytunnels @ £8/m^2	864	432	1,728	864
Bed feeding hopper system	1,000	800	1,200	1,000
Total costs of beds and housing	3,142	1,886	5,484	3,142

Appendix 3.1. Calculation of costs and benefits (£): pigs

Herd size	500		2,500		5,000	
Rate of conversion kg/m²/day	4	8	4	8	4	8
Costs						
Capital costs						
Beds and housing	3,597	2,115	15,243	7,938	29,717	15,243
Slurry separator	10,000	10,000	10,000	10,000	10,000	10,000
Total	13,597	12,115	25,243	17,938	39,717	25,243
Depreciation over 5 yrs (p.a.)	2,719	2,423	5,049	3,588	7,943	5,049
Interest @ 15% p.a.	1,020	909	1,893	1,345	2,979	1,893
Annual capital costs	3,739	3,332	6,942	4,933	10,922	6,942
Operating costs p.a.						
Energy	680	365	3,200	1,625	6,350	3,200
Labor	678	750	4,746	3,000	10,170	5,250
Maintenance and repairs	272	242	505	359	794	505
Total operating costs	1,630	1,357	8,451	4,984	17,314	8,955
Total annual costs	5,369	4,689	15,392	9,917	28,236	15,897
Benefits						
Output of beds p.a (m^3)	131.25		656.25		1,312.5	
Weight of output (t)	110.25		551.25		1,102.5	
– of which compost =	104.7		523.7		1,047.4	
– of which worms =	5.5		27.6		55.1	
Dry weight of worms (t)	1.1		5.5		11.0	
Value of worms @ £250/t	275		1,375		2,750	
Weight of dried compost (t)	52.35		261.85		523.7	
Assumed value of compost/t	15	40	15	40	15	40
Total value of compost	785	2,094	3,928	10,474	7,856	209
Total value of output	1,060	2,369	5,303	11,849	10,606	236

Appendix 3.2. Calculation of costs and benefits (£): dairy cattle

Herd size	100		200	
Rate of conversion kg/m²/day	4	8	4	8
Costs				
Capital costs				
Beds and housing	3,142	1,886	5,484	3,142
Slurry separator	10,000	10,000	10,000	10,000
Total	13,142	11,886	15,484	13,142
Depreciation over 5 yrs (p.a.)	2,628	2,377	3,097	2,628
Interest @ 15% p.a.	986	891	1,161	986
Annual capital costs	3,614	3,268	4,258	3,614
Operating costs p.a.				
Energy	568	309	1,087	568
Labor	678	750	1,356	750
Maintenance and repairs	263	238	310	263
Total operating costs	1,509	1,297	2,753	1,581
Total annual costs	5,123	4,565	7,011	5,195
Benefits				
Output of beds p.a (m³)	108		216	
Weight of output (t)	91		181	
– of which compost =	86.5		172	
– of which worms =	4.5		9	
Dry weight of worms (t)	0.9		1.8	
Value of worms @ £250/t	225		450	
Weight of dried compost (t)	43.25		86	
Assumed value of compost/t	15	40	15	40
Total value of compost	649	1,730	1,290	3,440
Total value of output	874	1,955	740	3,896

Appendix 3.3. Net annual returns of vermiculture on separated slurry under various assumptions (£ p.a.)

Assumed value of compost	£ 15/t		£ 40/t	
Assumed rate of conversion kg/m²/day	4	8	4	8
500 pig herd	–4,309	–3,629	–3,000	–2,320
2,500 pig herd	–10,059	–4,614	–3,543	+1,932
5,000 pig herd	–17,630	–5,291	–4,538	+7,801
100 cow herd	–4,249	–3,691	–3,168	–2,610
200 cow herd	–5,271	–3,455	–3,121	–1,305

DOMESTIC VERMICOMPOSTING SYSTEMS

MARY APPELHOF
Flowerfield Enterprises, Michigan, USA

Introduction

Earthworms have been used in a systematic way to convert household kitchen waste to compost for at least 30 years. Early systems were large and cumbersome, requiring a considerable volume of leaves, soil and manure to provide a suitable environment for the resident worm population. Organic waste burial in these units did, however, get rid of the waste, feed an increasing worm population, and provide additional nutrients in the compost produced. The compost enhanced plant growth in gardens. The worms were either sold for bait, used for fishing, or transplanted to the garden with the compost. When first mention is made of using worms inside the home to process organic kitchen waste, many are dubious. In 1981, a survey on vermicomposting of 800 individuals was taken; a third expressed a willingness to use worms to process food wastes; another third were willing to learn more about vermicomposting (Appelhof, 1981). Surprisingly, only a quarter could not stand the thought of using worms. Generally, most people had never heard of the concept. The more they learned about the technique, the more likely they were to want to try it.

A book describing how to set up and maintain home vermicomposting systems was published (Appelhof, 1982). Over 150 individuals and organizations have set up vermicomposting systems based on this book. The following survey of 126 users summarizes home vermicomposting as it is currently being utilized in the United States. Included are size, type and location of containers, beddings used, and kinds of garbage buried. Advantages, disadvantages and problems encountered are also addressed.

Methods

A three-page questionnaire was mailed to 126 individuals in 1983. The questionnaire had been previously developed with help from behavioral psychologists, Dr. Scott Geller and Dr. Cheryl Poche, who served as consultants on a study assessing attitudes towards vermicomposting. The 24 questions were intentionally open-ended to allow as much description as respondents were willing to provide. This format precludes a statistical treatment of the data, which is reflected in the results. It is estimated that it took about 30 min to complete the questionnaire.

Results

Response rate for this survey was slightly more than half (52%). Questionnaires varied in degree of completeness; some consisted of one and two word answers, others gave extensive details, dimensions, procedures and comments. Although the total $n = 65$, the number answering individual questions varied because certain questions were omitted, or, *e.g.*, a husband and wife filled out one questionnaire and gave

Earthworms in waste and environmental management
edited by C.A. Edwards and E.F. Neuhauser, pp. 157–161

demographic data for each. In eight instances, page 2 was not completed, presumably because it was on the back-side of page 1 and missed by the respondent.

Characteristics of the respondents

The age range for participation in home vermicomposting varied, from 2½ to 83 years. The majority were between 30–65 (73%), with 8% younger than 29, and 19% over 65. More women than men (58% to 41%) participated in this survey, and, presumably, the activity of vermicomposting itself. Almost half (47%) indicated an income level over $20,000 per year; 30% were between $10,000–20,000, 16% between $5,000–10,000, and 4% under $5,000. By far, the greatest number of respondents, lived in single family housing (89%), described variously as an old house, ranch, bungalow, mobile home, cabin, farmhouse, and rowhouse. Five respondents (8%) lived in apartments, with 2–4 units, and one with 60 units. The diversity of occupations was so great that to categorize them is not useful. There were several retirees, several homemakers, and a few teachers. Some of the other occupations were: author, engineer, attorney, dental hygienist, electrician, social workers, barber, clergyman, physician, auditor, student, handweaver, horticulturist and a director of a research institute.

Reasons for setting up a vermicomposting system

Of the many reasons given for setting up a vermicomposting system, about one third expressed a dual purpose: to dispose of garbage and make vermicompost. One-half did it 'to produce casting', 'potting soil', or 'compost'. About one third felt it would improve their composting efforts by reducing odor. Twelve (18%) did it 'to be able to compost organic materials through the winter months.' Four wanted to improve their garden soil. One particularly well-expressed statement was: 'To obtain rich soil for houseplants and garden and to learn a winter method in the cold north where my compost pile is inactive.'

How they learned about vermicomposting

The majority of correspondents (92%) indicated that they learned about vermicomposting through articles or my book. Twenty-seven (43%) specifically mentioned a feature article in Organic Gardening magazine. Other articles mentioned were Gardens for All newsletter (4), the Christian Science Monitor (3), and Prevention (2). The New York Post, Cincinnati Enquirer, Whole Earth Catalog (Co-Evolution Quarterly), and Mother Earth News were all mentioned once. Two learned about the technique from friends, two from a neighbor, and two from a unit set-up in an energy demonstration house.

Harvesting of worms and castings

Two-thirds (44%) indicated that they had harvested either the worms or castings. A variety of methods were used, most of which are described in my book. Eleven (25%) picked out worms by hand, seven 'dumped and sorted', six 'let the worms do the sorting', and six skimmed castings off the top. Several people experimented, trying one method, then another. Two let the worms die by drying out, then used the castings for their gardens. The most unique method mentioned was a plan to use an electric prod on a compost pile to get the worms out to re-establish the bin in the fall. Frequency of harvesting varied from once a week to once a year, with most harvesting between three and six months. Most keyed their harvesting to when they wanted to make use of the vermicompost and/or worms. Worms harvested were: used to start new bins (46%), put in the garden (15%), for fishing (15%), to feed robins or reptiles, or to put into compost bins or greenhouse beds. Several helped to set up family and friends with their excess of worms. Castings were used directly in gardens (42%), to make potting soil and mixes, to top dress houseplants, flowers, lawns, trees, shrubs, planters, and, to add to greenhouse beds, cold frames and compost bins. One used castings with a soil blocker to make soil blocks for starting seedlings.

Kinds of bedding used

Half of the correspondents used shredded newspaper for bedding, frequently mixed with peat moss, topsoil, or other materials. Shredded corrugated was used by 25%, shredded office paper by 7%. Six used manure, and two tore up corrugated boxes. Other materials included compost, composted oak leaves, oat straw, and ground leaves.

Types of materials used in vermicomposting systems

Organic materials used in the vermicomposting units primarily consisted of household garbage such as fruit and vegetable trimmings and table scraps. Items mentioned specifically were: citrus rinds, corn meal, egg, shells (preferably pulverized), coffee grounds, tea leaves, potato peels (which often sprouted), cucumbers, cantaloupe, melon, bananas, tomatoes, peppers, egg plants, pasta, oatmeal, broccoli, bone meal, human and dog hair, granite dust, rice and other grains, casseroles, onion skins, carrot tops and pulp, flour sweepings, spoiled food from the refrigerator, and even dryer lint. Meat wastes and bones were not generally used, although some used small amounts. A few ground up waste, making a slurry to add to the bin, but most did not. Few kept track of quantities.

Frequency and characteristics of depositing materials in the bin

Most correspondents (76%) deposited food wastes in the bin from 1–3 times per week. Three (6%) did so daily, one 'seldom', and another 'as often as box would take.' Site rotation was practiced by about two thirds initially. After they became more familiar with the system, many just put fresh wastes where none were recognizable from a previous burial. One correspondent had a chart under the hinged lid where the person depositing garbage could check off space used. Thirty-four correspondents (60%) buried wastes in the bin through a hole in the garbage, and covered it with bedding so no garbage was visible. Fewer (26%) did not bury the garbage, but did cover it. Only 14% threw the garbage on top. The use of a container to hold wastes until they were placed in the worm bin was preferred by 60%. Thirteen (23%)

deposited wastes directly in the bin without storing them any period of time. A few correspondents blended wastes, several cut larger pieces such as melon rinds into smaller pieces to make it easier for the worms to eat.

Location of vermicomposting units

The most common (54%) location for vermicomposting units was a basement or cellar. Of these, five specified the furnace room, six located their worm bins in a greenhouse, two each in a garage, porch or storage room. Other sites included a study, workshop, spare bedroom, bathroom, root cellar, or hallway inside the home. Outside locations were under a tree in backyard, under a deck, near the garden, or in a woodshed. Six kept their worm bins in the kitchen. Twelve (18%) changed locations of their bin seasonally, *e.g.*, from basement in winter to cellar in summer. In some climates it was necessary to move from outside to a basement during the summer when it was too hot. Temperatures for successful operation ranged from 40–100°F (4–38°C) with most ranging between 50–75°F (10–23°C). An outside 24°F (–4°C) was fatal to the worms. In most cases these temperatures were estimates of ambient temperature, and not based upon careful records of actual measurements. One correspondent, however, did record bed temperature of 62–95°F (17–49°C).

Problems experienced with vermicomposting

Correspondents were asked to specify whether they had encountered problems with fruit flies, worms crawling out, temperature fluctuations, structural problems with the box itself, odor, or other aspects. Thirty-four of the 60 responding (57%) had fruit flies at one time or another. Several users moved their worm bins outside when fruit flies became intolerable. Some changed bedding several times to reduce further populations, installed screened lids, and cooked food before placing in bin. Some placed insect strips above the bin to control fruit flies with apparent success.

Only four correspondents acknowledged any difficulties having worms crawl out of the bins, many did not know what was meant. Temperature fluctuations were rarely viewed as a problem, although some felt that worms weren't very active in cold weather even when bin was moved to the basement. One correspondent kept the bedding from getting too hot in summer by freezing water in plastic containers which were put in the box. Several considered it difficult to gauge the proper moisture content after the bins were first set up. In some cases drying was a problem, in others, too much moisture.

Conclusions

1. Vermicomposting is a technique for recycling organic wastes which can be successfully carried out by the layperson.
2. People most likely to be interested in home vermicomposting are those who have use for the two end products: (a) vermicompost for house plants and gardens; (b) worms for fishing or composting.
3. Vermicomposting can be carried out in temperate climates if necessary precautions

are taken to prevent freezing, overheating, or excessive drying.

4. Maintenance of a home vermicomposting system is minimal. However, some effort is required to set it up, monitor it, replenish bedding, and periodically harvest products.
5. Control of fruit flies, not odor, is the most serious unsolved problem in home vermicomposting systems.
6. Two motivators for setting up home vermicomposting systems are a strong environmental ethic and the desire to utilize resources wisely.

References

Appelhof, M. 1981. Biodegradable Solid Waste Conversion into Earthworms Castings. Grant Proposal No. ISP-800975, prepared for the National Science Foundation Appropriate Technology Program under the Small Business Innovation Research Program. 78 pp.

Appelhof, M. 1982. Worms Eat my Garbage. Flower Press, 10332 Shaver Road, Kalamazoo, MI 49002. 110 pp.

III. Earthworms as animal feed

CONTENTS

EARTHWORMS AS ANIMAL FEED: AN OVERVIEW

JOHN R. SABINE
University of Adelaide, Australia

Since the early 1970s, extensive research into biological waste management by earthworms, particularly in the United States, England, the Federal Republic of Germany and Australia, has indicated clearly the potential for three areas of significant financial return to be derived from an active vermiculture industry, namely, a reduction in the environmental pollution problems presented by the original waste, and its concurrent conversion to two useful products, *i.e.*, vermicompost, an enriched potting mix or plant-growth and soil-conditioning additive, and worm meal, a high-protein feed supplement for livestock rations.

Laboratory-scale studies have already established the biological suitability of earthworms, either fresh or processed, as a protein source for commercial livestock production. A list of published feeding trials is given in Table 1. Chemical analyses and dietary trials, both with laboratory animals and with pigs and poultry, have shown that earthworm meal is rich in protein, that this protein is of good biological value, and that it can effectively substitute for the more common protein sources, such as meatmeal and fishmeal, in the feeding of farm animals. Speciality livestock production, such as for *Tilapia*, or Japanese quail, or various shell-fish, may also be a substantial and profitable market for earthworms. Zoological gardens and aquatic parks are also likely customers for worms, and so is the recreational fish bait market, but any substantial vermiculture industry producing worm protein, must depend eventually upon supplying the large, intensive animal industries, *i.e.*, feed for pigs, poultry, dairy cattle.

As far as I am aware there is, as yet, no commercial production of earthworm protein. When this has been developed, then appropriate research in this area can move from the laboratory to the farm. Only then can the next most obvious research questions be tackled: What are the transfer rates of nutrients (energy, protein, minerals) from raw materials to wastes to vermicompost and worm meal? What is the fate of agricultural chemicals, industrial residues, heavy metals, along this food transfer pathway? What happens to actual or potentially pathogenic organisms in this food pathway? Will large-scale feeding trials confirm and extend laboratory findings of no deleterious effects and of increased food conversion efficiency in feeding worm meal to livestock? Then appropriate harvesting and processing technologies for worm meal can be developed. Full-scale vermiculture units will need to achieve the rates of protein production calculated as possible on the basis of results of small-scale trials.

Claimed rates of production, whether actually achieved or potentially possible (Table 2), are so large as undoubtedly to justify considerable further research. In the market place, however, the price for worm meal and thus the economics of its production, are dictated by the price, pro-rata for nutrient content, of the major, currently-available, alternative sources of high-quality protein, especially meat and fish meals. For commercial viability, the production of earthworms for protein must compete efficiently and effectively within this economic framework.

Earthworms in waste and environmental management
edited by C.A. Edwards and E.F. Neuhauser, pp. 165–167

Table 1. Dietary trials with earthworms

Worm species	In the diet of	Reference
Eisenia foetida	Mice	McInroy (1971)
Eisenia foetida	Rats	Schulz and Graff (1977)
Eisenia foetida	Poultry	Harwood (1976)
Eisenia foetida	Poultry	Harwood (1976)
Eisenia foetida	Poultry	Yoshida and Hoshii (1978)
Eisenia foetida	Poultry	Mekada *et al.* (1979)
Eisenia foetida	Pigs	Harwood and Sabine (1978)
Eisenia foetida	Young eels	Tacon (1981)
Eisenia foetida	Trout	Tacon *et al.* (1983)
E. foetida/L. rubellus	Poultry	Taboga (1980)
Perionyx excavatus	Japanese Quail	Guerrero (1983)
Perionyx excavatus	Tilapia	Guerrero (1983)
Eudrilus eugeniae	Trout	Hilton (1983)

Table 2. Rates of production of earthworm protein
(N.B. Estimates from small-scale experiments)

Reference	Protein	Substrate
McInroy (1971)	190 kg cow^{-1} yr^{-1}	Manure
Fosgate and Babb (1972)	150 kg cow^{-1} yr^{-1}	Manure
Hartenstein (1981)	6,685 kg ha^{-1} yr^{-1}	Sludge
Edwards (1983)	1,570 tn U.K.$^{-1}$ yr^{-1}	Potato waste
Walsingham (1981)	11,000 tn U.K.$^{-1}$ yr^{-1}	Vegetable processing waste
Compare: Milk (South Australian average) (Sabine, 1983)		140 kg cow^{-1} yr^{-1}
Fish farming		3,000 kg ha^{-1} yr^{-1}

References

Edwards, C.A. 1983. Production of earthworm protein for animal feed from potato waste. *In* Upgrading Waste for Feed and Food. pp. 153–162. Edited by D.A. Ledward, A.J. Taylor and R.A. Lawrie. Butterworths, London.

Fosgate, O.T. and Babb, M.R. 1972. Biodegradation of animal wastes by *Lumbricus terrestris*. J. Dairy Sci. 55: 870–872.

Guerrero, R.D. 1983. The culture and use of *Perionyx excavatus* as a protein resource in the Philippines. *In* Earthworm Ecology: From Darwin to Vermiculture. pp. 309–313. Edited by J.E. Satchell. Chapman and Hall, London.

Hartenstein, R. 1981 Use of *Eisenia foetida* in organic recycling based on laboratory experiments. *In* Workshop on the Role of Earthworms in the Stabilization of Organic Residues. pp. 155–156. Compiled by M. Appelhof. Beech Leaf Press, Kalamazoo, MI.

Harwood, M. 1976. Recovery of protein from poultry waste by earthworms. *In* Proceedings of First Australasian Poultry Stockfeed Convention Melbourne. pp. 138–143. Australian Poultry Industry Association, Sydney.

Harwood, M. and Sabine, J.R. 1978. The nutritive value of worm meal. *In* Proceedings of Second Australasian Poultry Stockfeed Convention Sydney. pp. 164–171. Australian Poultry Industry Association, Sydney.

Hilton, J.W. 1983. Potential of freeze-dried worm meal as a replacement for fish meal in trout diet formulations. Aquaculture 32: 277–283.

McInroy, D.M. 1971. Evaluation of the earthworm '*Eisenia foetida*' as food for man and domestic animals. Feedstuffs, Feb. 20th, pp. 36–47.

Mekada, H., Hayashi, N., Yokota, H. and Okumura, J. 1979. Performance of growing and laying chickens fed diets containing earthworms (*Eisenia foetida*). Jpn. Poult. Sci. 16: 293–297.
Sabine, J.R. 1983. Earthworms as a source of food and drugs. *In* Earthworm Ecology: From Darwin to Vermiculture. pp. 285–296. Edited by J.E. Satchell. Chapman and Hall, London.
Schulz, E. and Graff, O. 1977. Zur Bewertung von Regenwurmmehl aus *Eisenia foetida* (Savigny 1826) als Eiweissfuttermittel. Landb. Forschung-Völkenrode 27: 216–218.
Tacon, A.G.J. 1981. The possible substitution of fish meal in fish diets. Proc. Scott. Mar. Biol. Assoc. and Highlands and Islands Dev. Board Fish Farm. Meet., Oban, Scotland. pp. 46–56.
Tacon, A.G.J., Stafford, E.A. and Edwards, C.A. 1983. A preliminary investigation of the nutritive value of three terrestrial lumbricid worms for rainbow trout. Aquaculture 35: 187–199.
Taboga,L. 1980. The nutritional value of earthworms for chickens. Br. Poultry Science 21: 405–410.
Walsingham, J.M. 1981. The use of invertebrates in recycling. *In* Vegetable Productivity. pp. 239–246. Edited by C.R.W. Spedding. Institute of Biology Symposium 25. Macmillan, London.
Yoshida, M. and Hoshii, H. 1978. Nutritional value of earthworms for poultry feed. Jpn. Poult. Sci. 15: 308–311.

THE PRODUCTION AND PROCESSING OF EARTHWORM PROTEIN

CLIVE A. EDWARDS[1]* and ALAN NIEDERER[2]
[1]*The Ohio State University, Columbus, Ohio, U.S.A.;* [2]*Meat Research Institute, New Zealand*

Introduction

The large amounts of organic wastes produced by intensive animal production in Europe and USA cause major problems in disposal, odor and a considerable potential for pollution of waterways with nitrogen. Charles Darwin (1881) first emphasized the very important role that earthworms play in breaking down organic matter and incorporating it into soil. The importance of earthworms in processing organic matter in compost heaps is well-documented and there is little doubt of their importance in this role. Moreover, their contributions to the breakdown process in trickling sewage filters, which can be a major one, is not often recognized (Terry, 1951).

Work in recent years in the United States (Neuhauser *et al.*, 1979) (see also p. 19.) has shown that earthworms can be used to breakdown activated sewage sludges to finely divided materials. More recently, an extensive research program at Rothamsted Experimental Station has shown that earthworms can be used to breakdown not only animal but also vegetable and industrial organic wastes (Edwards, 1982, 1983a, 1983b, 1985).

The first workers to suggest that earthworms contained sufficient protein to be considered as animal food were Lawrence and Millar (1948), but it is only in the last 10 years that this has been confirmed (Fig. 1) and full analyses of the body tissues of earthworms have been available to show the kinds of amino acids they contain and the nature of the other constituents. The first successful animal feeding trial was by Sabine (1978); subsequently there have been a number of other such trials.

Food value of worms

The first analysis of the constituents of the tissues of different species of earthworms was by McInroy (1971), but there have been others since this (Schulz and Graff, 1977; Sabine, 1978; Yoshida and Hoshii, 1978; Mekada *et al.*, 1979; Toboga, 1980; Graff, 1982). The results of these studies have shown that the overall composition of earthworm tissues does not differ greatly from that of many vertebrate issues (Table 1).

The essential amino acid spectrum for earthworm tissues as reported by these different authors compares well with those from other currently used sources of feed protein. Clearly, the mean amounts of essential amino acids recorded are very adequate for a good animal feed when compared with the recommendations of FAO/WHO, particularly in terms of lysine and the combinations of methionine and cysteine, phenylaniline and tyrosine, all of which are important components of animal feeds. In addition, earthworm tissues contain a preponderance of long-chain

* Previously, Senior Principal Scientific Officer, Rothamsted Experimental Station, Harpenden, U.K.

Earthworms in waste and environmental management
edited by C.A. Edwards and E.F. Neuhauser, pp. 169–179

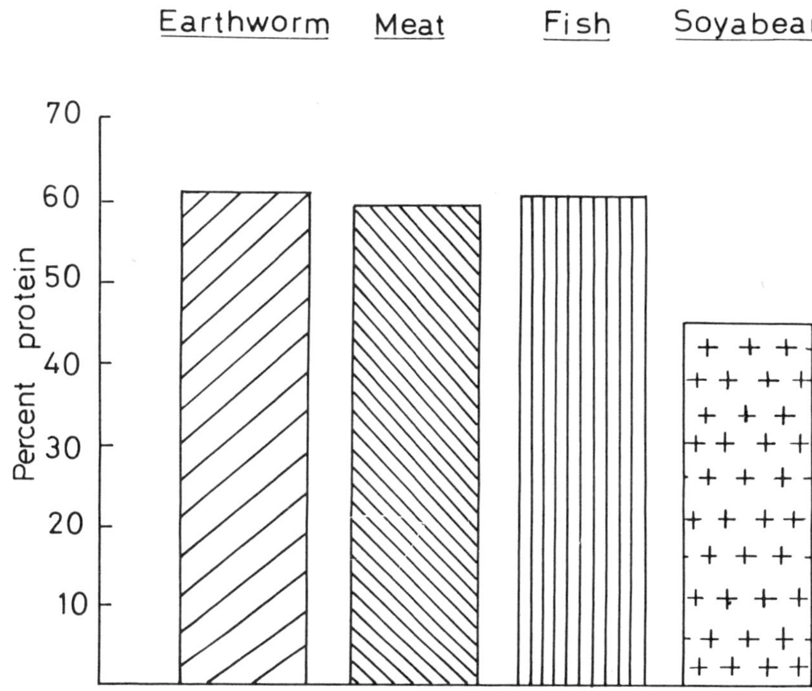

Fig. 1. Protein content of animal and vegetable feeds.

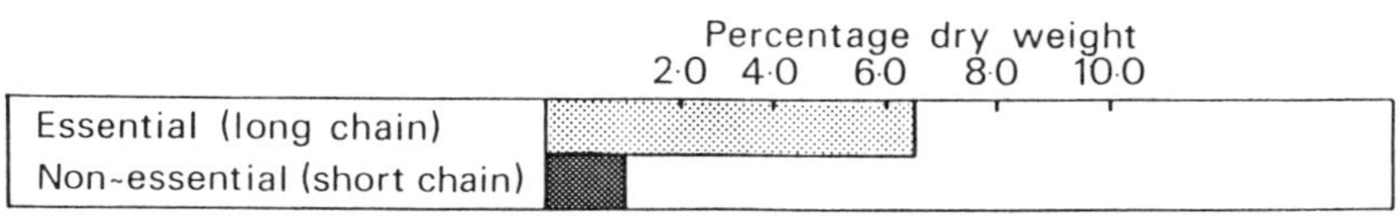

Fig. 2. Long and short chain fatty acid content of worms.

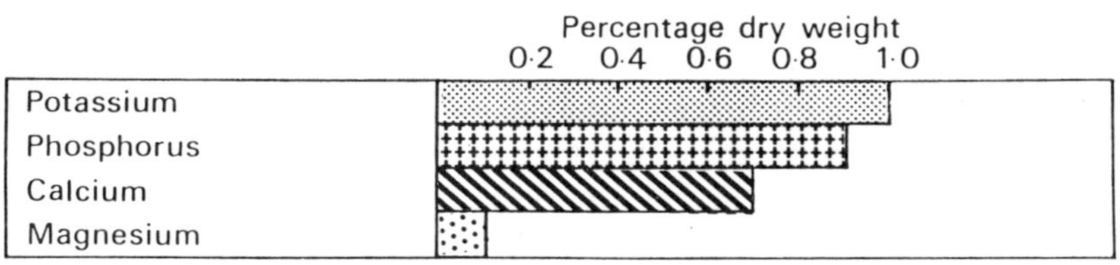

Fig. 3. Mineral content of worms.

fatty acids many of which non-ruminant animals cannot synthesize (Fig. 2) and an adequate mineral content (Fig. 3). They have an excellent range of vitamins and are rich in niacine which is a valuable component of animal feeds (Fig. 4) and are an unusual source of vitamin B_{12}. Overall, the nutrient spectrum of worm tissues has excellent potentials for animal feed.

Table 1. Essential amino acid contents of earthworm tissue (g/1000 g protein)

Author / Essential amino acid	WHO/FAO requirements of EAA's 1957–1965	1973	McInroy, 1971	Schulze and Graff, 1977	Sabine, 1978	Yoshida and Hoshii, 1978	Mekada *et al.*, 1979	Taboga, 1980	Graff, 1982 1982	Rothamsted, (unpublished)	Mean of all analyses
Arginine	—	—	6.1	6.1	4.2	6.9	4.5	7.3	6.1	6.5	6.0
Cysteine	2.0	*	1.8	1.4	2.3	0.8	—	1.8	1.4	0.7	1.5
Histidine	—	—	2.2	2.3	1.6	4.3	1.6	3.8	2.3	3.0	2.6
Isoleucine	4.2	4.0	4.6	4.7	2.6	4.7	4.3	5.3	4.7	4.1	4.3
Leucine	4.8	7.0	8.1	8.2	4.8	8.7	5.0	6.2	8.2	8.3	7.2
Lysine	4.2	6.5	6.6	7.5	4.3	8.7	5.9	7.3	7.5	6.5	6.8
Methionine	2.2	*	1.5	1.8	2.2	1.6	1.9	2.0	1.8	2.8	2.0
Phenylaniline	2.8	**	4.0	3.5	2.3	4.4	3.4	5.1	3.5	4.0	3.8
Threonine	2.8	4.0	5.3	4.7	3.0	5.2	6.9	6.0	4.7	5.6	5.2
Tryptophane	1.4	1.0	—	—	—	1.3	—	2.1	—	0.7	1.4
Tyrosine	2.8	**	—	3.0	1.4	4.4	2.7	4.6	3.0	3.5	3.2
Valine	4.2	5.0	5.1	5.2	3.0	5.1	5.1	4.4	5.2	4.7	4.7

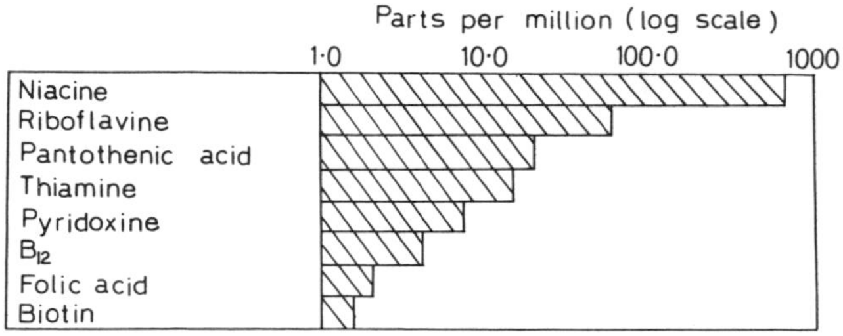

Fig. 4. Vitamin content of earthworm tissue.

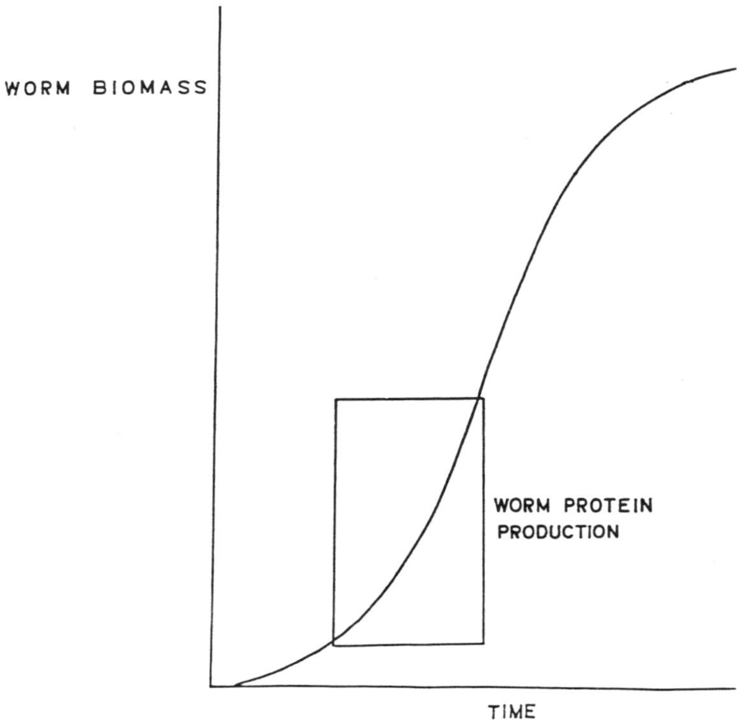

Fig. 5. Part of population curve of earthworm growth for maximum protein protection.

Growth of worms in animal, vegetable and industrial wastes

The biology of four species of worms that grow and reproduce well in organic wastes have been investigated thoroughly at Rothamsted. These are *Eisenia foetida* (Savigny), *Eudrilus eugeniae* (Kinberg), *Perionyx excavatus* (Gates), and *Dendrobaena veneta* (Rosa). The growth patterns of individual worms or whole populations follow classical sigmoid curves. The maximum protein production per unit time was achieved by inoculating large volumes of animal wastes with relatively small numbers of young worms (Fig. 5). Dry matter conversion ratios of waste to worms ranging from 10% for cattle and pig waste to 2% for duck waste, have been achieved readily in the laboratory and although rather less success has been attained in the field, such ratios should be feasible ultimately in large-scale production.

Methods of production

Decaying organic matter follows a natural microbial decomposition process with a succession of microorganisms and worms derive most of their nutrients from these microbes, particularly fungi. Worm colonization of decomposing organic matter in natural systems is limited by temperature, populations are localized and biomass productivity is low. To produce earthworms commercially in beds, boxes or continuous flow systems, as a source of protein, such systems must be controlled environmentally and waste added in small amounts regularly so that high temperatures do not develop and earthworms can colonize all of the organic matter and utilize the microorganisms growing on it to the maximum.

Animal and vegetable wastes range from almost liquid slurries through straw-based mixtures to relatively dry and finely dispersed materials such as those produced by laying chickens over deep litter systems. All these wastes must be brought to a suitable moisture content and temperature and the ammonia and salt contents reduced to acceptable levels by leaching, composting or some other method, before the worms can be grown successfully and productivity maximized.

Once suitable conditions are attained, worms may be grown by a variety of methods such as in ground-based beds of different sizes, in batch systems in boxes or crates or even in trickling filter systems similar to those used in sewage disposal (see p. 95). The choice of a system depends mainly on economic considerations; such as the cost of adding wastes to breeding containers, harvesting worms and handling the waste and worms.

It is outside the scope of the present paper to deal with the technology of worm production in detail, but it has been found that growing worms in beds, filled with animal waste to a depth of about 50 cm in successive shallow layers at regular intervals by automatically operated gantries, is an easy system and non-labor intensive. In colder climates, some form of insulated housing is necessary. Different forms of batch production in crates or boxes are useful and systems of continuous processing of wastes with automatic addition and removal of wastes have been developed. The technology to maximize worm and/or compost production is still evolving, but holds considerable promise for development into well-engineered highly productive non-laborious systems. A rapid rate of growth and multiplication of earthworms, and the efficient conversion of organic wastes into earthworm tissue protein would mean that systems could be developed using a minimum of labor or sophisticated technology. Moreover, earthworms have few diseases, predators or parasites that might hinder maximum protein production.

The efficient productivity of earthworm protein depends mainly upon detailed knowledge of the population dynamics of the appropriate species, maintenance of optimal environmental conditions and upon the engineering of suitable production and harvesting systems that involve small labor inputs. Our knowledge of these aspects is sufficient to make worm protein production an economic proposition if non labor-intensive harvesting methods are available.

Harvesting of earthworms from animal and vegetable wastes

Earthworms fragment animal wastes into finer and finer particles, producing eventually a peat-like material which has considerable potential in horticulture. However, worms grow best at relatively high moisture levels (80–90% M.C.), and this raises harvesting problems since it is not easy to separate worms mechanically from the finely divided organic matter at such high moisture contents, and some drying before harvest is usually necessary.

Machinery for separating worms from fully worked organic materials has been developed at Rothamsted and the National Institute for Agricultural Engineering, Silsoe (Phillips see p. 111), and patented. The efficiency of this machinery in terms of percentage recovery of worms is very high. The machine that has been developed currently, will separate worms from about one tonne of waste per hour and this machine can be automated and scaled up to increase throughput. However, the rate of separation of worms from waste is still too slow and labor-intensive to make the production of protein for animal feed from worms an economic process on its own. When combined with production of plant growth media (see p. 193) the economics become much more attractive.

Processing of worms for animal feed

After worms are collected from the separating machinery, they may have small particles of waste attached to their bodies and are likely to contain waste in their guts. Hence, the first stage of all the methods of processing tested is to wash the worms thoroughly and leave them standing in water for several hours, in order to evacuate the residual waste from their guts completely. A range of methods of processing the worms for animal feed have been developed at Rothamsted. Two of the methods produced a paste product and the other four a dry worm meal; all of the products acceptable for particular types of feed and the ultimate choice of a method of processing must depend upon: (i) type of feed required; (ii) cost of production of the protein; (iii) minimal loss of dry matter and (iv) minimal loss of nutrient value.

1. Incorporation with molasses

This method involves first blanching the worms in boiling water for one minute, then incorporating 30% molasses with them together with 0.3% potassium sorbate to produce a paste. The molasses effect a lowering of the water activity to about A_W 0.90 and the addition of potassium sorbate reduces this further to approximately A_W 0.65. At this level, the growth of yeasts and moulds is inhibited, permitting indefinite storage of the product.

2. Ensiling with formic acid

A second method tested was to incorporate 3% formic acid with the worms, thoroughly homogenize this then allow it to ensile until eventually a very stable paste or liquid product is produced.

3. Air drying

In this method a dry meal was produced by blanching worms in boiling water for one minute then air drying them and grinding. This needs high ambient temperatures.

4. Freeze drying

Another dry meal was produced by freezing worms quickly then freeze drying and grinding them. This is efficient but slow and expensive.

5. Acetone heat drying

A third type of dry meal was produced after killing worms by immersing them in acetone for one hour then air and later oven drying them in a hot oven at 95°C and finally grinding then into a pan. The acetone is evaporated off and reclaimed.

6. Oven drying

A dry meal was produced by killing and drying worms in trays in a large commercial oven at 95°C and then grinding them up into a dry powder.

All the methods tested produced a good material that could be used as animal feed but there were variations in the dry matter yield. Killing worms in boiling water, then drying them in an oven resulted in the lowest dry matter of 11.6%. Freeze drying produced a meal with 13.5% dry matter. Killing worms in acetone then drying them in air produced a dry matter of 14.5%, but after subsequent oven drying this fell to 12.8%. Killing and drying the worms in a hot air oven gave the greatest dry matter yield of 15.2%.

The different processing methods affected the amounts of essential amino acids slightly; the differences are summarized in Table 2. Most of the methods had relatively little effect on the amounts of amino acids in the product. However, lysine content was decreased slightly by ensiling with molasses using formic acid and by freeze drying, compared with the other methods. Clearly a stable protein feed can be produced by any of the methods and the choice of method must depend mainly on the use to which the protein is to be put and the cost of the processing method in relation to the value of the protein.

Assessment of the value of worm protein as animal feed

The main outlets suggested for utilization of earthworm protein have been in fish farming and as protein supplements in poultry and pig feeds.

1. Fish feeding trials

The first trials using earthworms as fish feed were by Tacon *et al.* (1983), in collaboration with Rothamsted. The growth of trout fed only on *E. foetida* (Savigny), *Allolobophora longa* (Ude) and *Lumbricus terrestris* L. was compared with that of fish fed on a commercial ration. Fish fed frozen *A. longa* and *L. terrestris* grew as well

Table 2. Amounts of essential amino acids in worm tissues after different processing methods (g/100g protein)

Processing method *Amino acid*	Molasses ensiling	Formic acid ensiling	Freeze dried	Acetone dried	Heat dried	Blanch and heat dried
Arginine	6.5	6.7	6.4	7.0	3.8	4.1
Cysteine	0.3	0.5	0.5	0.3	0.5	0.4
Histidine	2.8	2.7	2.7	2.5	2.3	2.3
Isolencine	4.2	4.3	4.0	4.3	4.4	4.3
Leucine	7.3	7.1	6.9	7.5	8.3	8.5
Lysine	5.4	5.9	5.7	6.1	6.2	6.5
Methionine	1.6	1.1	1.2	1.2	1.1	0.9
Phenylalinine	3.0	2.8	3.1	3.2	3.1	3.6
Threonine	4.8	5.1	8.5	5.1	5.6	5.6
Tyrosine	2.0	2.4	4.0	2.8	2.8	3.0
Valine	5.1	4.9	4.7	5.0	5.5	4.6

or better than fish fed on commercial trout pellets. Trout did not grow so well on a whole diet of freeze-dried *E. foetida*, although they grew much better on *E. foetida* that had been 'blanched' in boiling water before freezing (Stafford and Tacon, 1985) than on unsterilized worms. However, dried earthworm meal derived from *E. foetida* which had not been blanched could satisfactorily replace the fishmeal component of formulated trout pellets at levels of inclusion between 5% and 30% without affecting growth of trout adversely. The conclusion was that all earthworms have potential both as a complete feed or protein supplement for trout or other fish. Hilton (1983) reported that trout did not grow well on another earthworm species, *Eudrilus eugeniae*, but there are some doubts about his experimental techniques because other forms of protein supplement currently used commercially such as blood meal, would have also been unsuccessful if used in the same way as he used worm protein. Guerrero (1983) reported that *Tilapia* fish grew better on diets containing earthworm protein supplements from *Perionyx excavatus* (Perrier) than those with fish meal supplements.

2. Chicken feeding trials

The first trials which assessed the growth of chickens on earthworm protein were reported by Harwood and Sabine (1976) and Sabine (1978). They compared the use of worm meal as a protein supplement for chickens with meat meal and found no significant difference in growth and chickens between the two diets (Fig. 6). Similar results were reported by Taboga (1980) and Mekada *et al.* (1979), and the latter workers reported that when worms were fed to older birds, egg production was maintained. Jin-you *et al.* (1982a) reported that chickens fed on earthworms put on weight faster than those given other diets (including fish meal), they had more breast muscle and consumed less food.

These results have been confirmed by our collaborators at the Poultry Research Centre, Edinburgh (see p. 181), Fisher (1985). In his experiments, chickens grew well, had a good weight gain per unit of food and an excellent nitrogen retention when fed on diets with levels of worm meal from 72-215 g/kg.

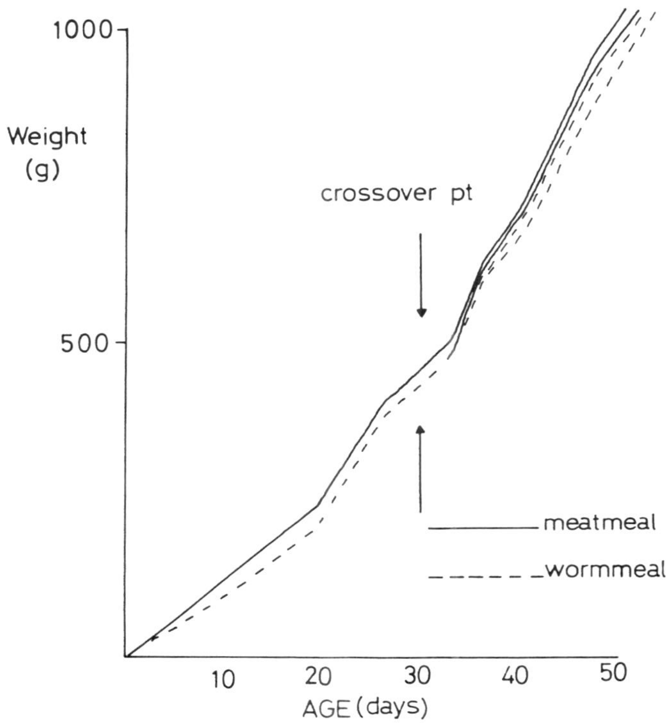

Fig. 6. Growth of chickens on meatmeal and wormmeal (from Sabine, 1978)

Table 3. Growth of pigs on worm protein supplements

	Ratios		
	Worm meal supplement	Meat meal supplement	Commercial protein supplement
Starter period (days 36–50)			
A. Mean weight gain (kg)	4.36	3.65	4.26
B. Mean % gain	48.9	41.2	48.9
C. Feed consumption (kg/pig/14days)	10.05	10.05	10.05
D. Feed conversion (C/A)	2.31	2.78	2.37
E. Mean rate of growth (kg/day)	0.31	0.26	0.30
Grower period (days 84–98)			
A. Mean weight gain	6.54	6.80	6.60
B. Mean % gain	28.1	28.2	27.0
C. Feed consumption (kg/pig/14days)	16.78	16.78	16.78
D. Feed conversion (C/A)	2.47	2.48	0.46

(From Sabine, 1978)

Table 4. Value of earthworm meal for different animal feeds

Animal	Maximum value/tonne
Cows	£ 130
Ducks	£ 183–218
Broiler chickens	£ 220–249
Turkey (starter)	£ 1011
Turkey (finisher)	£ 338
Trout	£ 404
Eels	£ 2000

Based on computer programme using nutrient composition

3. Pig feeding trials

There have been only two trials reporting the growth of pigs on earthworm protein supplements. Harwood and Sabine (1978) and Sabine (1978) showed that in feeding trials with both starter and grower pigs, young pigs fed on an earthworm protein supplement grew equally well and had similar feed conversion ratios to those grown on commercial rations (Table 3). Jin-you (1982b) reported that piglets grew better on earthworm protein supplements than on others, had accelerated weaning, produced earlier oestrus in sows, increased disease resistance and a decreased incidence of white diarrhoea.

Economics of production of earthworm protein

A study of the economics of production of earthworm protein is presented elsewhere (p. 145) (Fieldson, 1984). His general conclusions were that earthworm production had the best prospects of good profits when done by larger farmers with considerable amounts of animal wastes. The most important criterion is that the worm meal must be produced at an economic price, although the value of the compost can also be taken into account. Currently, the only labor-intensive part of worm protein production is the harvesting process and this remains the main barrier to commercial production, but seems likely to be relieved by engineering of better harvesting methods.

In a computer analysis of the economic value of earthworm protein based on its amino acid, fatty acid, mineral and vitamin content, it emerged that it is extremely valuable as feed for certain animals (Table 4) particularly eels, young turkeys and has the same value for fish, pig and poultry feed as fish or meat meal.

References

Darwin, C. 1981. The Formation of Vegetable Mould Through the Action of Worms. Murray, London. 326 pp.

Edwards, C.A. 1983a. Earthworms, organic waste and food. Span. 26 Mar. 83.: 106–108.

Edwards, C.A. 1983b. Production of earthworm protein for animal feed from potato waste. *In* Upgrading Wastes for Feed and Food. pp. 153–162. Edited by D.A. Ledward, A.J. Taylor and R.A. Lawrie. Butterworth's, London.

Edwards, C.A. 1985. Production of feed protein from animal wastes by earthworms. Phil. Trans. Roy. Soc. London. 310: 299–307.

Fieldson, R.S. 1987. The Economic Viability of Earthworm Culture on Animal Wastes. This volume (See p. 145).

Fisher, C. 1985. The Nutritional Value of Earthworm Meal for Poultry. This volume (See p. 181ff).

Guerrero, R.D. 1983. The culture and use of *Perionyx excavatus* as a protein resource in the Philippines. *In* Earthworm Ecology. pp. 309–313. Edited by J.E. Satchell. Chapman and Hall, London.

Graff, O. 1982. Vergleich der Regenwurmarten *Eisenia foetida* and *Eudrilus eugeniae* hinsichtlich ihrer Eignung zur Proteingewinnung aus Abfallstoffen. Pedobiologia 23: 277–282.

Hartenstein, R., Neuhauser, E.F. and Kaplan, D.L. 1979. A Progress Report on the Potential Use of Earthworms in Sludge Management. Proceedings of the 8th National Sludge Conference, Florida, March 1979. pp. 238–241. Transfer Inc.

Harwood, M. and Sabine, J. 1978. The Nutritive Value of Worm Meal. Proceedings of the Australian Poultry and Stock Feed Conference, Sydney, March 1978. pp. 164–171. Australian Poultry Industry Association, Sydney.

Hilton, J.W. 1983. Potential for freeze-dried worm meal as a replacement for fish meal in trout diet formultions. Aquaculture 32: 277–283.

Jin-you, X., Xian-Kuan, Z., Zhi-ren, P., Zhen-yong, J., Yan-hua, G., Hong-bo, I., Xue-yan, H., Qiao-ping, X. 1982a. Experimental research on the substitution of earthworm for fish meal in feeding broilers. J. South China Norm. Coll. 1: 88–94.

Jin-you, X., Xi-cong, H. and Wen-xi, L. 1982b. An observation on the results of using earthworms as supplementary food for suckling pigs. J. South China Norm. Coll. 1: 1–8.

Lawrence, R.D.and Miller, R.H. 1948. Protein Content of Earthworms, Nature (London) 3939. April 28, 517.

McInroy, D.M. 1971. Evaluation of the earthworm *Eisenia foetida* as food for man and domestic animals. Feedstuffs 43: 46–47.

Mekada, H., Hayashi, N., Yokota, H. and Okomura, J. 1979. Performance of growing and laying chickens fed diets containing earthworms. Jpn. Poult. Sci. 16: 293–297.

Neuhauser, E.F., Hartenstein, R. and Kaplan, D.L. 1979. Second Progress Report on Potential Use of Earthworms in Sludge Management. Proceedings of the National Conference on Sludge Composting. 9 pp. Inf. Transf. Inc.

Sabine, J. 1978. The nutritive value of earthworm meal. *In* Proceedings of the Conference on Utilization of Soil Organisms in Sludge Management. Syracuse, N.Y. pp. 122–130. Edited by R. Hartenstein. S.U.N.Y., Syracuse.

Schulz, E. and Graff, O. 1977. Zur Bewertung von Regenwurmmehl aus *Eisenia foetida* (Savigny) als Eiweissfuttermittel. Landb. Forsch. Volkr. 27: 216–218.

Stafford, E.A. and Tacon, A.G.J. 1985. The Use of Earthworms as Food for Rainbow Trout *Salmo gairdneri* (This volume p. 193–208).

Taboga, L. 1980. The nutritional value of earthworms for chickens. Brit. Poult. Sci. 21: 405–410.

Tacon, A.G.J., Stafford, E.A. and Edwards, C.A. 1983. A preliminary investigation of the nutritive value of three terrestrial lumbricid worms for rainbow trout. Aquaculture 35: 187–199.

Terry, R.J. 1951. The behavior and distribution of the larger worms in trickling filters. J. Proc. Inst. Sewage Purif. 16–25.

Yoshida, M. and Hoshii, H. 1978. Nutritional value of earthworms for poultry feed. Jpn. Poult. Sci. 15(6): 308–311.

THE NUTRITIONAL VALUE OF EARTHWORM MEAL FOR POULTRY

C. FISHER*

Agricultural and Food Research Council's Poultry Research Centre, Roslin, Midlothian EH25 9PS, Scotland

Introduction

Of the nutrients given to poultry, 60–70% of the crude protein, 20% of the gross energy, 50% of the phosphorus and similar proportions of other nutrients are voided in the excreta. Systems of agriculture which conserve and recycle these excretory products effectively are obviously required to increase efficiency and, in the long run, their development is essential to the future of intensive poultry production. Traditionally, animal excreta has been re-cycled by its use as a fertilizer and this remains a major route for the poultry industry. However, the industrial nature of poultry production in many countries leads to handling, organizational and environmental problems and potential manurial values are often not realized. Some poultry litter (a mixture of excreta and a cellulosic or lignified material) is used as food for ruminants but this has a poor public image and several problems which limit its use. Some re-cycling of litter through poultry is also possible but this is obviously unaesthetic, restricted by toxin, residue or pathogenic organism accumulations and does not lead to large increases in nutrient utilization.

Such considerations, and the composition of poultry excreta and litter, lead to the conclusion that the material should be used as a medium for culturing some other organism before re-cycling the nutrients to poultry or other animals. Suggestions have been made for growing house-fly larvae (Calvert, 1977), algae (Priestley, 1976) and aerobic bacteria (Schuler *et al.*, 1979) on poultry waste. A practical technology for adding earthworms to this list would be of considerable interest to the poultry industry.

The use of earthworms in poultry feeds should be uncontroversial in view of the natural feeding habits of poultry and other birds. However, nutrient content and availability need to be defined accurately to facilitate feed formulation. In industrial countries, the worms would probably be dried to a meal although presentation of raw worms has been clearly shown to be satisfactory (Taboga, 1980). In the experimental work described here, dried meals have been used to study nutrient content, nutrient availability and feeding value. These are preliminary results owing to the small amounts of material so far available.

Worm meals

All meals were prepared at Rothamsted Experimental Station (Dr C.A. Edwards). The worms were separated manually from compost, immersed in water for three hours, blanched in water for five minutes at 100°C and then either freeze-dried or oven-dried (80°C for 2–4 hours). Oven-dried worms were ground to a fine meal before incorporating them into feeds.

* *Present address:* AFRC Institute for Grassland and Animal Production, Poultry Department, Roslin, Midlothian EH 25 9PS, Scotland.

Earthworms in waste and environmental management
edited by C.A. Edwards and E.F. Neuhauser, pp. 181–192

Table 1. Gross composition of earthworm meals

	Unpublished					Published[a]				
Meal no.	1	2	3	4	5	6	7	8	9	10
Species[b]	Ef	Ef	Lt	Lt	Lt	—	Ef	Ef	Ef	Ef
Drying method[c]	OD	FD	OD	FD	OD	ND	FD	AD	FD	OD
Dry matter g/kg	926	899	938	934	966	—	926	—	928	912
Ash g/kg	108	62	212	245	55	—	88	80–100*	56	78
Crude protein										
g/kg DM	655	627	594	551	723	620–710	610	620–640	642	657
g/kg OM	741	674	768	746	767		674		683	718
Ether extract										
g/kg DM	75	96	43	37	79	23–45	85	70–100	122	120
g/kg OM	84	103	55	51	83		94	—	130	131
Gross energy										
kJ/g DM	20.7	20.1	18.3	17.4	—		—	16.3–17.2	—	—
kJ/g OM	23.4	21.6	23.7	23.6	—		—	—	—	—

[a] Meal 6 – Taboga (1980) – range of values given: 7 – Yoshida and Hoshii (1978); 8 – Sabine (1981) – range of values given; 9 and 10 – Sugimura *et al.* (1974).
[b] Ef – *Eisenia foetida*; Lt – *Lumbricus terrestris.*
[c] OD – oven dried; FD – freeze dried; ND – not dried; AD – air dried.
* Dry-matter basis.

Table 2. Mineral composition of earthworm meals (air dry basis)

	Unpublished[a]					Published[a]	
Meal no.	1	2	3	4	5	7	8
Species[a]	Ef	Ef	Lt	Lt	Lt	Ef	Ef
Drying method[a]	OD	FD	OD	FD	OD	FD	OD
Mineral content g/kg							
Ash	108	62	212	245	55	88	80–100*
Insoluble ash[b]	62	6	154	166	—	—	—
Soluble ash[b]	46	56	58	79	—	—	—
Calcium	6.1	3.9	6.7	5.9	5.8	4.8	5.5*
Phosphorus	—	7.9	—	—	7.9	8.7	10.0*
Sodium	6.6	6.2	2.5	4.0	1.1	6.9	
Potassium	3.3	9.0	3.5	7.7	1.6	8.9	
Magnesium	1.2	0.8	0.8	1.0	—		
Chloride	—	3.8	—	—	0.6		

[a] See Table 1.
[b] Solubility in concentrated HCl; insoluble material is largely silica.
* Dry-matter basis.

The gross composition of the five meals used in the present experiments are summarized in Table 1 together with published values from papers on poultry feeding (the general literature on the composition of worms has not been reviewed in the preparation of this paper).

Meals 1–4 were atypically high in ash content and meals prepared from *Lumbricus terrestris* worms contained twice as much ash as those prepared from *Eisenia foetida* worms. This excess ash is almost entirely insoluble in acid (mainly silica) and since the worms were well-washed, presumably represents retained intestinal contents. Minimum clearance times for *L. terrestris* worms may be longer than the three hours used. A typical ash content appears to be 50–100 g/kg for worm meals.

The crude protein (N × 6.25) content of the worm meals is about 700–750 g/kg organic matter (OM). The variation is of a degree which would be of importance economically. The apparent loss of nitrogen on freeze-drying (meals 1–4) is curious and merits further investigation if freeze-drying is to be used as a standard in experimental studies. Sugimura *et al.* (1984) also compared freeze- and air-dried worm meals and found no difference in protein content (meals 9 and 10). In this case the air-drying treatment was very prolonged; 60°C for seven days.

The meals prepared from *E. foetida* worms consistently contained about 90 g/kg ether extractable fat, but this may be higher if acid hydrolysis was used prior to extraction (Sugimura *et al.*, 1984). With meals 1 and 3 acid ether extract values were 15 g/kg higher than ether extract. Meals from *L. terrestris* contain lower levels of fat. No studies on the composition of body lipids of earthworms have been found.

Protein and fat together account for about 800 g/kg of the organic matter. Again no description of the remainder has been found but it seems unlikely to be of great nutritional significance. The mineral composition of the worm meals is summarized in Table 2. The data were rather variable, which is not surprising if part of the ash is intestinal contents. The results for meals 1–4 suggest that removal of silica by acid

Table 3. Amino acid composition of earthworm meals (g/16gN)

	Unpublished				Published						
Meal no.[a]	1	2	3	4	6	7	8	9	10		
Species[a]	Ef	Ef	Lt	Lt	—	Ef	Ef	Ef	Ef	Typical[b]	Rel. to
Drying method[a]	OD	FD	OD	FD	—	FD	OD	FD	OD	value	chick requ.
Crude protein g/kg DM	741	674	768	746	620–710	610	620–640	642	657		
ARG	6.91	6.67	7.08	6.41	7.3	6.9	6.8	6.7	(3.5)	6.8	88
GLY	4.17	4.40	4.10	4.01	4.3	5.5	4.8	4.0	4.9	4.4	98[c]
HIS	3.21	3.31	2.85	2.89	3.8	4.3	2.6	2.6	(2.1)	3.2	106
i-LEU	4.71	4.56	4.60	4.31	5.3	4.7	4.2	3.2	4.5	4.5	85
LEU	8.50	8.15	8.33	7.80	(6.2)	8.7	7.9	(5.4)	7.2	8.1	89
LYS	7.41	6.85	7.28	6.73	7.3	8.7	7.1	7.3	6.8	7.3	(100)
MET	2.10	2.10	2.15	2.03	2.0	1.6	3.6	1.3	1.6	2.1	64
PHE	4.48	4.35	4.38	4.05	5.1	4.4	3.7	3.2	3.5	4.1	90
THR	4.96	5.06	4.88	4.66	6.0	5.2	4.8	(3.6)	(3.3)	5.1	109
TRP	1.50	0.99	1.45	0.90	2.1	—	—	1.3	1.3	1.4	111
VAL	5.76	5.72	5.67	5.23	4.4	5.1	4.9	(3.5)	5.2	5.3	92
CYS	1.58	1.78	1.45	1.56	1.8	(0.8)	(3.8)	1.0	1.0	1.5	63[d]
SER	5.01	5.18	5.14	5.17	5.8	5.3	4.7	3.7	(3.0)	5.0	—
TYR	3.94	3.99	3.81	3.56	4.6	4.4	(2.2)	2.7	3.1	3.9	87[e]
ALA	5.41	5.72	5.55	5.31	5.4	5.8	—	4.3	6.8	5.4	—
ASP	11.00	10.62	10.40	9.86	10.5	11.4	—	(7.5)	(8.1)	10.6	—
GLU	13.50	13.50	13.75	13.14	13.2	14.5	—	(9.7)	13.0	13.5	—
PRO	4.03	3.95	3.76	3.45	5.3	3.8	—	(2.8)	(1.9)	4.1	—
Total	98.2	96.9	96.6	91.07	100.4	101.1	—	73.8	80.8	(96.3)	—
Reactive lysine (FDNB)	7.2	6.6	7.1	6.5	—	—	—	—	—	—	—

[a] See Table 1; [b] Average excluding values in parenthesis; [c] GLY + SER; [d] MET + CYS; [e] PHE + TYR.

leaves a roughly constant level of 'soluble' ash of 60 g/kg. There appears to be no danger of causing of dietary imbalance by feeding worms to chickens. Phosphorus is the only mineral of great economic significance and the level of about 8.0 g/kg is fairly well-defined.

The amino acid content of the worm meal is summarized in Table 3. In most of the analyses, the sum of those amino acids commonly recovered accounts for all but a small proportion of the total nitrogen. The analyses of Sugimura *et al.* (1984) must be suspect on this basis. In addition to those listed about 0.5 g/16gN hydroxyproline has been recovered (unpublished results).

There is reasonable agreement between the different studies and a 'typical' value can be derived from the mean in most cases. The results for meals 1–4 reveal few effects of either species or drying method except for the lower histidine content of *L. terrestris* protein and small overall losses on freeze drying. Sugimura *et al.* (1984) report relatively large effects of drying method but in view of the extreme nature of their air-drying procedure and the poor recovery of amino acids these results must be suspect. Studies of reactive lysine, determined by the FDNB-method (Carpenter, 1960) revealed 'availabilities' ranging from 0.956–0.979.

When the typical amino acid composition of worm meals is expressed relative to the amino acid requirements of the chick (lysine = 100) then, apart from the sulphur amino acids which have scores of 63–64, the balance of the protein is excellent. The addition of 2 g methionine per 16 gN, or approximately 12 g per kg worm meal, should be sufficient to maximize its value as a protein source.

Nutrient availability

The availability to chickens of energy, fat, protein, amino acids and phosphorus in worm meals has been investigated or reported. The oven-dried samples of *E. foetida* and *L. terrestris* (meals 1 and 3 in Tables 1, 2 and 3) were used in experiments to determine energy and amino acid availability by methods similar to those described by Sibbald (1983). Mature adult cockerels (ISA Brown), weighing about 3.5 kg were starved for 48 hours and then given 20 or 40 g of the test food by intubation into the crop. Excreta were collected for exactly 48 hours, freeze-dried and equilibrated prior to analysis for gross energy, nitrogen and amino acids. The true metabolizable energy value (TME_N) was estimated from the increment of excreted energy, corrected to zero N-balance at the rate of 34.4 kJ/g N, at the two levels of intake. An apparent metabolizable energy (AME) value for full fed birds, also corrected to zero N-retention, was calculated according to McNab and Fisher (1981). True amino acid retentions were determined only in birds given 40 g test feed and were corrected for endogenous losses estimated from birds given 40 g granulated glucose in the same procedure. The results of these experiments are summarized in Table 4.

The metabolizable energy values of the two meals differ widely when expressed on a dry-matter basis but represent very similar proportions of the gross energy, about 60%. Thus, the differences are in organic matter content (Table 1); AME_N (100) values per gram organic matter were 13.75 kJ/g for *E. foetida* and 13.67 kJ/g for *L. terrestris*. Yoshida and Hoshii (1978) reported data on freeze-dried *E. foetida* (meal 7) equivalent to 14.57 kJ/g organic matter. Calculation from the results of Sugimura *et al.* (1984) yields 18.04 and 12.09 kJ/g organic matter for freeze- (meal

9) and oven-dried (meal 10) *E. foetida*. It is suggested on the basis of these figures, that 14 kJ/g organic matter be used a 'typical' AME_N value for the evaluation of worm meals although the considerably higher value, and the effect of drying, implied by the experiments of Sugimura *et al.* (1984) both merit further investigation.

If the protein and fat in the meals are assumed to have digestibilities of 0.90 and 0.88 respectively (see below), and to be the only sources of ME, then the calculated AME_N values of the *E. foetida* and *L. terrestris* meals (1 and 3) are 15.18 and 14.65 kJ/g organic matter respectively. These are slightly higher than the observed values, but tend to confirm the small difference between the species and suggest that the other organic components of the meals are of low digestibility. Sugimura *et al.* (1984) found that the 'crude fiber' component had a digestibility of about 0.3 and the residual organic matter (NFE) of 0.46 in their freeze-dried meal (no. 9) and zero in the oven-dried meal (no. 10). It should be noted, however, that the determination of the digestibility of these components is subject to many technical difficulties. In these same studies fat digestibility was 0.88 in the freeze-dried meal (no. 9) and 0.59 in the oven-dried meal (no. 10).

The methods used for measuring amino acid availability do not yield a result for glycine and give unreliably low values for cystine (due to feather and scurf loss) and histidine (due to the presence of methyl histidines in the excreta). Apart from these amino acids and the values around 1.0 for tryptophan the true retention of all other amino acids varied little from the mean values of 0.93 for *E. foetida* (meal 1) and 0.92 for *L. terrestris* (meal 3). Sugimura *et al.* (1984) found crude protein digestibility of 0.873 in their freeze-dried meal (no. 9) and 0.745 after oven drying (no. 10). It appears that an average digestibility of 0.9 can be assumed for the protein components of well prepared worm meals.

The evidence concerning the susceptibility of earthworm protein to damage in drying is inconclusive. The extreme effects found by Sugimura *et al.* (1984) may be attributable to the very prolonged heating that was used, 60°C for seven days, and contrasts sharply with the values for reactive lysine of meals 1–4 in Table 3. The stability of nutrients in earthworms during drying requires further investigation. Yoshida and Hoshii (1978) found the phosphorus in worm meal to be completely available to young chicks.

Feeding experiments with worm meals

It is not possible in a feeding experiment to test the 'effect' of a feed ingredient on animal performance. The ingredient must be substituted for one or more other components of the diet and it is the effect of these substitutions that is measured experimentally.

Meal number 5 (Tables 1 and 2) was substituted for a mixture of herring meal and soyabean meal which, by calculation, had an amino acid balance similar to the worm meal. The mixture was prepared and analyzed and the following substitutions made to balance protein, energy and some minerals; 100 parts of worm meal were replaced by adding 125.6 parts herring/soyabean meal mix (42:58 w/w) and removing 6.5 starch, 15.9 cellulose, 1.9 limestone, 0.9 dicalcium phosphate and 0.4 methionine. Feeds containing 0, 72, 144 and 215 g/kg worm meal were prepared with the compositions shown in Table 5. The biological performance of chicks given these feeds from

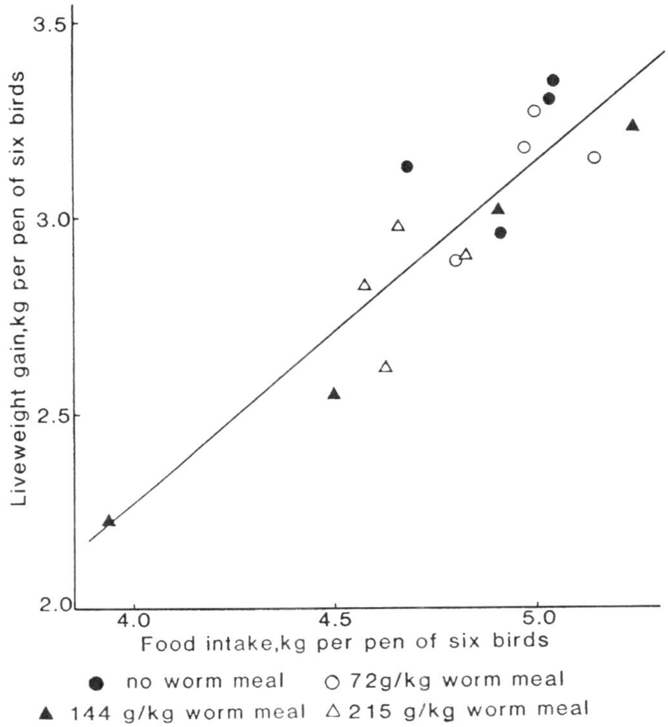

Fig. 1. Food intake and gain of pens of six birds given diets containing 0 (●), 72 (○), 144 (▲) or 215 (▿) g/kg worm meal. The pooled regression line has the equation $y = -1.22 + 0.875\ x$ and is not significantly different from separate lines fitted for each treatment.

two to four weeks of age is summarized in Table 6.

When the worm meal formed 72 g/kg of the diet and provided about a quarter of the dietary protein there was virtually no effect on the growth and feed intake of the birds nor on dietary ME or N-retention. However, at higher levels of inclusion, there is a small but significant depression in growth and feed conversion efficiency. Food consumption tended to decline but the overall effect of the treatments was not significant ($p < 0.05$).

Possible explanations for this growth depression are not immediately obvious. The diets may not have provided identical levels of nutrients but the substitutions were calculated from actual analyses of the worm meal and herring/soyabean mix and should therefore have been accurate with respect to protein and amino acids. This was borne out by the protein analyses. There is some indication that dietary energy level declined as the worm meal increased but this was a small effect and should, in theory, have affected food conversion and not growth.

The other main possibility is that the birds tended to reject the feeds containing worm meal. Although the overall differences in food consumption were not significant when weight gain is plotted against food intake for individual hens (Fig. 1) it is found that all the data lie around a single line. This indicates that the efficiency of food utilization above maintenance level is not affected by diet and this is consistent with the hypothesis that the poorer growth obtained with the diets containing worm meal is due to reduced food intake. This requires direct investigation, but it appears that some worm meals may have physical or chemical characteristics which tend to make them relatively unpalatable for chickens.

It is difficult to compare these results critically with those in the literature because of the wide variety of dietary and other conditions involved. In general, however, most investigators have not found growth depressions on feeding earthworms although reduced food intake has been widely commented on. Most experiments have, however, used raw worms or lower levels of worm meals than in this study.

Taboga (1980) mainly fed live worms with maize and, provided vitamin supplements were used, obtained growth rates equivalent to those on a proprietary feed. In one experiment a worm 'powder' (dessicated naturally for 24 hours) was fed at 350 g/kg with maize and vitamins; food intake was normal but growth slightly depressed.

Sabine (1981) reports the feeding of 123.5 g/kg earthworm meal (air dried, 60°C for 4–6 hours) from 0–52 days of age with growth equal to a control feed. Low protein diets were used but the exact substitutions are not known. Food intake was not reported but the author comments 'food consumption was consistently lower, by some 15% in the latter stages of the trial'.

Mekada *et al.* (1979) made extremely complex substitutions when using 50 g/kg worm meal in diets for chickens. No effect on growth rate was observed but there was a trend to reduced food consumption. Those authors also report successful experiments with laying hens in which raw worms were fed.

Practical trials in China showed that worms (boiled and chopped) could replace fish meal and give improved growth performance (Jin-you *et al.*, 1982) whilst Sugimura *et al.* (1984) substituted 60 g/kg air- or freeze-dried worm meal for fish meal with no effects on growth or food consumption. However, this latter experiment failed to distinguish between the two earthworm meals (9 and 10) which had very different available nutrient contents.

Discussion

As a component of poultry feeds earthworms and meals prepared from them have an excellent chemical composition. The composition of the organic matter appears to be fairly constant but methods of maximizing the yield of organic matter, a desirable objective, need more investigation. Earthworms do not seem to be particularly sensitive to heat damage in drying although prolonged air-drying at 60°C can reduce the availability of nutrients considerably (Sugimura *et al.*, 1984). More critical evaluation of drying conditions is requried.

The feeding trials with poultry are rather inconclusive. It is clear that low levels of earthworms, about 50 g/kg, can be used but feed consumption may be reduced and this merits further investigation. Species of worm and, in particular, method of meal preparation may be important variables in this respect. Sensitivity to freeze-drying was unexpected and should be studied further if this method is to be used as a control in research work.

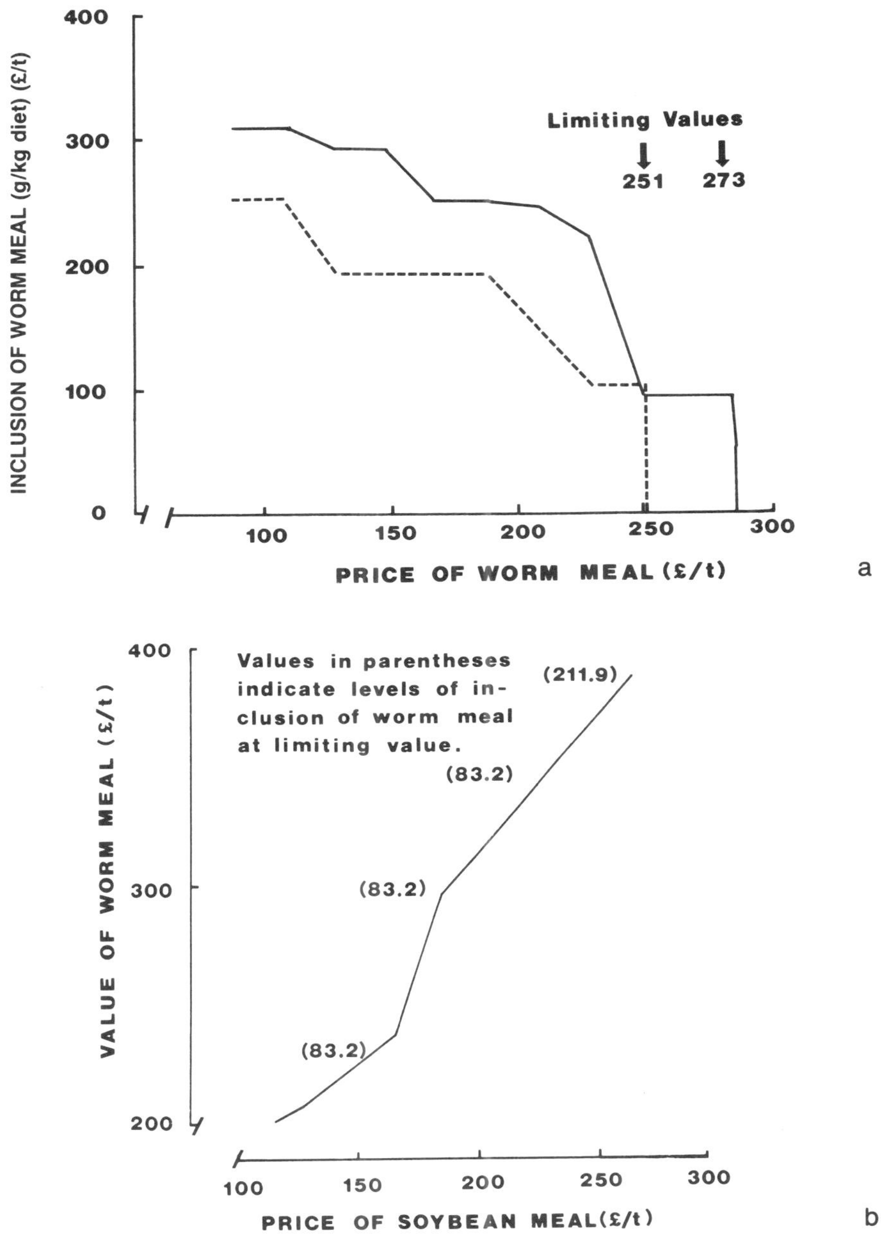

Fig. 2. Evaluation of worm meal by linear programming. Broiler starter diets were formulated to normal standards of energy, amino acids and minerals with feedstuffs at typical prices (July 1984). Figure 2a shows the inclusion of worm meal at different prices, (——) when soyabean meal costs £195/t and other animal proteins are not available, (------) with herring meal fixed at 50 g/kg. Figure 2b shows the limiting value of worm meal at different soyabean prices, when worm meal is the only animal protein source available.

Table 4. Availability of energy and amino acids in oven dried earthworm meals prepared from *E. foetida* and *L. terrestris*

	E. foetida	*L. terrestris*
Gross energy	20.7[a]	18.3[a]
True metabolizable energy at zero N-balance, TME_N, kJg^{-1}	12.467[a]	10.909[a]
TME_N/GE	.602	.596
Apparent metabolizable energy at zero N-balance calculated for a food intake of 100g, AME_N (100), kJg^{-1}	12.262[a]	10.772[a]
$AMAE_N$ (100)/GE	.592	.589
True retention of amino acids		
Lysine	.93	.91
Methionine	.94	.92
Cystine	(.83)	(.76)
Tryptophan	1.03	.99
Mean of 9 essential amino acids[b]	.931 ± .01	.916 ± .01
Mean of 15 amino acids[c]	.937 ± .03	.920 ± .02

[a] Dry-matter basis.
[b] MET, THR, SER, VAL, i-LEU, LEU, PHE, LYS, ARG.
[c] As [b] plus ASP, GLU, PRO, ALA, TYR, TRP.

Table 5. Substitution of earthworm meal (No. 5) into a chick diet

	Level of earthworm meal (g/kg)[a]	
	0	215
	g/kg diet	
Starch	46	60
Cellulose	20	54.2
Maize oil	40	40
Maize	590	590
Earthworm meal	—	215
Herring/soya meal mixture	270	—
D.L. Methionine	—	0.8
Mineral and vitamin supplements	34	40
	1000	1000
Calculated composition		
Crude protein (N × 6.25), g/kg	199.7	200.0
Metabolizable energy, kJ/g	12.90	12.89
Calcium, g/kg	10.0	10.0
Available phosphorus, g/kg	4.5	4.5
Sodium g/kg	1.6	2.3
Potassium g/kg	6.5	3.9
Chloride g/kg	3.2	3.1
Lysine g/kg	12.6	12.9
Methionine g/kg	5.4	5.4
Meth + Cystine g/kg	8.2	8.9
Tryptophan g/kg	2.1	2.7
Threonine g/kg	8.4	9.4

Table 5. Cont.

	Level of earthworm meal (g/kg)[a]	
	0	215
	g/kg diet	
Determined composition		
Crude protein, g/kg	206.9	209.4
Ether extract, g/kg	73	80
Ash, g·kg	55	48

[a] Diets with 72 and 144 g/kg earthworm meal were obtained by interpolation.

Table 6. Performance of chicks given feeds containing different levels of worm meals[a]

	Level of worm meal, g/kg				
	0	72	144	215	SEM[b]
Proportion of protein from worm meal	0	.24	.48	.71	
Initial liveweight, g	203	201	198	201	
Final liveweight, g	735	722	677	674	
Weight gain, g	532	521	479	473	16.4*
Food consumption, g	820	830	774	778	19.9[NS]
Gain/food	0.649	0.628	0.619	0.608	0.009**
Diet GE, kJ/g	17.752	18.016	18.020	18.154	
ME/GE	0.738	0.730	0.713	0.710	0.003***
N-retention, g/g diet N	0.588	0.573	0.569	0.599	0.011[NS]

[a] Each feed was given to four replicate groups of six broiler chicks (Ross I) from 14–28 days of age. No birds died. Diet ME and N-retention were determined at 25–27 days of age by incorporation of an indigestible marker (TiO_2) in the feed.
[b] Standard error of mean and significance of diet effect; NS – not significant, * $p < 0.05$, ** $p < 0.01$; *** $p < 0.001$.

In the final analysis the evaluation of a potential feedstuff must be in economic terms. Linear programming provides a means by which this can be done. Of course the value can be defined only relatively and is influenced both by the cost of alternative feedstuffs and by many other factors in the programming. In the present exercise a typical worm meal has been used in the formulation of a broiler starter diet. Typical current (July 1984) feed costs were used, the value of protein sources being largely determined in these conditions by the price of soyabean meals.

The worm meal was assumed to have the following composition (g/kg); moisture 50, ash 100, crude protein 600, amino acids calculated from Table 3, calcium 5.8, available phosphorus 80, sodium 1.1, potassium 1.6, chloride 0.6 and ME 11.5 kJ/g. The other assumptions made in this analysis, and some of the results are summarized in Fig. 2.

When ordinary soyabean meal (440 g/kg crude protein) costs £165/t the limiting value of the worm meal was £266. By comparison, meat and bone meal was worth

£273 and herring meal £406. Forcing 50 g/kg herring meal into the mix, a common practice, only reduced the value of worm meal to £251. The relationship between the price of worm meal below these limiting values and its use in broiler diets is illustrated in Fig. 2. It will be observed that once the meal is economic it rapidly becomes used at high levels. Also shown in Fig. 2 is the relationship between soyabean meal prices and the limiting value of worm meal. When soyabean is very cheap the differential is about £80–100/t rising, at higher soyabean prices, to £120–140/t. For the present time this simple evaluation suggests that the value of a typical earthworm meal is about £100/t more than soyabean meal and roughly equivalent to meat and bone meal.

Acknowledgements

The experiments were carried out by S. White, L. Broadbent and Mrs. W. Smith. A first draft of this paper was prepared by Mrs. J. Arnott.

References

Calvert, C.C. 1977. Systems for the indirect recycling by using animal and municipal wastes as a substrate for protein production. Proc. Tech. Consult., Nov. 22–24, 1976, FAO, Rome. New Feed Res. 4: 245–264.

Carpenter, K.J. 1960. The estimation of lysine in animal protein foods. Biochem. J. 77: 604–610.

Jin-you, X., Xian-Kuan, Z., Zhi-ren, P., Zhen-yong, H., Yan-hua, G., Hong-bo, T., Xye-yan, H., and Qiao-Ping, X. 1982. An experimental research on the substitution of earthworm for fish meal in feeding meat chicken. J. South China Norm. Coll. 1: 88–94.

McNab, J.M. and Fisher, C. 1982. The choice between apparent and true metabolizable energy systems – recent evidence. Proc. III Europ. Poult. Nutr. Symp., Peebles 1981, pp. 45–55. Edited by D.W.F. Shannon and I.E. Wallace. World's Poultry Science Association.

Mekada, H., Hayashi, N., Yokota, H. and Okumura, J. 1979. Performance of growing and laying chickens fed diets containing earthworms (*Eisenia foetida*). Jpn. Poult. Sci. 16: 293–297.

Priestley, G. 1976. Algal proteins. *In* Food From Waste. p. 114. Edited by G.G. Birch, K.J. Parker and J.T. Worgan. Applied Science Publishers Ltd, London.

Sabine, J.R. 1981. Vermiculture as an option for resource recovery in the intensive animal industries. *In* Workshop on the Role of Earthworms in the Stabilization of Organic Residues. Vol. 1. proceedings. 241 pp. Compiled by M. Appelhof. Beechleaf Press, Kalamazoo, Michigan.

Schuler, M.L., Roberts, E.D., Mitchell, D.W., Kargi, F., Austic, R.E., Henry, A., Vashon, R. and Seely, H.W. Jr. 1979. Process for the aerobic conversion of poultry manure into high-protein feedstuff. Biotechn. and Bioengin. 21: 19–38.

Sibbald, I.R. 1983. The T.M.E. System of Feed Evaluation. Animal Research Centre Contribution 83-1, Animal Research Centre, Ottawa, Ontario, Canada.

Sugimura, K., Hori, E., Kurihara, Y. and Itoh, S. 1984. Nutritional value of earthworms and grasshoppers as poultry feed. Jpn. Poult. Sci. 21: 1–7.

Taboga, L. 1980. The nutritional value of earthworms for chickens. Br. Poult. Sci. 21: 405–410.

Yoshida, M. and Hoshii, H. 1978. Nutritional value of earthworms for poultry feed. Jpn. Poult. Sci. 15: 308–311.

THE USE OF EARTHWORMS AS A FOOD FOR RAINBOW TROUT *SALMO GAIRDNERI*

E.A. STAFFORD and A.G.J. TACON
Institute of Aquaculture, University of Stirling, Stirling, Scotland, FK9 4LA

Introduction

Large scale vermiculture has been suggested as an alternative to conventional methods of waste disposal (Hartenstein *et al.*, 1979). As a by-product of this process, earthworm biomass represents a novel protein source with potential for use in animal feeds, including fish feeds.

At present, fish meal is the major source of dietary protein used in commercial salmonid rations. Increasing costs of fish meal and a decline in the availability of good quality fish meal has resulted in high cost fish feed which may account for between 40–60% of the running costs of an intensive fish farm. Alternative sources of protein are being sought to alleviate the dependence of the fish feed industry on a single commodity and to reduce the cost of intensive aquaculture through the use of less expensive ingredients.

Preliminary investigations into the nutritional quality of earthworms have suggested that earthworms or dried earthworm meal may be used to supply successfully at least part of the protein fraction in diets for chickens, (Sabine, 1978; Taboga, 1980) pigs, (Sabine, 1978) mice (McInroy, 1971) and fish (Guererro, 1983; Hilton, 1983; Tacon *et al.*, 1983; Stafford and Tacon, 1984). However, it has also been suggested that at high levels of earthworm meal inclusion a dietary deficiency or imbalance may result (Hilton, 1983; Stafford and Tacon, 1984).

Experimental growth trials described here aimed to nutritionally evaluate five selected species of earthworm as a feed for rainbow trout, *Salmo gairdneri*.

The nutritional composition of earthworms

Five species of earthworm: *Allolobophora longa* (Ude, 1985), *Lumbricus terrestris* Linnaeus, 1758), *Eisenia foetida* (Savigny, 1826), *Dendrobaena veneta* (Rosa, 1886) var. *zebra* (Michaelsen, 1902) and *Dendrodrilus subrubicundus* (Eisen, 1886) were selected for nutritional evaluation as food for rainbow trout.

The gross nutritional composition of these five earthworm species compared favorably with that of a commercially available trout pellet (Table 1). This suggested that earthworms may have potential for use as a complete diet or supplementary feed for rainbow trout. Comparison of the essential amino acid profile of each of the five species of earthworm with that of the conventional sources of protein – herring meal, meat meal and soybean meal indicated that earthworm protein could potentially replace these dietary protein sources (Table 2). For each amino acid, with the possible exception of the sulphur amino acids (methionine and cystine), the level of amino acid in earthworm protein appeared to satisfy the dietary amino acid requirement of the rainbow trout (Table 2).

Earthworms in waste and environmental management
edited by C.A. Edwards and E.F. Neuhauser, pp. 193–208

Table 1. Nutritional composition of five earthworm species and a fingerling trout ration

	A. longa	*L. terrestris*	*E. foetida*	*D. veneta*	*D. subrubicundus*	Commercial trout[a] pellet
Component						
Moisture (%)	78.29	81.09	83.26	81.67	84.26	10
Moisture-free basis						
Crude protein (6.25 × N) (%)	50.43	56.10	58.78	56.98	67.79	45
Lipid (%)	1.44	2.13	9.04	19.15	12.41	10
Ash (%)	35.20	28.72	17.24	4.24	6.66	14
Nitrogen free extractives (%)[b]	12.93	13.05	14.94	19.63	13.14	31

[a] Mean values (to nearest whole figure) for a range of commercially available fingerling trout rations.
[b] Nitrogen free extractives (NFE) = 100 − (crude protein (%) + lipid (%) + ash (%))

Table 2. Essential amino acid composition of five earthworm species, three dietary protein sources and the essential amino acid requirement of rainbow trout. Amino acids are expressed as g amino acid/100 g recovered amino acid

	A. longa	*L. terrestris*	*E. foetida*	*D. veneta*	*D. subrubicundus*	Herring meal[a]	Meat meal[a]	Soybean[a] meal	R. trout[b] requirement
Amino acid									
Arginine	7.17	6.54	5.62	6.31	6.56	7.8	6.7	7.7	3.5
Histidine	2.30	2.85	2.96	2.68	2.93	2.6	2.0	2.5	1.6
Isoleucine	5.10	4.53	4.14	4.34	4.48	4.2	3.5	5.7	2.4
Leucine	8.13	8.48	8.29	7.46	7.90	7.1	6.4	7.7	4.4
Lysine	7.81	7.26	6.52	6.77	7.45	7.9	5.5	6.6	5.3
Methionine	1.14	2.29	2.80	2.78	2.58	3.1	1.4	1.5	1.8
(Cystine)[c]	(0.68)	(0.66)	(0.70)	(3.15)	(1.95)	(1.0)	(1.2)	(1.5)	(0.90)
Phenylalanine	6.04	4.17	3.97	4.54	5.04	3.6	3.5	5.0	3.1
(Tyrosine)	(4.53)	(3.67)	(3.46)	(4.54)	(3.81)	(3.4)	(1.6)	(2.9)	(2.1)
Threonine	4.80	5.12	5.60	4.67	5.41	4.0	3.3	3.9	3.4
Valine	5.60	4.75	4.65	6.57	6.51	7.9	4.7	5.4	3.1

[a] Source: Tacon, 1982.
[b] Source: Ogino, 1980 (expressed as a percentage of protein).
[c] These values may be low due to oxidation during acid hydrolysis.

Extensive analysis of the lipid fraction of earthworms, conducted by Hansen and Czochanska (1975) revealed a high proportion of polyunsaturated fatty acids, including fatty acids of the linolenic series, at sufficient level to satisfy the dietary requirement of rainbow trout. A considerable variation was observed in the ash content of the five earthworm species studied, ranging from 4–35% (Table 1). Concentrations of selected mineral elements within the earthworms also showed a wide variation between species (Table 3). Prior to analysis, the earthworms were not held on an inert substrate to void their gut contents and therefore this variation may be a reflection of the earthworm gut contents and substrate media. For comparison Table 3 also gives the range of concentrations reported in commercial fingerling trout rations available in Europe (Tacon and De Silva, 1983).

The nutritional evaluation of earthworms in rainbow trout feeding trials

In three experimental feeding trials conducted with rainbow trout, the performance of fish fed diets comprised wholly or partially of earthworms was compared with the performance of fish fed on a control fishmeal-based trout pellet. Assessment of fish performance was based upon measurements of fish growth, feed utilization efficiency and whole fish carcass composition. A similar experimental procedure was employed for each feeding trial and has been described in detail elsewhere (Tacon *et al.*, 1983). Groups of experimental fish were held in 60 liter tanks supplied with mains water at a rate of 2 litre/min in a through-flow system. Each group of fish was allocated one of the experimental diets. Fish were fed twice daily 'to appetite' and an accurate record of the bi-weekly feed intake was kept. Fish in each group were individually weighed at the start of the experiment and at bi-weekly intervals throughout the experimental period. A representative sample group of fish was killed at the start of each feeding trial (initial fish sample) and fish fed each of the experimental diets were killed at the end of the experimental period. All fish were subject to whole carcass proximate analysis.

Fish for which the whole carcass mineral composition was required were dissected and the gastro-intestinal tract washed free of undigested feed using de-ionized water. Analytical methods have been described previously (Tacon *et al.*, 1983; Stafford and Tacon, 1984). The livers of five fish from each group were removed and weighed for the determination of liver somatic index (LSI = Liver weight/Fish weight × 100). Tissue slices of gill, kidney, liver and spleen were removed from five fish in each group and prepared for histopathological examination (Tacon and Cooke, 1980).

Statistically significant difference between means was determined using Duncan's multiple range test (Duncan, 1955).

Trial 1

Over an experimental period of 70 days, rainbow trout were maintained on diets which consisted solely of frozen slices of each of the earthworm species: *A. longa, L. terrestris* or *E. foetida*. A commercially available trout pellet served for comparison as a control diet.

Table 3. Mineral composition of the five species of earthworm nutritionally evaluated with rainbow trout and the range of concentrations reported for commercially available fingerling trout rations in Europe. (All values expressed on a dry weight basis)

	A. longa	*L. terrestris*	*E. foetida*	*D. veneta*	*D. subrubicundus*	Commercial[a] trout pellet
Macro-elements (g/kg)						
Ca	3.40	6.96	6.13	1.48	1.94	14–32
P	8.95	8.10	9.07	16.45	14.15	11–22
Mg	0.73	0.82	0.63	0.57	0.68	2–22
Micro-elements (mg/kg)						
Fe	5280.69	4014.09	1228.49	320.33	370.43	80–270
Zn	300.99	304.49	252.94	111.31	204.87	55–150
Pb	24.62	23.40	20.85	7.85	15.28	2–7
Cu	12.25	13.34	25.79	16.59	21.28	7–34
Cd	4.06	3.30	4.90	3.35	1.07	0.1–1.0

[a] Source: Tacon and De Silva, 1983.

Materials and methods

Twenty fingerling rainbow trout (ca. 30g in weight) were alloted dietary treatments, as follows:

Diet 1 Frozen slices of *A. longa*
Diet 2 Frozen slices of *L. terrestris*
Diet 3 Frozen slices of *E. foetida*
Diet 4 Control – Commercial trout pellet

The nutritional composition of each of these three earthworm species employed as fish diets in this trial is given in Table 1. The control commercial trout pellet (trout grower, Edward Baker Ltd., Sudbury, Suffolk) comprised 46.30% crude protein, 7.66% lipid, 13.58% ash and 32.46% nitrogen free extract (NFE). The water temperature during the experimental period was 7.5°C ± 1°C and fish were subject to natural photoperiod.

Results

Fish fed diets consisting solely of frozen slices of *A. longa* or *L. terrestris* achieved similar growth rates to fish fed the control diet (Fig. 1, Table 4). There was no significant difference ($p < 0.05$) between the mean final body weights of fish fed these three diets (Table 4). The feed utilization efficiency, expressed in terms of food conversion ratio (FCR = Feed intake, dry weight/Live fish weight gain), protein efficiency ratio (PER = Protein intake/Live fish weight gain) and apparent net nitrogen utilization (ANNU = N retention in the fish carcass/N intake over the experimental period × 100) was similar for those fish which achieved similar rates of growth over the experimental period (Table 4).

A significant difference was observed in the carcass composition of fish fed diets

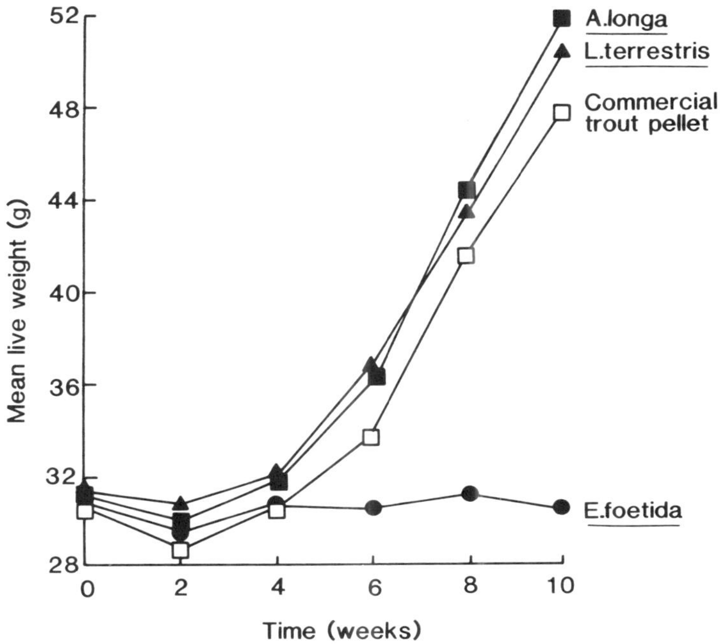

Fig. 1. Trial 1: the growth response of rainbow trout, fed only on earthworms – compared with a commercial diet.

Table 4. Trial 1: growth performance and feed utilization efficiency of rainbow trout fed the experimental diets for 70 days

Diet	*A. longa*	*L. terrestris*	*E. foetida*	Commercial trout pellet	± SE*
Mean initial weight (g)	31.10[a]	30.90[a]	30.80[a]	30.40[a]	0.992
Mean final weight (g)	50.24[b]	51.72[b]	30.89[a]	47.70[b]	2.160
Food conversion ratio	1.37	1.23	—	1.48	
Protein efficiency ratio	1.45	1.45	—	1.46	
Apparent net N utilization	24.34	24.39	—	27.20	
Liver somatic index	1.83[b]	1.84[b]	1.17[a]	1.68[b]	0.129

* Standard error – calculated from residual mean square in the analysis of variance.
[ab] Mean values for components with the same superscript are not significantly different ($p < 0.05$).

consisting of frozen earthworms compared with fish fed the control fish pellet (Table 5). The carcass lipid content of fish fed frozen slices of earthworm was significantly lower ($p < 0.05$) than fish fed the control diet. Conversely, the carcass moisture content of fish fed frozen slices of earthworm was significantly higher ($p < 0.05$) than fish fed the control diet. A similar inverse relationship between the whole fish carcass moisture and lipid content has been reported previously (Cowey and Sargent, 1979).

Histopathological examination of tissue slices of gill, kidney, liver and spleen showed no evidence of any toxic effect of the earthworms to the fish.

Table 5. Trial 1: proximate carcass composition of rainbow trout fed the experimental diets for 70 days

Diet	Initial fish	*A. longa*	*L. terrestris*	*E. foetida*	Commercial trout pellet	± SE*
		After 70 days				
Component (% wet weight)						
Moisture	74.03	76.45[b]	75.45[a]	77.49[c]	75.20[a]	0.280
Crude protein	15.37	15.89[a]	15.99[a]	15.50[a]	15.75[a]	0.376
Lipid	7.19	4.50[a]	5.60[b]	3.99[a]	6.45[c]	0.231
Ash	2.37	2.69[a]	2.63[a]	2.47[a]	2.47[a]	0.042

* Standard error – calculated from residual mean square in the analysis of variance
[abc] Mean values for components with the same superscript are not significantly different

Discussion

The results of this experiment suggest that, for a period of 70 days, the earthworm species *A. longa* and *L. terrestris* replaced the commercial trout pellet as a sole source of food for rainbow trout adequately, without adverse effect on the health, growth rate or feed utilization efficiency of the fish. However, the earthworm *E. foetida* did not support fish growth and without some form of pretreatment this species can serve little or no purpose as a complete diet or feed supplement for rainbow trout. This result was surprising in view of the recommended use of *E. foetida* as a feeding stimulant for fish (Nightingale, 1980) and its sale for use as fish bait. The European eel, *Anguilla anguilla*, has also been observed to find a diet consisting of frozen slices of *E. foetida* to be unpalatable (Aston *et al.*, 1982).

The banded or striped appearance of *E. foetida* distinguishes it from the majority of other terrestrial lumbricids. This may be a warning to predators of its unpalatable nature. As a defense mechanism *E. foetida* ejects a foul smelling yellow coelomic fluid from the dorsal pores in response to unfavorable stimulii (Gerard, 1964; Edwards and Lofty, 1977) and this may cause it to be unpalatable to the rainbow trout.

In other experiments, (Stafford, 1984) several pre-treatment processes have been applied to live *E. foetida* in an attempt to improve the acceptability of these worms to the fish. A significant improvement was achieved by 'blanching' the live *E. foetida* before use (dropping the live earthworms into boiling water for five minutes before freezing for subsequent use as a fish diet). After a six-week growth period, there was no significant difference ($p < 0.05$) between the mean final body weight of fish fed frozen 'blanched' *E. foetida* and fish fed a control fish meal based pellet. However, the extent to which the observed improvement was due to the heat treatment applied to the worms in this process or to the elimination of a large proportion of the noxious yellow coelomic fluid of *E. foetida* achieved by the 'blanching' process was not established.

The reduced carcass lipid content of fish fed *A. longa* and fish fed *L. terrestris* may have been a reflection of the low lipid content of these diets (Table 1). Fish fed *E. foetida* consumed only sufficient feed to maintain their body weight and their low carcass lipid content may have resulted through the utilization of body lipid reserves. Fish fed *E. foetida* also displayed the lowest liver somatic index which may also be a reflection of insufficient feed intake (Takeuchi and Watanabe, 1982).

Trial 2

The earthworm species *D. veneta* and *D. subrubicundus* were evaluated nutritionally as a frozen complete diet for rainbow trout over an experimental period of 60 days. For comparison a commercially available trout pellet served as a control diet.

Material and methods

Three experimental diets were each fed to groups of 15 fingerling rainbow trout (ca. 24 g in weight) as follows:

Diet 1 Frozen slices of *D. veneta*
Diet 2 Frozen slices of *D. subrubicundus*
Diet 3 Control – Commercial trout pellet

The nutritional composition of the earthworms employed as fish diets for this trial is given in Table 1. The control commercial trout pellet comprised: 52.12% crude protein, 13.25% lipid, 12.24% ash and 22.39% NFE. Fish were subject to natural photoperiod and a water temperature of 14.5°C ± 3.5°C. At the end of the experimental period all fish were subject to whole carcass proximate and mineral analysis.

Results

After a 60-day experimental feeding period, there was no significant difference ($p < 0.05$) between the mean final body weights of fish fed each of the three experimental diets (Fig. 2, Table 6). Fish fed frozen *D. subrubicundus*, and fish fed the control diet, exhibited a more efficient FCR compared with fish fed frozen *D. veneta* and a higher PER and ANNU was achieved by fish fed the control diet compared with fish fed the frozen earthworm diets (Table 6).

Maintaining rainbow trout on a diet of only frozen slices of earthworm for a period of 60 days affected the carcass composition of the fish significantly (Table 7). A similar result was reported in Trial 1. However, while in this experiment the carcass lipid content was significantly lower ($p < 0.05$) the carcass crude protein content was significantly higher ($p < 0.05$) in fish fed frozen earthworms compared with fish fed the control diet.

The mineral composition of the *D. veneta* and *D. subrubicundus* employed as fish diets in Trial 2 is given in Table 3. The control commercial trout pellet contained 17.27% Ca; 17.96% P; 2.13% Mg; 301.73% Fe; 183.79% Zn; 4.62% Pb; 15.31% Cu and 0.52% Cd. Despite low levels of Ca and Mg in diets consisting of frozen earthworms (Table 3), at the end of the experimental period similar levels of these elements were present in the carcass of fish fed all experimental diets (Table 7). Rainbow trout have been shown to compensate for low levels of these elements in the diet by uptake through the gills and fins from the surrounding water (Ichikawa *et al.*, 1962).

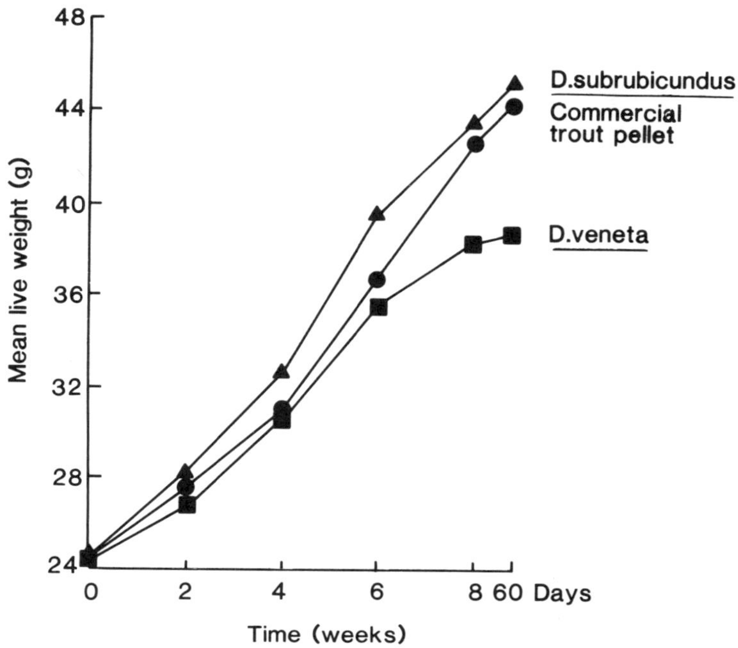

Fig. 2. Trial 2: the growth response of rainbow trout.

Table 6. Trial 2: growth performance and feed utilization efficiency of rainbow trout fed the experimental diets for 60 days

	D. veneta	*D. subrubicundus*	Commercial trout pellet	± SE*
Mean initial weight (g)	24.43[a]	24.51[a]	24.16[a]	0.602
Mean final weight (g)	38.37[a]	45.02[a]	43.81[a]	2.812
Food conversion ratio	1.53	1.07	1.12	
Protein efficiency ratio	1.15	1.37	1.72	
Apparent net N utilization	24.78	27.72	30.22	
Liver somatic index	1.42[b]	1.55[b]	1.18[a]	0.072

* Standard error – calculated from the residual mean square in the analysis of variance
[ab] Mean values for components with the same superscipts are not significantly different ($p < 0.05$)

Concentrations of Pb, Cu and Cd were higher in the diets consisting of frozen earthworms compared with the control diet (Table 3). Fish fed frozen *D. veneta* exhibited a significantly higher ($p < 0.05$) concentration of Zn and fish fed frozen *D. subrubicundus* a significantly higher ($p < 0.05$) concentration of Pb compared with fish fed the control diet (Table 7). Although higher levels of Cu and Cd were present in the frozen earthworm diets compared with the control diet, there was no significant increase ($p < 0.05$) in the carcass Cu or Cd concentration of fish fed *D. veneta* or *D. subrubicundus* (Table 7).

Discussion

Maintaining rainbow trout on a diet comprised solely of frozen *D. veneta* or frozen *D. subrubicundus* caused no significant decline in the growth rate or health of these fish compared with fish fed the control commercial trout pellet. The observed decline in efficiency of protein utilization may have been a result of the high crude protein content of frozen *D. veneta* (56.98%) and frozen *D. subrubicundus* (67.79%) which is in excess of the dietary protein requirement of rainbow trout (40–50% protein on a dry weight basis). Under these circumstances, excess protein may be catabolized as an energy source (Cowey and Sargent, 1979).

The proximate carcass composition of rainbow trout fed frozen earthworm diets was significantly altered compared to fish fed a control fish-meal based diet in both Trial 1 and Trial 2. In trial 1, an inverse relationship between carcass lipid and moisture content was observed. However, in Trial 2, the significantly reduced carcass lipid content varied inversely with the carcass crude protein content. In contrast to the nutritional composition of *A. longa* and *L. terrestris*, both species of earthworm employed as fish diets in Trial 2 had lipid contents similar to the control commercial trout pellet. Further experimentation is necessary to discover the factors causing alteration to the carcass composition of fish fed frozen earthworm diets.

Some evidence emerged in the results of Trial 2 that Zn and Pb may accumulate in the carcass of fish fed frozen slices of earthworms (Table 7). Both Zn and Pb have been shown to accumulate in the tissues of earthworms (Czarnowska and Jopkiewicz, 1978; Hartenstein *et al.*, 1980) and the passage of these elements through the food chain should be closely monitored if earthworms are used in animal feeds. Cd has been demonstrated to accumulate within earthworms to a greater extent than other elements (Van Hook, 1974; Wade *et al.*, 1982). However, in Trial 2 no evidence emerged that Cd accumulated within the carcass of fish fed *D. veneta* or *D. subrubicundus* compared with fish fed the control diet.

Trial 3

The chemical composition of earthworms bears important similarities to the dietary protein source which is herring meal (Table 2), and dried earthworm meal may therefore have potential for use as a direct replacement for the herring meal ingredient in trout diets.

For the purposes of Trial 3, *E. foetida* was selected, since this is the earthworm species most widely cultured in commercial vermiculture and dried *E. foetida* worm meal was evaluated nutritionally in diets for rainbow trout. Five semi-production trout diets were formulated to include dried *E. foetida* meal at levels of 0% (control diet), 5%, 10%, 20% and 30% of the diet, directly replacing the herring meal component.

Materials and methods

E. foetida were freeze-dried and ground to produce a free flowing meal. This 'worm meal' comprised 64.48% crude protein and 11.17% lipid and was used to directly replace herring meal (68.03% crude protein and 7.59% lipid) in five experimental

Table 7. Trial 2: proximate and mineral carcass composition of rainbow trout fed the experimental diets for 60 days

	Initial fish	*D. veneta*	*D. subrubicundus*	Commercial trout pellet	± SE*
Proximate composition (% wet weight)					
Moisture	74.28	75.29[a]	73.94[a]	73.94[a]	0.765
Crude protein	15.95	17.91[b]	17.70[b]	16.70[a]	0.212
Lipid	7.25	5.05[a]	6.24[b]	7.41[c]	0.318
Ash	2.45	2.79[a]	2.45[a]	2.32[a]	0.065
Mineral composition (expressed on a wet weight basis)					
Macro-element (g/kg)					
Ca	3.27	3.75[a]	3.08[a]	2.94[a]	0.305
P	2.16	2.50[c]	1.92[a]	2.25[b]	0.069
Mg	0.15	0.23[a]	0.21[a]	0.21[a]	0.009
Micro-elements (mg/kg wet weight)					
Fe	12.11	16.18[a]	15.39[a]	15.52[a]	1.627
Zn	14.81	26.61[b]	23.47[ab]	18.54[a]	1.826
Pb	1.14	0.96[ab]	1.33[b]	0.83[a]	0.161
Cu	1.02	0.88[a]	1.28[b]	1.14[ab]	0.114
Cd	0.11	0.17[a]	0.12[a]	0.15[a]	0.018

* Standard error – calculated from residual mean square in the analysis of variance
[abc] Mean values for components with the same superscripts are not significantly different

diets (Table 8). The inert marker Cr_2O_3 was included within the pelleted diets for the determination of digestibility co-efficients. All diets were formulated to contain 45% crude protein and 12% lipid and the proximate and mineral composition of the five diets is given in Table 8.

Each of the five diets was fed to duplicate groups of 15 rainbow trout (ca. 7 g in weight). The water temperature during the experimental period of 84 days was 13°C ± 3.5°C and fish were subject to natural photoperiod. During the final week of Trial 3, fecal samples were collected, by hand stripping, from fish fed each of the experimental diets (Austreng, 1978). Fecal samples were dried at 105°C to constant weight and used for the determination of digestibility co-efficients (Furukawa and Tsukahara, 1966). At the end of the experimental period, fish fed each of the experimental diets and fish sacrificed at the start of the experiment were subject to whole carcass proximate and mineral analysis.

Results

All fish achieved similar growth rates over the experimental feeding period (Fig. 3) and no significant difference ($p < 0.05$) emerged between the mean final body weights of fish fed each of the experimental diets (Table 9). Similar efficiency of feed utilization was observed for all experimental diets (Table 9). Apparent dry matter digestibility and apparent nitrogen digestibility increased slightly with increasing dietary inclusion of dried *E. foetida* meal (Table 9).

Fish fed 5%, 10% and 20% by weight dried *E. foetida* meal in the diet exhibited a significantly higher ($p < 0.05$) carcass lipid content and fish fed the diet containing

Table 8. Trial 3: dietary formulation and proximate and mineral composition of the five experimental diets

Diet no.	1	2	3	4	5
% inclusion of *E. foetida* meal	0	5	10	20	30
Component (% by weight)					
Herring meal	40	35	30	20	10
E. foetida meal	—	5	10	20	30
Soybean meal	10	10	10	10	10
Meat and bone meal	15	15	15	15	15
Blood meal	5	5	5	5	5
Wheat meal	15	15	15	15	15
Corn oil	3	3	3	2.5	2
Cod liver oil	7	7	7	7.5	8
Carboxymethylcellulose	0.5	0.5	0.5	0.5	0.5
Vitamin mix*	2	2	2	2	2
Mineral mix*	2	2	2	2	2
Chromic oxide	0.5	0.5	0.5	0.5	0.5
Proximate composition (% wet weight)					
Moisture	7.56	7.33	7.92	7.90	8.15
Crude protein	44.86	45.24	44.88	43.89	43.46
Lipid	12.24	12.97	12.25	10.93	10.37
Ash	15.40	15.14	14.04	14.85	12.75
Nitrogen free extractives	19.94	19.32	20.91	22.43	25.27
Mineral composition (mg·kg dry weight)					
Fe	221.39	317.35	207.34	119.92	175.06
Zn	125.08	119.84	134.09	145.39	147.81
Pb	12.26	11.23	11.67	11.54	10.37
Cu	17.28	16.56	18.34	23.59	20.63
Cd	1.14	1.00	1.12	1.12	1.01

* Composition of mineral and vitamin mix given in Tacon *et al.*, 1983

Table 9. Trial 3: growth performance and feed utilization efficiency of rainbow trout fed the five experimental diets for 84 days

Diet no.	1	2	3	4	5	± SE*
% inclusion of *E. foetida* meal	0	5	10	20	30	
Mean initial weight (g)	7.25	7.28	7.28	7.69	7.46	
Mean final weight (g)	24.91[a]	23.57[a]	27.97[a]	26.34[a]	22.81[a]	1.774
Feed conversion ratio	1.55	1.61	1.37	1.58	1.63	
Protein efficiency ratio	1.45	1.38	1.63	1.45	1.42	
Apparent net N utilization	24.94	24.08	25.82	23.88	23.29	
Apparent dry matter digestibility	45.79	53.24	50.94	59.51	59.03	
Apparent N digestibility	93.05	92.65	94.51	94.31	95.26	
Liver somatic index	1.69[a]	1.63[a]	1.55[a]	1.58[a]	1.69[a]	0.074

* Standard error – calculated from residual mean square in the analysis of variance

[abc] Mean values for components with the same superscripts are not significantly different.

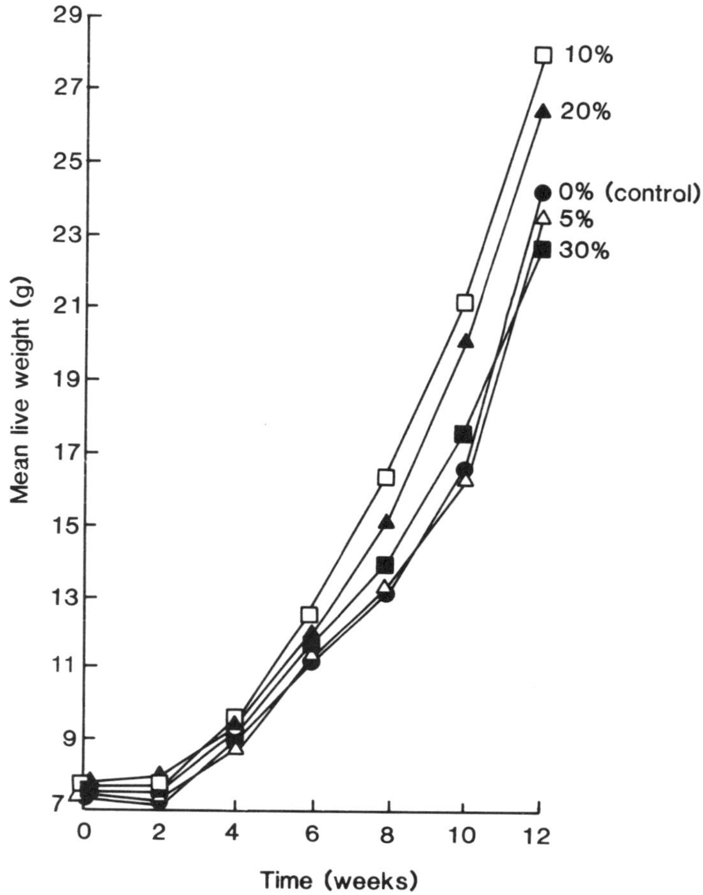

Fig. 3. Trial 3: the growth response of rainbow trout fed production diets containing low levels of *E. foetida* meal.

5% by weight of *E. foetida* meal exhibited a significantly reduced ($p < 0.05$) carcass moisture content compared with fish fed the control diet without dried *E. foetida* meal inclusion (Table 10). Five micro-elements: Fe, Zn, Pb, Cu and Cd were monitored in the experimental diets and the whole carcass of fish fed these diets (Table 8 and Table 10). Linear regression analysis revealed no significant linear relationship ($p < 0.05$) between increasing dried *E. foetida* meal inclusion and the concentration of each element within the fish carcass (Table 11).

Discussion

Dried *E. foetida* meal replaced the herring meal component of a production trout diet, up to a dietary inclusion level of 30% by weight, without detrimental effect to the growth rate or feed utilization efficiency of the fish. Previous results had demonstrated that rainbow trout, maintained on a semi-synthetic ration in which dried

Table 10. Trial 3: proximate and mineral carcass composition of rainbow trout fed the experimental diets for 60 days

Diet no.		1	2	3	4	5	± SE*
% inclusion of *E. foetida* meal		0	5	10	20	30	
	Initial fish	After 84 days					
Proximate composition (% wet weight)							
Moisture	77.96	73.34[b]	72.50[a]	72.66[ab]	72.66[ab]	73.79[b]	0.241
Crude protein	15.07	16.60[a]	16.70[a]	16.39[a]	16.08[a]	16.15[a]	0.207
Lipid	4.68	7.83[a]	8.64[bc]	8.97[c]	8.95[c]	8.10[ab]	0.207
Ash	2.38	2.58[a]	2.52[a]	2.43[a]	2.42[a]	2.48[a]	0.056
Mineral composition (mg/kg wet weight)							
Fe		10.21	12.26	11.27	11.13	11.48	
Zn		25.19	30.86	29.12	22.19	27.35	
Pb		1.29	1.41	1.29	1.59	1.40	
Cu		1.37	1.26	1.09	1.50	1.32	
Cd		0.17	0.18	0.10	0.17	0.16	

* Standard error – calculated from residual mean square in the analysis of variance
[abc] Mean values for components with the same superscripts are not significantly different

Table 11. Trial 3: regression equations relating the level of each element in the fish carcass (Y) to the level of dietary inclusion of dried *E. foetida* meal (X)

Fe:	$Y = 11.1 + 0.0146X$	$r^2 = 5.7\%$ n.s.[a]
Zn:	$Y = 27.9 - 0.0710X$	$r^2 = 6.4\%$ n.s.
Pb:	$Y = 1.33 - 0.0052X$	$r^2 = 25.9\%$ n.s.
Cu:	$Y = 1.27 + 0.0031X$	$r^2 = 6.0\%$ n.s.
Cd:	$Y = 0.16 - 0.0001X$	$r^2 = 0.1\%$ n.s.

[a] n.s. = not significantly related ($p < 0.05$) as ascertained by F test

E. foetida meal was used exclusively to supply dietary protein exhibited significantly reduced growth rates, feed utilization efficiency and a significantly altered carcass composition (Tacon *et al.*, 1983). In the same experiment, a significant improvement was observed in fish performance when fish were fed semi-synthetic rations containing only 50% dried *E. foetida* meal and 50% herring meal protein. The results of Trial 3 confirmed this trend demonstrating that dried *E. foetida* meal may be used to replace fish meal at lower levels of dietary inclusion without significantly affecting the growth rate and feed utilization efficiency of rainbow trout.

Although a significant difference was observed in the carcass lipid content of fish fed diets containing 5%, 10% and 20% by weight of *E. foetida* compared with fish fed the control diet, carcass lipid levels reported in this experiment (7.83%–8.97%) remained within the acceptable limits for rainbow trout: 3.7%–14.2%, wet weight (Pieper and Pfeffer, 1979).

Concentrations of trace elements measured in the experimental diets (Table 8), fell below the maximum levels permitted in salmonid feedstuffs (Edwards and Densem, 1980) and, with the exception of Pb, also fell within the range of concentrations measured in commercially available feedstuffs in Europe (Table 3 and Table 8). Over the 84-day feeding period no significant accumulation of Fe, Zn, Pb, Cu or Cd in the carcass of rainbow trout was evident as the dietary inclusion levels of *E. foetida* meal increased from 0–30% by weight. The high dietary Ca level present in these diets (due to the meat and bone meal inclusion), but not present in the frozen earthworm diets in Trial 2 (Table 3) may afford additional protection against heavy metal uptake by the fish (Hsu *et al.*, 1975; Six and Goyer, 1970). Evidence also exists that fish may be able to regulate tissue levels by excreting excess amounts of heavy metals including non-essential elements such as Cd (Cearly and Coleman, 1974; Miettinen, 1975).

Conclusions

Five species of earthworm – *L. terrestris*, *A. longa*, *E. foetida*, *D. veneta* and *D. subrubicundus* were nutritionally suitable as a feed for rainbow trout. Fish fed solely on frozen slices of earthworm (with the exception of fish fed *E. foetida*), achieved growth rates and feed utilization efficiency comparable to fish fed a control, fish meal based diet. Fish fed solely on frozen slices of *E. foetida* achieved little or no growth over the experimental period. The proximate carcass composition of rainbow trout fed the frozen earthworm diets differed significantly compared with fish fed on a control diet.

A dried 'worm meal' derived from *E. foetida* adequately replaced the dietary protein source herring meal at levels of up to 30% inclusion by weight without loss in fish growth and feed utilization efficiency. A significant effect was observed on the proximate carcass composition of fish fed diets containing dried *E. foetida* meal.

Some evidence emerged that levels of Pb and Zn increased in the carcass of fish fed only on frozen slices of earthworm compared with fish fed a control diet. However, at lower levels in inclusion (up ot 30% of dried *E. foetida* meal by weight in the diet) no significant linear relationship emerged between levels of heavy metals in the fish carcass and increasing dietary inclusion of dried *E. foetida* meal. In no instance did the results indicate a toxic effect of the earthworm diets to rainbow trout.

References

Aston, R.J., Morgan, N. and Abrams, F.S.H. 1982. The growth and heavy metal content of eels (*Anguilla anguilla* L.) fed on oligochaete worms. Central Electricity Generating Board, Technology, Planning and Research Division, Central Electricity Research Laboratories TPRD/L/2356/N82.

Austreng, G. 1978. Digestibility determination in fish using chromic oxide marking and analysis of contents from different segments of the gastro-intestinal tract. Aquaculture 13: 265–272.

Cearly, J.E. and Coleman, R.L. 1974. Cadmium toxicity and bioconcentration in largemouth bass and bluegill. Bull. Environ. Contam. Toxicol. 11: 146–151.

Cowey, C.B. and Sargent, J.R. 1979. Nutrition. *In* Fish Physiology. Vol. VIII. pp. 1–69. Edited by W.S. Hoar and D.J. Randall. Academic Press.

Czarnowska, K. and Jopkiewicz, K. 1978. Heavy metals in earthworms as an index of soil contamination. Pol. J. Soil. Sci. 11(1): 57–62.

Duncan, D.B. 1955. Multiple range and multiple F-tests. Biometrics 11: 1–42.

Edwards, C.A. and Lofty, J.R. 1977. The Biology of Earthworms, 2nd Ed. Chapman and Hall, London. 283 pp.
Edwards, R.W. and Densem, J.W. 1980. Fish from sewage. Appl. Biol. 5: 221–269.
Furukawa, A. and Tsukahara, H. 1966. On the acid digestion method for determination of chromic oxide as an indicator substance in the study of digestibility in fish. Bull. Jap. Soc. Sci. Fish. 32: 502–504.
Gerard, B.M. 1964. Synopses of the British fauna (6), Lumbricidae. The Linnean Society of London. 58 pp.
Guererro, R.D. 1983. The culture and use of *Perionyx excavatus* as a protein resource in the Philippines. *In* Earthworm Ecology from Darwin to Vermiculture. pp. 309–313. Edited by J.E. Satchell. Chapman and Hall, London.
Hansen, R.P. and Czochanska, Z. 1975. The fatty acid composition of the lipids of earthworms. J. Sci. Food Agric. 26: 961–971.
Hartenstein, R., Neuhauser, E.F. and Kaplan, D.L. 1979. A progress report on the potential use of earthworms in sludge management. Proceedings of the 8th National Sludge Conference, Municipal Sludge Management 19–21 March 1979. Hazardous Matter Control Res. Inst. & Sludge Mag. Inf. Transfer Inc. Silver Spring M.D. pp. 238–241.
Hartenstein, R., Neuhauser, E.F. and Collier, J. 1980. Accumulation of heavy metals in the earthworm *Eisenia foetida*. J. Environ. Qual. 9(1): 23–26.
Hilton, J.W. 1983. Potential of freeze dried worm meal as a replacement for fish meal in trout diet formulations. Aquaculture 32: 277–283.
Hori, E., Terauchi, S., Seki, Ts., Kurihara, Y., Itoh, S. and Sugimara, K. 1983. Nutritional evaluation of earthworms for growing chicken. Proceedings of the 5th World Conference on Animal Production, Tokyo, Japan. 14–19 August 1983 (Abstract only).
Hsu, F.S., Krook, L., Pond, W.G. and Duncan, J.R. 1975. Interaction of dietary calcium with toxic levels of lead and zinc in pigs. J. Nutr. 105: 112–118.
Ichikawa, R., Oguri, M. and Takada, N. 1962. Metabolism of radionuclides in fish. III Comparative behaviour of strontium and calcium at gill absorption in environmental water with different amounts of calcium. Bull. Jap. Soc. Sci. Fish. 28: 1160–1163.
McInroy, D. 1971. Evaluation of the earthworm *Eisenia foetida* as a food for man and domestic animals. Feedstuffs 43: 37 and 46–47.
Miettinen, J.K. 1975. The accumulation and excretion of heavy metals in organisms. *In* Ecological Toxicology Research Effects of Heavy Metal and Organohalogen Compounds. pp. 215–229. Edited by A.D. McIntyre and C.F. Mills. Plenum Press, London.
Nightingale, P. 1980. Worms and shelled eggs – appetizers and body builders. Fish Farmer 3(3): 47–49.
Ogino, C. 1980. Requirements of carp and rainbow trout for essential amino acids. Bull. Jap. Soc. Sci. Fish 46(2): 171–174.
Pieper, A. and Pfeffer, E. 1979. Carbohydrates as possible sources of dietary energy for rainbow trout (*Salmo gairdneri* Richardson). *In* Finfish Nutrition and Fishfeed Technology. Vol. I. pp. 209–220. Edited by J.E. Halver and K. Tiews. I.H. Heenemann Gumblt and Co. Berlin.
Sabine, J.R. 1978. The nutritive value of earthworm meal. *In* Utilisation of Soil Organisms in Sludge Management pp. 122–130. Edited by R. Hartenstein. National Technical Information Services, Springfield VA. No. PB 286932.
Six, K.M. and Goyer, R.A. 1970. Experimental enhancement of lead toxicity by low dietary calcium. J. Lab. Clin. Med. 76(6): 933–942.
Stafford, E.A. 1984. The Use of Earthworms as a Feed for Rainbow Trout, (*Salmo gairdneri*) Ph.D. Thesis, Stirling University. 242 pp.
Stafford, E.A. and Tacon, A.G.J. 1984. Nutritive value of the earthworm, *Dendrodrilus subrubicundus*, grown on domestic sewage, in trout diets. *Agric. Wastes* 9: 249–266.
Taboga, L. 1980. The nutritional value of earthworms for chickens. Br. Poultry Sci. 21(5): 405–410.
Tacon, A.G.J. 1982. Utilisation of chick hatchery waste: the nutritional characteristics of day old chick and egg shells. Agric. Wastes 4: 335–343.
Tacon, A.G.J. and Cooke, D.J. 1980. Nutritional value of dietary nucleic acids to trout. Nut. Rep. Int. 22(5): 631–640.
Tacon, A.G.J. and De Silva. 1983. Mineral composition of some commercial fish feeds available in Europe. Aquaculture 31: 11–30.
Tacon, A.G.J., Stafford, E.A. and Edwards, C.A. 1983. A preliminary investigation of the nutritive value of three terrestrial worms for rainbow trout. Aquaculture 35: 187–199.
Takeuchi, T. and Watanabe, T. 1982. Effects of various polyunsaturated fatty acids on growth and fatty

acid compositions of rainbow trout, *Salmo gairdneri*, coho salmon, *Oncorhynchus kisutch* and chum salmon *Oncorhynchus keta*. Bull. Jap. Soc. Sci. Fish. 48(12): 1745–1752.

Van Hook, R.I. 1974. Cadmium, lead and zinc distributions between earthworms and soils: Potentials for biological accumulation. Bull. Environ. Contam. Toxicol. 12(4): 509–512.

Wade, S.E., Bache, C.A. and Lisk, D.J. 1982. Cadmium accumulation by earthworms inhabiting municipal sludge ammended soil. Bull. Environ. Contam. Toxicol. 28: 557–560.

IV. Earthworms in production of plant growth media

CONTENTS

THE POTENTIAL OF EARTHWORM COMPOSTS AS PLANT GROWTH MEDIA

CLIVE A. EDWARDS[1]* and IAN BURROWS[2]**
[1]*The Ohio State University, Columbus Ohio, U.S.A.;* [2]*Rothamsted Experimental Station, Harpenden, Herts, U.K.*

Introduction

Charles Darwin first demonstrated the ability of earthworms to break down vegetable matter and incorporate it into the soil (Darwin 1881), and since then there has been an increasing body of information on the importance of earthworms in soil formation and in increasing soil fertility (Edwards and Lofty 1977; Stockli 1928; Swaby 1949). Earthworms can comminute or breakdown mineral particles to smaller units (Teotia *et al.* 1950; Joshi and Kelkan 1952) and form stable aggregates (Gurianova, 1970; Hopp and Hopkins 1946; Dawson 1947; Guild 1955), particularly when vegetable matter is added to the soil (Dutt 1948; Dawson 1948), suggesting that after plant remains have passed through earthworms they reinforce and hold the aggregates together. Earthworms are important agents in breaking down plant remains into smaller pieces which provide a greater surface area for further microbial activity. They also turnover large amounts of soil and mix this intimately with decomposed plant material. Estimates of soil turnover by earthworms range up to 268 tonnes/ha (Beauge 1912), to a typical figure for an English grassland with a good worm population of 22 tonnes/ha (Evans and Guild 1947). It has been demonstrated clearly that earthworms improve soil aeration, porosity, waterholding capacity and drainage (Evans 1948; Teotia *et al.* 1950) so that soils with large populations of worms not only drain well but can also retain moisture better in dry conditions.

There is also good evidence that earthworms have a strong influence on the availability of mineral nutrients in soils. The carbon to nitrogen ratio (C:N) has been reported by several workers to be higher in worm casts than in the soil in which they live (Satchell 1963; Graff 1971). There is much more nitrate nitrogen in worm casts than in the parent soil, moreover, the casts have a higher base-exchange capacity, and more exchangeable calcium magnesium and potassium and available phosphorus than the soil in which the worms live (Puh 1941; Lunt and Jacobson 1944; Nijhawan and Kanwar 1952; Graff 1971).

There is a strong interaction between earthworm feeding and micro-organism populations. Many workers have demonstrated dramatic increases in numbers of microorganisms after passage of soil through the earthworm gut (Bassalik 1913; Dawson 1948; Teotia *et al.* 1950; Parle 1963; Ghilarov 1963; Went 1963; Atlavinyte and Lugauskas 1971). However, most of these micro-organisms have been identified only to major groups and their function is obscure; some may be beneficial in breaking down organic matter or making nutrients available and others harmful because they are plant or animal pathogens.

There have been many conclusive demonstrations that earthworms can increase the

* Previously Senior Principal Scientific Officer, Rothamsted Experimental Station, U.K.
** *Present address:* University of Papua New Guinea.

Earthworms in waste and environmental management
edited by C.A. Edwards and E.F. Neuhauser, pp. 211–219

growth of plants in soils. These results have used differing experimental methods to demonstrate increased growth of plants after adding earthworms to soil in which was grown clover, rye and spinach (Russell, 1910; van Rhee, 1965), peas and oats (Kahnitz, 1922), barley (Uhlen, 1953), oats (Aldag and Graff, 1974) and wheat (Edwards and Lofty, 1978). All these studies, summarized by Edwards and Lofty (1977), demonstrate the considerable value of earthworms in improving the structure and fertility of soil.

However, more recently, there has been considerable interest in using species which live in decaying organic matter, such as the brandling or tiger worm *Eisenia foetida* (Savigny), to break down various human, animal, vegetable and industrial organic wastes and turn them into composts. Such materials have been marketed in U.S.A., Italy, the Philippines and elsewhere, often with extravagant claims on their usefulness as plant growth media. There is little scientific substantiation of such claims and the scientific literature on the subject is extremely sparse (Edwards, 1983; Edwards *et al.*, 1985).

An extensive research program at Rothamsted which began in 1980, was aimed at developing methods of using earthworms to break down animal, vegetable and industrial wastes into useful composts. In this program, the suitability of composts produced by earthworms from cattle, pig, chicken, duck, turkey, potato, brewery, paper and spent mushroom wastes as plant growth media has been tested for a wide range of plants.

Production of plant growth media

Most animal and vegetable based organic wastes can be broken down by appropriate species of earthworms, the one most commonly used being *Eisenia foetida*. Methods range from using simple windrows, heaps, beds or boxes, to complex commercial continuous flow systems. These are discussed elsewhere in this book (pp. 21, 65, 111, 157). Such systems turnover wastes extremely rapidly and efficiently, and produce finely divided materials that not only have a good structure but contain inorganic nutrients. However, organic wastes differ greatly in their nutrient content and this governs the spectrum of nutrients that plant growth media contain. Earthworms can change the form of the nutrient compounds but have little effect on the total amounts contained.

Characteristics of plant growth media produced by earthworms from organic wastes

The final physical structure of the plant growth media produced from organic wastes depend very much on the original material. Some, such as cattle, pig and poultry manures from indoor systems and spent mushroom compost contain straw, which takes longer for the earthworms to fragment than more particulate materials such as animal solids and slurries, brewery waste, paper pulp and similar materials. However, the final product is usually a finely-divided peat-like material with excellent structure, porosity, aeration, drainage and moisture-holding capacity (Edwards, 1982, 1983). It has the appearance and many of the characteristics of peat.

The nutrient content differs greatly depending on the parent material. However, when the nutrient content is compared with that of a commercial plant growth medium to which inorganic nutrients have been added, such as Levington compost, it is

Table 1. Major plant nutrient elements in worm processed animal wastes

Waste material	Element content (% dry wt.)					
	N	P	K	Ca	Mg	Mn
Separated cattle solids	2.20	0.40	0.90	1.20	0.25	0.02
Separated pig solids	2.60	1.70	1.40	3.40	0.55	0.03
Cattle solids on straw	2.50	0.50	2.50	1.55	0.30	0.05
Pig solids on straw	3.00	1.60	2.40	4.00	0.60	0.05
Duck solids on straw	2.60	2.90	1.70	9.50	1.00	0.10
Chicken solids on shavings	2.75	2.70	2.10	4.80	0.70	0.08
Levington compost (Commercial plant growth medium)	1.80	0.21	0.48	0.94	2.20	0.02

usually higher for most elements, although there is often a deficiency of magnesium (Table 1). An important feature is that during the processing of the wastes by earthworms, many of the nutrients they contain are changed to forms more readily taken up by plants, such as nitrate, nitrogen, exchangeable phosphorus and soluble potassium, calcium and magnesium (Table 2).

Moreover, organic wastes tend to be on the alkaline side of neutral > 7.0, whereas, most plants prefer a growth medium on the acid side of neutral, *e.g.*, a pH of 6.0. The processing by worms does not change the pH appreciably, so for ideal plant growth, worm-worked wastes benefit from some acidification or by mixing with an acid medium such as peat.

Growth of plants in worm-worked organic wastes

The research program at Rothamsted Experimental Station involved an extensive series of trials testing the growth of a wide range of plants in a variety of worm-worked wastes. As well as the extensive trials at Rothamsted, other tests were made at the National Vegetable Research Station, the Glasshouse Crops Research Station, Oaklands Agricultural College (St. Albans), the Agricultural Developmental Stations at Lea Valley and Bedford (see p. 221), and by a fertilizer company in Barcelona, Spain. There is no space here to report these extensive trials in detail, but it would be appropriate to give some general data and results on particular aspects and uses of worm-worked composts.

A wide range of plants have been grown successfully in both undiluted wastes or a number of mixes including 3:1 or 1:1 ratios of worm-worked wastes to peat, pine bark or Kettering loam. Growth has usually been better than in recommended commercial growing media such as Levington compost, and seed germination has been found to be more rapid for most species of plant in worm-worked wastes.

Plants tested include vegetables such as: aubergine, cabbage, capsicum, cucumber, lettuce, radish and tomato, bedding plants such as alyssum, antirrhinum, aster, campanula, calceolaria, cineraria, coleus, French marigold, plumose asparagus, polyanthus, salvia and sweet pea, and ornamental shrubs such as *Eleagnus pungens, Cotoneaster conspicua, Pyracantha, Viburnum bodnantense, Chaemaecyparis lawsonia, Cupressocyparis leylandii* and *Juniperus communis*.

Table 2. Effect of worm activity on nutrients in organic wastes

Organic waste	Ammonium nitrogen (ppm)	Nitrate nitrogen (ppm)	Readily soluble P. (% d.m.)	Exchangeable (% d.m.)		
				K	Ca	Mg
Cattle (unworked)	117.1	8.8	0.11	0.19	0.35	0.05
Cattle (worm-worked)	141.5	259.4	0.18	0.41	0.59	0.08
Pig (unworked)	858.4	31.6	1.05	1.49	1.56	0.45
Pig (worm-worked)	2040.0	110.3	1.64	1.76	2.27	0.72
Potato (unworked)	1982.5	74.6	0.19	1.94	0.91	0.24
Potato (worm-worked)	681.8	1428.0	0.22	3.09	1.37	0.34

Table 3. Emergence of seedlings in worm-worked and composted animal wastes compared with a commercial growth medium

Medium	% Emergence (after 14 days)		
	Tomato	Cabbage	Radish
Composted cattle	15.0	30.0	76.7
Composted pig	11.7	66.7	75.0
Worm-worked cattle	91.7	88.3	90.0
Worm-worked pig	93.3	83.3	95.0
50% worm-worked cattle/50% pig	98.3	30.0	96.7
50% worm-worked pig/50% peat	93.3	86.7	91.7
Recommended commercial Levington potting compost	86.7	86.7	90.0

The waste, after working by the worms, is at about 75% M.C. Some form of partial sterilization, such as by heating to between 60° and 80°C, may be used to kill residual earthworms and their cocoons, insects and to lessen pathogen problems. Usually, a magnesium sulphate supplement is used to rectify the usual deficiency in magnesium, and the pH is adjusted in some way.

a. Effects on seedling emergence

A wide range of tests of seedling emergence of pea, lettuce, wheat, cabbage, tomato and radish were made in small pots and trays using the standard E.E.C. recommended seedling emergence test. The results of typical seedling emergence and growth tests are given in Tables 3 and 4. Clearly, seedling emergence of tomatoes, cabbage and radish tended to be as good and usually better in worm-worked animal wastes than in a commercial medium, and much better than in composted animal wastes with no earthworms. Similarly, early growth of seedlings of ornamentals up to the stage they

Table 4. The growth status of seedlings in worm-worked waste/peat mixtures compared with a commercial plant growth medium at 'potting-on'

Plant	Cabbage	Tomato	Dahlia	Petunia	Capsicum	Salvia	Aubergine	Coleus	Cineraria	Calceolaria
Growth medium										
50% worm-worked pig/50% peat	5	4	4	4	4	3	5	2	5	3
50% worm-worked cattle/50% peat	5	3	4	5	5	3	5	2	5	5
Levington commercial plant growth medium	4	4	4	5	5	4	5	2	4	3

Growth index 1 (poorest) → 5 (best)

Table 5. Growth of ornamentals in worm-worked animal wastes

Growth medium	Aubergine	Dahlia	Calceolaria	Coleus	Capsicum	Marjoram	Salvia	Polyanthus	Cineraria	Petunia
Recommended commercial compost (Levington)	7	9	6	8	7	5	3	6	8	7
Worm-worked cattle	8	9	6	8	9	2	3	6	8	7
Worm-worked pig	9	9	8	8	9	9	6	8	8	7
25% worm-worked cattle/75% peat	6	6	8	8	4	8	7	8	8	6
25% worm-worked pig/75% peat	8	8	8	6	8	7	7	6	9	7

Growth index: 1 (poorest) → 10 (best).

Table 6. Growth of tomatoes and cabbages in worm-worked paper waste

Plant	Tomatoes		Cabbages	
	% emergence	Mean dry weight (g)	% emergence	Mean dry (weight (g)
Growth medium				
Recommended commercial compost (Levington)	98	3.66	90	11.9
100% worm-worked paper	97	3.67	90	16.6
100% peat	98	3.38	85	9.9
75% peat/25% worm-worked paper	93	4.36	93	12.6
50% peat/50% worm-worked paper	97	4.29	97	15.1
100% peat/sand	100	3.26	93	10.9
75% peat/sand/25% worm-worked paper	98	3.96	88	15.3
50% peat/sand 50% worm-worked paper	97	5.11	92	11.5

were transplanted into larger pots or outdoors was as good or better in the worm-worked animal wastes mixed with peat as in the commercial plant growth medium.

b. Growth of young plants

After transplanting into pots, the growth of ornamentals was also good in worm-worked animal wastes and mixtures of these with peat (Table 5) and was often better than in the commercial medium. Aubergines, dahlias, coleus, capsicum and polyanthus did particularly well in such worm-worked waste/peat mixtures. Some of the ornamentals, particularly chrysanthemums, salvias and petunias flowered much earlier in worm-worked waste mixtures and this could possibly be due to a hormonal effect. A number of other species of plants followed significantly different growth patterns, another indication of a possible hormonal effect.

c. Growth of tomatoes and cabbage in worm-worked paper waste

When the solids in the liquid effluent from paper manufacturing plants (about 0.6% solids), were separated and concentrated to about 35% dry matter and used in worm processing systems, *E. foetida* grew in them extremely well and processed the solids into an excellently-textured growth material very quickly. Although this medium was lower in nutrients than materials produced from animal wastes, it was still excellent for seedling germination and growth (Table 6). Tomatoes and cabbages germinated well and grew better in mixtures of worm-worked paper waste with peat and sand than in a commercial plant growth medium or in 100% peat or a peat/sand mixture. It was one of the best materials tested and there were no problems in processing or producing a standardized material.

Table 7. Effects of a range of mixtures between an Efford commercial plant growth medium and worm-worked pig and cattle waste on growth of ornamentals

Efford (%)		100	95	90	75	50	25	0
Medium (%)	Worm-worked cattle	0	2.5	5.0	12.5	25	37.5	50
	Worm-worked pig	0	2.5	5.0	12.5	25	37.5	50
Plant growth	*Chaemaecyparis lawsoniana*	6	8	9	9	9	10	9
	Pyracantha sp.	5	9	10	10	9	9	9
	Viburnum bodnantense	7	8	8	9	9	10	8

Growth index: 1 – (poorest); 10 – (best).

d. Effects of dilution of worm-worked animal wastes with a commercial plant growth medium (Efford) on growth of ornamentals

In trials of three ornamental plants, when a 50/50 mixture of pig and cattle worm-worked animal wastes were diluted at a range of levels with a commercial plant growth medium (Efford), all of the mixtures grew *Chaemaecypanis lawsoniana Pyracomtha* sp. and *Viburnum bodnantense* better than the recommended medium (Table 7). The most surprising result was that even 5% of worm-worked animal waste in the worm-worked waste/commercial mixture had a significant effect on the growth of the plants. All dilutions of the different mixtures also tended to grow better plants than did the 100% worm-worked wastes, which had a tendency to dry out more rapidly than the different mixtures.

e. Use of worm-worked animal waste as a 'blocking' material for growth of seedlings to transplant into the field

Cabbages were grown in machine-compressed blocks made from either a commercial material or worm-worked cattle or pig waste. The seedlings were transplanted into the field in the blocks and their subsequent growth followed and measured. Those transplants grown in worm-worked pig waste were significantly larger and more mature at harvest than those grown in the commercial blocking material (Table 8).

Thus, there is evidence that good worm-worked organic wastes mixed with peat and other materials make excellent plant growth media for a variety of purposes. They need standardization to ensure an adequate nutrient status, adjustment of pH and sterilization to kill insects etc., but have considerable commercial potential.

Commercial markets for worm-worked animal wastes as plant-growth media

The potential commercial outlets for such materials vary very much between countries, and so do the economic returns on the sale of these media. In general, for the high value market, the product must be produced to within relatively small tolerances, as a standard material varying little in consistency or nutrient content. For such a product, uniform sources of organic wastes must be available and if more than one kind of waste is used, the mixture and additives must be in constant proportions.

Table 8. Growth of cabbages sown in seedling blocks and transplanted to the field

Blocking medium	No. of plants	Mean weight of cabbage head (g)
Worm-worked cattle	16	267 ± 59
Worm-worked pig	15	493 ± 60
Commercial blocking material	16	299 ± 96

Table 9. Markets for plant growth media produced from worm-worked organic wastes

Market	Product	Value	Tolerance
Home and gardens	Grow-bags	High	Narrow
Horticulture	Growing media	Moderate	Narrow
Amerity/Parks/ Horticulture	Tree planting compost	Moderate	Moderate
Horticulture/ gardens	General soil	Moderate	Moderate

Additionally, there must be batch analyses to ensure the standardization of the product (Table 9). As the product is produced with lower technology and more variable sources of wastes, so its value decreases but so does the cost of production processing and packaging.

If a large market is to be reached, the production and packaging must be mechanized and the wholesale cost be very competitive with that of existing products.

References

Adag, R. and Graff, O. 1974. Einfluss der Regenwurm tätigkeit auf proteingehalt und proteinqualität junger Haferpflanzen. Zeitschr. Lebensm. Forsch. 31(11): 277–284.

Atlavinyte, O. and Luganskans, A. 1971. The effect of Lumbricidae on soil microorganisms. Ann. Anim. Zool. Ecol. (Spec. publ.) 4: 73–80.

Bassalik, K. 1913. On silicate decomposition by soil bacteria. Zeitschr. Garungs-phys. 2: 1–52.

Beauge, A. 1912. Les vers de terre et la fertilité du sol. J. Agric. Pract. Paris 23: 506–507.

Dawson, R.C. 1947. Earthworm microbiology and the formation of waterstable aggregates. Soil Sci. 69: 175–184.

Dawson, R.C. 1948. Earthworm microbiology and the formation of waterstable aggregates. Proc. Am. Soil Sci. Soc. 12: 512–516.

Darwin, C. 1881. The Formation of Vegetable Mould through the Action of Worms. Murray, London. 326 pp.

Dutt, A.K. 1948. Earthworms and soil aggregation. J. Am. Soc. Agron. 40: 407.

Edwards, C.A. and Lofty, J.R. 1977. Biology of Earthworms. Chapman and Hall, London. 333 pp.

Edwards, C.A. and Lofty, J.R. 1978. The influence of arthropods and earthworms upon root growth of direct drilled cereals. J. Appl. Ecol. 15: 789–795.

Edwards, C.A. 1982. The use of earthworms in waste disposal and protein production. Rothamsted Experimental Station Report for 1981. Part 1, 103–105.

Edwards, C.A. 1983. The use of earthworms in waste disposal and protein production. Rothamsted Experimental Station Report for 1982. Part 2, 102–105.

Edwards, C.A. 1985. Utilization of earthworm composts as plant growth media. *In* International Symposium on Agricultural and Environmental Prospects in Earthworm Farminag. Rome. 183 pp. Edited by U. Tomaki and A. Grappelli. Consortium of Humus Producers, Rome.

Edwards, C.A. 1983. Production of earthworm protein for animal feed from potato waste. *In* Upgrading Waste for Feeds and Food. pp. 153–162. Edited by D.A. Ledward, A.J. Taylor and R.A. Lowrie. Butterworths, London.

Edwards, C.A., Burrows, I., Fletcher, K.E. and Jones, B.A. 1985. The use of earthworms for composting farm wastes. *In* Composting of Agricultural and Other Wastes. pp. 229–242. Edited by J.K.R. Gasser. Elsevier Publ. Co.

Evans, A.C. 1948. Relations of worms to soil fertility. Discovery. Norwich, England, 9(3): 83–86.

Evans, A.C. and Guild, W.J. Mcl. 1947. Studies on the relationships between earthworms and soil fertility. I. Biological studies in the field. Ann. Appl. Biol. 34: 307–330.

Ghilarov, M.S. 1963. On the interrelations between soil dwelling invertebrates and soil microorganisms. *In* Soil Organisms. pp. 255–259. Edited by J. Doeksen and J. van der Drift. North Holland Publishing Co. Amsterdam.

Graff, O. 1971. Stikstoff, Phosphor und Kalium in der Regenwurmlösung auf der Wiesenversuchflache des Solling projecktes. Ann. Anim. Zool. Ecol. (Spec. publ.) 4: 503–512.

Guild, W.J.Mcl. 1955. Earthworms and Soil Structure. *In* Soil Zoology. pp. 83–98. Edited by D.K. McE. Devan. Butterworths, London.

Gurianova, O.Z. 1940. Effect of earthworms and organic fertilizers on structure formation in chernozem soils. Pedology 4: 99–108.

Hopp, H. and Hopkins, H.T. 1946. Earthworms as a factor in the formation of water-stable aggregates. J. Soil Water Cons. 1: 11–13.

Joshi, N.V. and Kelkav, B.V. 1952. The role of earthworms in soil fertility. Ind. J. Agric. Sci. 22: 189–196.

Kahsuitz, H.G. 1922. Investigations on the influence of earthworms on soil and plant. Botan. Arch. 1: 315–351.

Lunt, H.A. and Jacobson, G.M. 1944. The chemical composition of earthworm casts. Soil Sci. 58: 367.

Nijhawan, S.D. and Kanwar, J.S. 1952. Physiochemical properties of earthworm castings and their effect on the productivity of soil. Ind. J. Agric. Sci. 22: 357–373.

Parle, J.N. 1963. Microbiological study of earthworm casts. J. Gen. Microbiol. 13: 13–23.

Puh, P.C. 1941. Beneficial influence of earthworms on some chemical properties of the soil. Contrib. Biol. Lab. Sci. Soc. China 15: 147–155.

Russell, E.J. 1910. The effect of earthworms on soil productiveness. J. Agric. Sci. 3: 246–257.

Rhee, J.A. van 1965. Earthworm activity and plant growth in artificial cultures. Plant Soil 22: 45–48.

Satchell, J.E. 1963. Nitrogen turnover by a woodland population of *Lumbricus terrestris*. *In* Soil Organisms. pp. 60–66. Edited by J. Doeksen and J. van den Drift. North Holland Publishing Co., Amsterdam.

Stockli, A. 1928. Studien über den Einfluss der Regenwurmer auf die Beschaffenheit des Bodens. Landwirtsch. Jahrb. Schweiz 42 (1).

Swaby, 1949. The influence of earthworms on soil aggregation. J. Soil Sci. 1(2): 195–197.

Teotia, S.P., Duley, F.L. and McCalla, T.M. 1950. Effect of stubble mulching on numbers and activity of earthworms. Bull. Nebraska Agric. Exp. Station 165: 20.

Uhlen, G. 1953. Preliminary experiments with earthworms. Landbruklogskolans. Institute Jordkulture. Meld. 37: 161–183.

Went, J.C. 1963. Influence of earthworms on the number of bacteria in the soil. *In* Soil Organisms. pp. 260–265. Edited by J. Doeksen and J. van der Drift. North Holland Publishing Co., Amsterdam.

THE USE OF WORM-DIGESTED ANIMAL WASTE AS A SUPPLEMENT TO PEAT IN LOAMLESS COMPOSTS FOR HARDY NURSERY STOCK

MARGARET A. SCOTT
Agricultural Development and Advisory Service, Efford Experimental Horticulture Station, Lymington, Hampshire, England

Introduction

Efford Experimental Horticulture Station is part of the Ministry of Agriculture Fisheries and Food's Agricultural Development and Advisory Service (ADAS) and one of its sectors of responsibility, is for work with propagation and container production of hardy nursery stock. With nearly 70 million container plants being produced annually (MAFF, 1983) compost forms the basic raw material of the industry and the limited volume available to each plant has to give support, and have a structure which provides adequate drainage to achieve the correct air/water ratio, as well as containing a balance of nutrients, both micro and macro elements. Early work concentrated on developing a loamless compost mix based on peat:sand, and more recently on peat:granulated pine bark (Scott, 1984), with particular attention to structure (Scott, 1983) and nutrition using long term controlled release fertilizers. (Efford, 1982a, 1983). Since 1980, there has been considerable interest from the livestock industry as to the potential of animal waste products for use in horticultural composts. Initial work at Efford EHS was in co-operation with ADAS Farm Waste Unit and Microbiology Department at Reading together with the National Institute for Research in Dairying (NIRD), and examined the use of composted separated solids ('Crumb') from cattle and pig slurry pits as a supplement with peat and bark based mixes. (MAFF/ADAS, 1981). The inclusion of 25% cow crumb produced satisfactory growth, but results with pig crumb were poor. This work also underlined the basic problems of variability of these materials, particularly nutrient levels, depending on source, type of animal, and the bedding used, making them less easy to manage than the standard peat:sand, peat:bark mixes.

Another development was the research project at Rothamsted Experimental Station which looked at the use of earthworms in breaking down and converting animal waste products into material which appeared to have greater potential as a compost supplement, since the worm activity changed most of the nitrogen from the ammonium to the nitrate form (Edwards, 1981, 1982). Co-operative trials between Rothamsted and Efford began in 1983 with the objective of assessing plant response in composts containing a proportion of worm digested animal waste. Similar work with bedding and pot plants was undertaken at the Lee Valley Experimental Horticulture Station (Lee Valley, 1983).

The scope of this paper is to review the two years work at Efford EHS where worm digested animal waste have been used as a supplement to peat in loamless composts for hardy nursery stock.

Earthworms in waste and environmental management
edited by C.A. Edwards and E.F. Neuhauser, pp. 221–229

Table 1. Nutrient analyses of animal wastes from different sources

Waste	pH	P ppm	K ppm	Mg ppm	NO_3 ppm	NH_4 ppm	Conductivity index (ADAS Scale 0–9)
Duck (worm digested)	8.0	414	3300	110	264	< 6	(9)
Pig (worm digested)	6.5	258	1132	366	1020	< 6	(9)
Pig (composted)	6.7	72	1092	96	354	174	(9)
Cattle (worm digested)							
1983	7.0	70	306	24	120	< 6	(2)
1984	6.5	252	1512	338	900	< 6	(9)
Cattle (composted)							
1982	7.1	52	612	24	13	8	(3)
1984	7.9	330	1794	115	174	156	(9)

Animal wastes available

A range of materials have been received during the two years and analyses indicated considerable variation between products. (Table 1)

The worm-digested duck waste, as well as having an excessively high pH, also had extremely high phosphate and potash levels together with high nitrate resulting in very high conductivity.

The composted and worm-digested pig wastes cannot be compared directly as they were from different sources. Both had very high conductivity values and potash and nitrates, but the ammonium element, present in the composted product was virtually absent in the worm digested material, albeit that nitrate levels were very high, increased no doubt by the conversion of ammonia to nitrate by the worm digestion process.

The cattle wastes were also from different sources, and seasons, making direct comparison difficult, but again all the nitrogen source was in the form of nitrates in the worm digested product. There was a large variation in soluble salt levels in both sources of cattle material between the two seasons, underlining the problems which arose in calculation of base dressings.

Depending on the analysis of each batch of animal waste received the following adjustments had to be considered.

1. *pH*

Ideally, a loamless compost for hardy nursery stock should be in the region of pH 5.0–5.6 (4.5–5.0 for Ericaceae), particularly if the water supply is hard. The higher the pH of the waste, the less Magnesian lime required, and in extreme cases would be replaced by Kieserite (Mg SO_4) as the magnesium source. Table 2 was originally prepared by soil science colleagues for pH adjustment of bark-based mixes, but was successfully used in the animal waste trials.

2. *Phosphate*

The importance of phosphate in the production of quality container grown nursery

Table 2. Rates (kg/m³) of magnesian lime addition in the base dressing according to pH of the animal waste

pH of waste	Up to 30% waste		Above 30% waste	
	General	Ericacieae	General	Ericacieae
4.5	2.4	1.2	2.4	1.2
4.5–5.0	1.8	1.0	1.2	0.6*
5.0–6.5	1.2	0.6[a]	0.6[a]	Nil[b]
6.5+	0.6[a]	Nil[b]	Nil[b]	Nil[b]

[a] Add 0.6 kg/m³ Kieserite (Magnesium Sulphate) with Mg Lime.
[b] Add 1.2 kg/m³ Kieserite as a replacement for magnesium lime.

stock had become increasingly apparent from trials results over four years, where it was shown that many species were sensitive to phosphate and too much applied in the base dressing could be damaging (Scott, 1982). Hence there was concern as to the relatively high levels of phosphate in the animal wastes, particularly as the standard fertilizer used in the base dressing, Osmocote 18:11:10, had sufficient soluble phosphate in its formulation to supply the plants requirements. Other species, *i.e.*, Tomatoes, Chrysanthemums, Carnations, appear less sensitive to phosphate, growing well in rates which would have caused severe damage to the P-sensitive nursery stock species. The adverse effects of phosphate increase as pH increases.

3. *Fertilizer addition*

Rate of controlled release fertilizer used may need adjustment depending on the soluble salt levels of the waste material.

1983 Trial

The first trial looked at the incorporation of 25% or 50% worm-digested cattle waste as a supplement with sphagnum moss peat (medium, Irish). At the same time, the amount of additional nutrition required was monitored by using nil, half or 'standard' rates (2 kg/m³) of the controlled release fertilizer Osmocote 18:11:10, a single application of this fertilizer in the base dressing normally being sufficient to last the season without the need for supplementary feeding. The mixes were compared with a 'standard' peat:sand mix (75:25), a peat: granulated pine bark (Cambark fine) mix (60:40) and a peat: pine bark: worm waste (35:40:25).

Base dressing: 2 kg/m³ Osmocote 18:11:10 (unless reduced in experimental treatment) 2.4 kg/m³ Magnesian limestone (reduced as necessary, see Tables 1&2) 0.3 kg/m³ Fritted Trace Elements WM 255

0.4 kg/m³ Ammonium nitrate (Nitram) was included in bark mixes to counteract initial N lock up by Cambark

Method: Rooted cuttings, potted direct into 1 liter rigid pots in early July 1983, were grown on outdoor low level irrigated drained sand beds and included a range of nine hardy nursery stock species of differing vigor and habit (Table 4).

Table 3. Compost analyses at mixing

% worm waste	kg/m³ rate osmocote	pH	ppm P	K	Mg	NO_3	Cond. index
Nil (Peat:Sand)	2 kg	5.7	12	30	20	40	(2)
25%	Nil	5.6	23	119	61	87	(1)
	1 kg	5.7	27	142	70	86	(2)
	2 kg	5.6	64	234	74	150	(3)
50%	Nil	6.0	47	228	71	132	(2)
	1 kg	6.1	74	318	67	153	(3)
	2 kg	6.1	62	270	71	150	(3)

Results

This was a preliminary investigation without replication, apart from species, with five plants per treatment, and caution is needed in interpreting the results. The main points can be summarized as follows.

Compost analyses

1. Adjustment of the Magnesian lime/Kieserite in the base dressing successfully produced an initial compost pH at mixing similar to the peat:sand control where 25% worm waste was used (Table 3), but inclusion of 50% waste increased pH by around 0.4 units.
2. As was expected, the inclusion of waste increased N:P:K levels in the mix, especially at 50% plus the full rate of Osmocote 18:11:10, but conductivities were still within an acceptable range for hardy nursery stock.

Plant growth

1. Overall 25–50% worm-digested waste in the compost produced as good or better results than the peat:sand control (Table 4). An exception was *Chamaecyparis lawsoniana* 'Stardust' where 50% inclusion produced poorer results, though 25% gave equally good results as the control.
2. The response to fertilizer (Osmocote 18:11:10) where worm waste was used appeared to be related to the vigor of the species. Slower growing species produced equally good results at 1 kg/m³ while the more vigorous species (except *X Cupressocyparis leylandii*) responded to 2 kg/m³. However, this was a late potted trial and whether similar results would have been observed over a longer growing season needs confirming.
3. The peat:pine bark:worm waste and peat:pine bark mixes produced equal or better results than the peat:sand control.
4. The percentage phosphate uptake by the plant, while increasing as proportion of waste and Osmocote increased, did not reach levels which had proved damaging in earlier trials (Scott, 1982).

1984 Trial

Results of the 1983 trial were sufficiently encouraging to extend the work in 1984. While basically a repeat of the 1983 work comparisons included worm-digested pig

Table 4. Effects of treatments on plant dry weight and foliage phosphate by December 1983

Rate osmocote (kg/m³)	Peat sand	25% worm waste			50% worm waste			Peat bark worm	Peat bark
	2 kg	Nil	1 kg	2 kg	Nil	1 kg	2 kg	2 kg	2 kg
Slower growing species									
Chamaecyparis lawsonian 'Ellwoods Gold'									
Dry weight (g)	5.4	4.0	5.3	5.3	3.7	4.9	6.1	4.8	6.5
Elaeagnus pungens 'Maculata'									
Dry weight (g)	1.4	1.2	2.1	1.8	1.2	1.5	1.6	1.4	1.9
% P in foliage	0.20	0.38	0.42	0.44	0.47	0.41	0.40	0.44	0.25
Juniperus communis 'Repanda'									
Dry weight (g)	0.5	1.3	1.3	1.4	0.9	1.0	1.6	1.3	1.1
% P in foliage	0.27	0.25	0.26	0.28	0.25	0.29	0.28	0.28	0.24
Vigorous growing species									
Chamaecyparis lawsoniana 'Columnaris Glauca'									
Dry weight (g)	8.1	5.9	7.4	9.7	6.0	7.5	8.7	8.0	8.4
% P in foliage	0.17	0.26	0.26	0.30	0.28	0.31	0.34	0.30	0.19
C.l 'Stardust'									
Dry weight (g)	5.6	4.6	5.9	6.8	3.5	4.8	2.8	5.3	6.7
X Cupressocyparis leylandi									
Dry weight (g)	3.3	5.2	7.5	5.5	6.4	6.2	4.5	6.1	7.9
% P in foliage	0.15	0.21	0.24	0.30	0.27	0.29	0.29	0.30	0.21
Cotoneaster conspicus									
Dry weight (g)	3.4	3.2	3.5	5.4	3.4	3.8	5.4	4.0	4.4
% P in foliage	0.17	0.26	0.26	0.33	0.33	0.32	0.31	0.28	0.23
Pyracantha × Orange Glow									
Dry weight (g)	3.6	3.5	6.5	7.8	4.4	5.7	6.6	6.8	7.2
Viburnum bodnantense									
Dry weight (g)	3.4	2.1	5.6	7.4	3.8	6.0	5.7	6.3	5.5
% P in foliage	0.13	0.29	0.24	0.28	0.22	0.24	0.22	0.23	0.13

waste as well as cattle, and in addition a composted cattle waste from a commercial enterprise was included. As previously 25 and 50% of each waste was used to supplement medium Irish sphagnum moss peat with nil, half or 'standard' rate Osmocote 18:11:10 (2 kg/m³). Bark was not included in 1984, the control being the 75;25 peat:sand mix with 2 kg/m³ Osmocote 18:11:10. Other base dressings were modified as necessary. (see 1983 trial and Tables 1 and 2).

Method: Rooted cuttings were potted into 2 liter polybags in early May 1984 and as in the earlier trial grown on outdoors on irrigated sand beds. Four species were included this season (see Table 6), to enable two replicates to be included, again with five plants per plot.

Table 5. Compost analyses at mixing and after six weeks

% Waste	Rate osmocote	P (ppm)		NO_3 (ppm)		NH_4 (ppm)		Cond. at mix.	Index after 6 wks.
		at mix.	after	at mix.	after	at mix.	after		
Nil	2 kg	18	17	84	162	45	57	(1)	(2)
Worm worked pig									
25%	Nil	151	198	228	276	10	19	(3)	(4)
	1 kg	148	212	198	372	22	38	(3)	(6)
	2 kg	180	317	156	402	39	44	(3)	(6)
50%	Nil	169	238	330	708	14	21	(5)	(8)
	1 kg	169	202	318	510	25	40	(6)	(6)
	2 kg	223	295	384	720	33	72	(7)	(8)
Worm worked cattle									
25%	Nil	162	166	189	420	17	<6	(4)	(6)
	1 kg	144	216	183	378	17	41	(3)	(6)
	2 kg	130	173	246	300	36	59	(2)	(5)
50%	Nil	180	198	390	504	26	10	(5)	(7)
1 kg	176	184	432	498	33	32	(5)	(6)	
	2 kg	216	227	588	594	30	66	(6)	(8)
Composted cattle									
25%	Nil	122	119	77	114	66	72	(3)	(4)
	1 kg	79	140	54	156	84	102	(2)	(5)
	2 kg	115	148	71	174	150	162	(3)	(5)
50%	Nil	97	101	35	40	132	84	(3)	(4)
	1 kg	112	108	32	42	138	72	(3)	(4)
	2 kg	97	158	32	114	186	192	(3)	(6)

(figures are an average for the conifer species)

Results

Compost analyses

1. As in the previous trial, adjustment of the Magnesian lime in the base dressing according to the pH of the waste used (Table 2) produced a satisfactory pH in the mixed compost (Table 5).
2. The high initial soluble salts in the worm-worked waste materials (Table 1) increased the conductivity in the mixed composts, particularly where 50% was incorporated, and there was a correspondingly high early release of nutrients as monitored in the six week sample. Part of this would have been due to fertilizer, but when compared with the peat:sand control there was obviously considerable release of nitrates from the worm worked waste. The level of soluble salts was lower in the composted cattle waste, but this mix had a higher concentration of ammonia present.

Plant growth

Response varied with species and appears to be related to its vigor of growth (Table 6).

Table 6. Effects of treatments on plant dry weight (g) by November 1984

Rate of Osmocote (kg/m^3)	Peat:sand 75:25	Worm-worked pig		Worm-worked cattle		Composted cattle	
		25%	50%	25%	50%	25%	50%
Juniperus sabina tamariscifolia							
Nil	—	6.6	5.1	3.1	5.7	6.3	1.9
1 kg	—	4.2	4.5	5.9	5.0	5.2	5.3
2 kg	6.6	6.1	2.3	6.1	6.9	4.9	4.3
Chamaecyparis lawsoniana – 'Allumii'							
Nil	—	14.5	9.8	16.1	11.1	12.4	5.9
1 kg	—	16.6	10.9	14.5	11.5	15.8	13.1
2 kg	19.3	17.1	7.4	13.3	10.6	14.1	7.0
Pyracantha – Orange Glow							
Nil	—	25.4	26.3	28.7	33.7	20.6	13.9
1 kg	—	36.6	30.8	34.2	40.9	23.1	24.1
2 kg	30.2	28.6	26.6	39.4	43.3	31.3	29.6

1. *Juniperus sabina tamariscifolia* (Slower growing)
25% addition of all three types of animal waste could produce similar results to the control, but whereas with worm-digested cattle waste the best results were with 1–2 kg/m^3 Osmocote, with the worm-digested pig and composted cattle materials the best growth was obtained without controlled release fertilizer application in the base dressing.

Apart from the worm-digested cattle mix increasing the proportion of waste up to 50% produced poorer results, particularly at the higher rates of fertilizer where conductivities were high.

2. *Chamaecyparis lawsoniana* 'Allumii' (moderate vigor)
Results were similar for all three types of animal wastes. While the addition of 25% waste gave satisfactory growth it was not quite as good as the peat sand control, but increasing the proportion to 50% produced a marked reduction in growth.

3. *Pyracantha* Orange Glow (vigorous)
This vigorous growing species responded well to the increased level of nutrients available in the mixes containing the wastes and results were as good (composted cattle) or better (worm digested material) than the control. Overall, the worm-digested cattle material produced better results than pig, particularly as proportion and rate of fertilizer increased.

These results have yet to be statistically analysed, and foliage analyses will not be available until later.

Discussion

Results of the two seasons work have shown that worm-digested animal wastes can be used as a supplement with peat for hardy nursery stock grown in loamless composts. While the data was from a small plot size, the plants were of clonal origin and therefore relatively uniform, and some valuable guidelines have been obtained as to the potential of the waste materials for container nursery stock.

Overall, the worm-digested cattle material gave the most consistent result across all species, and the lower ammonia levels in this material compared with the composted product may be part of the reason for this. The poorer results in the worm-digested pig waste could not be accounted for entirely by high salt levels, but other factors may have been involved. Facilities were not available for trace element analyses, but an analysis of composted pig waste in 1982 gave a high copper value (57 ppm) as opposed to 4.1 ppm Cu in composted cattle waste. This aspect needs further investigation.

The trials suggest that up to 25% of the compost can be supplemented by the animal waste, but 50% produced variable results depending on species. The major problem, however, is variability of the product, not only between wastes of different animals, but from the same source in different seasons. The variable and high soluble salts present in the waste made calculation of fertilizer requirements difficult and an analysis of each batch would be necessary. A large proportion of the hardy nursery stock industry uses controlled release fertilizers as a single application in the base dressing at each potting. Results suggest that this will still be required even though the waste will supply some nutrients. This also appeared to be the case with bedding plants in the Lee Valley EHS work (Lee Valley, 1983). Analysis of the basic material prior to use is also necessary for pH assessment, though once available the lime adjustments used in Table 2 worked well. The high levels of phosphate in the waste were also of concern for P-sensitive nursery stock species, and it remains to be seen what foliage uptake was in 1984.

The objective of the work was to monitor plant performance where animal wastes, particularly worm-digested materials from Rothamsted, were included in the mix. More detailed work is required on nutritional requirements but composts containing up to 25% waste performed well. The worm-worked cattle waste looked particularly promising but the variability of the product is a major draw back to its wider use on a commercial scale at present.

References

Edwards, C.A. 1981. Rothamsted Expt Stn. The Use of Earthworms in Waste Disposal and Protein Production.

Edwards, C.A. 1982. Report Rothamsted Expt. Stn. 1981 (1982) pp. 103–106.

Edwards, C.A. 1983. Earthworms, Organic Waste and Food. Span. Publ. Shell Chemical Co. 26(3): 106–108.

Efford EHS. 1981. Use of Organic Waste (Cow Crumb) to Supplement Peat and Bark in Composts (FL02/07697). MAFF/ADAS Reference Book 235, Hardy Ornamental Nursery Stock pp. 110–112.

Efford EHS. 1982a. Comparison of Slow Release Fertilizers (FL02/09009). MAFF/ADAS Reference Book 235, Hardy Ornamental Nursery Stock, pp. 64–67.

Efford EHS. 1982b. Use of Organic Waste (Crumb to Supplement Peat and Bark (FL02/09014). MAFF/ADAS Reference Book 235 Hardy Ornamental Nursery Stock. pp. 88–90.

Efford EHS. 1983. Comparison of Slow Release Fertilizers (FL02/10003). MAFF/ASAS Reference Book 235, Hardy Ornamental Nursery Stock.

Lee Valley EHS 1983. Bedding Plants, Compost Additives (GP38/09235). MAFF/ADAS Reference Booklet 236, Protected Crops Ornamentals. pp. 3–4.

MAFF 1983, Statistics for 1983.

Scott, M.A. 1982. The Role of Phosphate in the Production of Quality Container Grown Nursery Stock. Report Efford Expl. Hort Stn. for 1981. pp. 17–24.

Scott, M.A. 1983. Composts for Container Nursery Stock Production Proc. 18th Annual Pershore Conference.

Scott, M.A. 1984. Hardy Nursery Stock: The Use of Bark in Composts – A Review of Work at Efford EHS 1983–83. Rev. Efford Expl. Hort. Stn. for 1983 pp. 23–34.

CHEMICAL CHARACTERIZATION AND NITROGEN MINERALIZATION POTENTIALS OF VERMICOMPOSTS DERIVED FROM DIFFERING ORGANIC WASTES

M.A. BUCHANAN, G. RUSSELL and S.D. BLOCK
University of California, Santa Cruz, California 95060

Introduction

The potential for utilizing earthworms in the digestion of organic wastes and the corresponding production of composts for agri-horticultural industries has been established by many researchers and limited-scale projects (see earlier chapters). While the immediate potential for earthworm composts (vermicomposts) lies mainly in horticultural use, particularly as a potted-plant medium, there is interest in improving and extending their application as a fertilizer or soil amendment. However, there seems to be no data available on how individual crops respond to earthworm composts or how vermicomposts compare with other composts.

Knowledge of how earthworm composts perform in a variety of applications will be critical to understanding and selecting appropriate end-uses. In the U.S., compost marketing has often been hindered by lack of such analysis (Buchanan and Gersper, 1984). Thus, it follows that appropriate use of composts should be determined, in part, by the composition (physical-chemical) of the material, as well as, the short and long-term soil productivity benefits, including availability of nutrients to crop plants. It should be noted that chemical analysis of waste materials and composts is not always performed prior to application or, if obtained, is rarely a reliable predictive assay for nutrient availability and crop response.

It is acknowledged widely that many compositional parameters can affect the release of nitrogen (N) from organic forms in composts, crop residues, and soils (Tinsley, 1969; Stevenson, 1982). For organic amendments, total organic carbon (C) and N content, as well as the character of carbon compounds may often be the prime determinants of the quantity and rate of N mineralization following soil incorporation (Marumoto *et al.*, 1980).

Waring and Bremner (1964) proposed that incubating soils under waterlogged conditions gave a reliable measure of inorganic-N production. Smith and Stanford (1968) compared anaerobic and aerobic incubations on 39 different soils and found that N mineralization in both procedures was significantly correlated (r = .93). Advantages and accuracy of a biological index (*e.g.*, as determined by incubation) over direct chemical analyses and in correlation with crop uptake have been reported (Stanford and Legg, 1968; Castellanos and Pratt, 1981; Griffin and Laine, 1983). Attempts to correlate N mineralization with organic C or N content, C:N ratio, or CO_2 evolution from incubated soils have shown no consistent results (Iratini *et al.*, 1960; Smith and Stanford, 1971; Castellanos and Pratt, 1981).

This work addresses a key point in earthworm compost utilization, by evaluating materials produced from differing waste materials by the earthworm, *Eisenia foetida*. In the study, utilizing chemical analysis and a modified anaerobic incubation procedure, we examined the influence of chemical-physical composition of vermicomposts on the mineralization of nitrogen from amended soil.

Earthworms in waste and environmental management
edited by C.A. Edwards and E.F. Neuhauser, pp. 231–239

Materials and methods

Laboratory analysis

All composts were primarily earthworm casts produced by cultures of *E. foetida* from a commercial source (Collier's Compost Systems, Santa Cruz, CA) and from a research project at the University of California-Santa Cruz Experimental Farm. The five materials from which earthworm composts were made (5 is a finished product) are:

1. Watsonville sludge (WS) – a municipal sewage sludge;
2. Tannery sludge-horse manure mix (TSM) – tannery sludge waste mixed 1:1 with manure;
3. Household refuse (HR) – vegetable and yard wastes;
4. Horse Manure (HM) – both fresh and aged manure;
5. Commercial mix (CM) – a blend of horse and rabbit manure castings (35%) and pulverized fir bark (65%)

Samples of the finished earthworm composts were air-dried and ground to pass a 0.5 mm sieve. The soil used in the mineralization incubation was an Elkhorn sandy loam sampled from a plot which had recently been cropped. The soil was air-dried and ground to pass a 2 mm sieve. A subsample from all materials was oven-dried for 24 hours (at 70°C for composts and 105°C for soil to determine air-dry moisture content.

The pH was determined from a saturation paste using a glass electrode with a Beckman 2000 ion analyzer. Organic-C was determined using a modified Walkley-Black procedure (Nelson and Summers, 1982). Organic-N was extracted using a micro-Kjeldahl digestion analysis completed using an ammonium electrode with the Beckman 2000 ion analyzer. Ammonium (NH_4) was also assayed with the ammonia electrode after the samples were shaken for one hour with a 2N KCl extractant. Nitrate (NO_3) was extracted in a 0.001% HCl solution and measured using a nitrate electrode with a double junction glass reference.

Total phosophorus (P) was determined with a modification of an $HClO_3$-HNO_3 digestion procedure (World Health Organization, 1978). An aliquot of the digest was complexed with ammonium molybdate and measured by absorbance at 660 nm using a Bausch and Lomb Spec 20 spectrophotometer. Soluble or available P was extracted by shaking 5 mg samples in 100 ml of a 0.001N HCl solution for 30 minutes and assayed by the same procedure used for total-P. Total-K, Ca, and Mg were assayed from the total-P digest, while available content was obtained by shaking samples with 1N NH_4OAc at a 10:1 extractant to sample ratio. All analysis of cations was performed by atomic absorbtion spectrophotometry using a Varian Techtron, Model-70.

N-mineralization experiment

Nitrogen mineralization was determined over a seven-week-period. The soil used was the Elkhorn sandy loam previously mentioned. The vermicompost treatments were prepared from the air-dried samples used for analysis. Five replications per treatment

Table 1. Chemical composition of soil and earthworm composts

Properties	Samples					
	Soil	WS	TSM	HR	HM	CM
pH	6.1	6.9	7.2	6.8	7.5	7.6
org.C %	1.77	7.47	8.43	8.83	9.03	9.22
org.N %	.124	1.05	1.31	.74	.75	.80
NH_4 ppm	1.3	3.7	4.7	1.1	1.5	1.7
NO_3 ppm	4.3	47.5	129.6	48.5	91.0	133.7
Total-P %	.05	1.38	.33	.48	.47	.92
Avail.-P ppm	37.5	11.4	9.7	18.4	11.1	16.9
Avail.-K ppm	1830	2442	1029	887	2129	2636
Total-Ca ppm	—	23895	9240	—	15717	24541
Avail.-Ca ppm	1907	4354	2140	2548	4543	5506
Total-Mg ppm	—	8762	4910	—	2535	5489
Avail.-Mg ppm	1145	1858	846	1443	1825	3841

were prepared with amendments added at a 1:100 ratio (0.1 gm vermicompost to 9.9 gms soil). The control was a 10 gm soil only treatment. The samples were mixed in 50 ml centrifuge tubes, then twice washed by centrifugation with 25 ml aliquots of 0.01M $CaCl_2$. These washes were stored at $-4°C$ and later analyzed for NH_4 and NO_3.

The sample tubes were then placed in an incubator at $30 \pm 2°C$ following a 25 ml addition of a 'nutrient' solution consisting of 0.001M $MgSO_4$ and 0.001M K_2HPO_4. Extracts were taken at one, three, five and seven weeks by the described washing procedure. After the second wash 25 ml of the 'nutrient' solution was again added to each sample tube, together with 1 ml of a soil suspension (microbial innoculant) consisting of 5 gm soil:250 ml distilled water. Ammonium-N in the washes was determined with the ammonia electrode.

Results and discussion

Results of the analyses are summarized in Table 1. Although supportive data on the composition of the raw wastes was not available, the range of these macronutrients most probably reflects both the initial composition of the wastes and the production method. In bin or windrow digestion a varying proportion of the N contained in the material may be lost through volatilization, denitrification, or leachates. While the volume reduction obtained during digestion may partially balance this loss, the net effect will be a lower N fraction in the converted material.

Data for N content supports the assumption that N loss may occur, and be substantial, during digestion and stabilization. Raw sewage sludges are usually higher in total-N and, in particular, typical tannery wastes may contain up to 7% N (Cooke, 1963). In addition, nitrate data indicate a high nitrification rate potential which may greatly affect total-N. Williams *et al.* (1980) noted this in a comparison of a fresh and highly-weathered sewage sludge, where nitrate-N levels were five times higher in the latter material. Significant leachate losses may occur during a long earthworm stabilization process.

Table 2. Nitrogen mineralization during incubation

Treatment	N mineralized in ppm (time in weeks)				Total	
	0–1	1–3	3–5	5–7	ppm	kg/ha
Soil	11.9	2.3	1.0	1.5	16.7	40.3
HR	11.6	2.9	1.2	2.4[a]	18.1	43.4
TSM	13.7[b]	5.3[b]	2.4[b]	3.2[b]	24.6[b]	59.1
WS	12.1	3.6[b]	2.0[b]	2.8[b]	20.5[b]	49.2
HM	11.0	2.4	1.5[a]	3.3[b]	18.2	42.7
CM	10.6[a]	2.0[a]	1.1	1.6	15.3[a]	36.6

Data significance versus control (soil only).
[a] $p < .05$.
[b] $p < .01$.

The carbon content of the earthworm composts seemed consistent with their physical appearances with the exception of the commercial mix. The commercial mix contained a large amount (15–20%) of gravel and sand which reduced the C-percentage in analysis. Only the Watsonville sludge-derived earthworm compost had no visible fiber content. All others contained a varying amount of undigested material (*i.e.*, resistant cellulose-lignins).

The pH reaction showed a trend towards a generally neutral status. It is interesting to note the slightly alkaline characteristics of the manure-based and commercial mix composts. The neutral to alkaline pH buffering capacity may be attributable to $CaCO_3$ secretions from the earthworm gut and, in this case, a substantial sodium content in the relatively poor quality irrigation water used in producing the vermicomposts. Analysis for calcium reveals this strong buffering potential. This potential has also been observed in a field trial which monitored pH changes in the 0–10 cm soil depth of control treatments as compared with a number of application rates of vermicomposts, composts, and inorganic-N fertilizer (Buchanan, unpublished).

Larger amounts of N mineralized in the first week of incubation than in successive weeks. Other workers have observed similar patterns (see Table 2). Stanford *et al.* (1972) and Griffin and Laine (1983) speculated that this may be due to either excess exchangeable-NH_4 which is not removed by the 0.01M $CaCl_2$ washes or, by the mineralization patterns inherent to the soils examined. It has been shown that air-dried soils can have increased mineralization rates upon subsequent rewetting (Eagle, 1969). Our research leads us to suggest that much of the increase may be due labile portions of the desiccated soil microbial biomass along with residual soil enzyme activity. Eagle (1969) states that only freshly air-dried soils (within two days) could be used for mineralization incubations as proportionate increase in mineralization potential is far from consistent in soils.

Another factor affecting the measured mineralization rates in soil relates to field conditions at the time of sampling. The soil used in this study was taken in early spring from a continuously-cropped plot, which received undetermined additions of manures and composted crop residues, the previous fall. Much of easily mineralizable-N may have been exhausted during the previous cropping. This along with the relatively low C:N ratio (14.3), may explain the relatively low release as com-

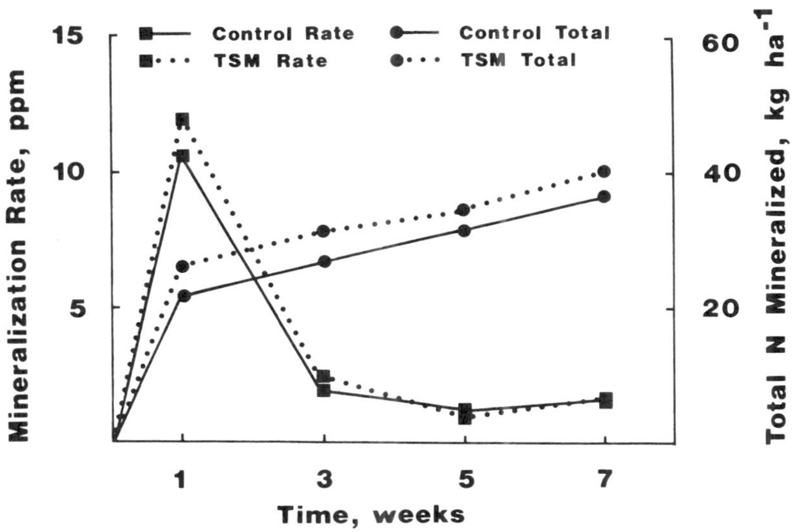

Fig. 1. Tannery sludge mix v.s. control.

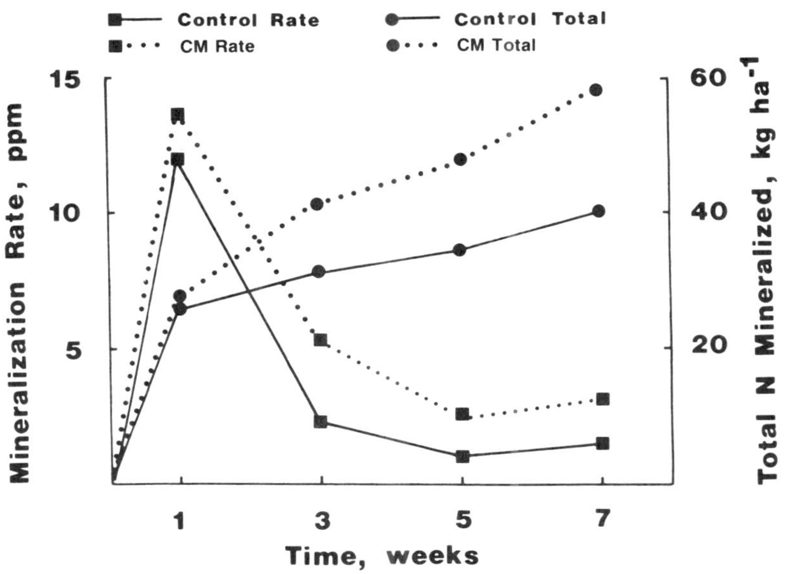

Fig. 2. Commercial mix vs control.

pared to data reported by other workers. Griffin and Laine (1983) found a negative correlation between organic C content and measured mineralization rates in soils amended for a long period with organic wastes.

Data for the mineralization rates and total-N accumulated over the seven-week-period are given in Table 2. These data give a strong indication of the N supplying power of the soil alone and when amended with earthworm composts. The sludge-

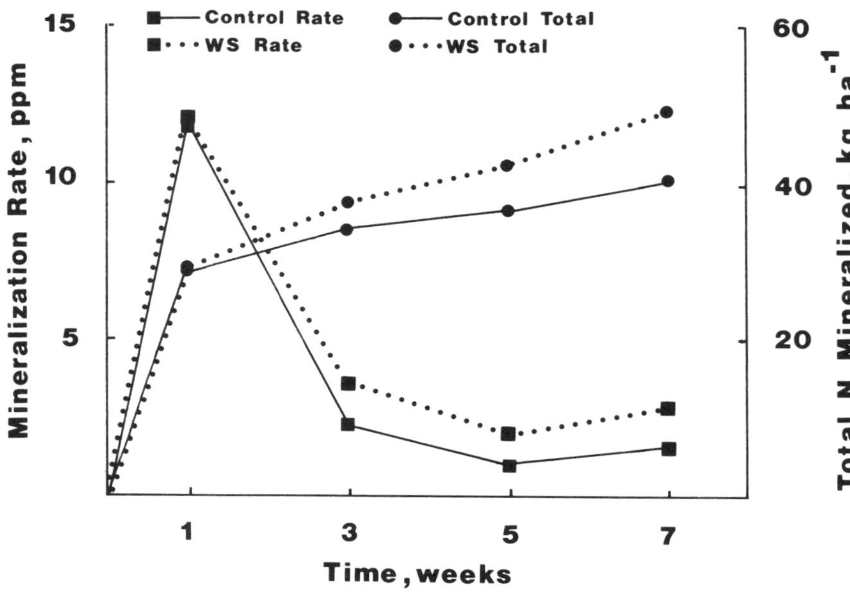

Fig. 3. Watsonville sludge vs. control.

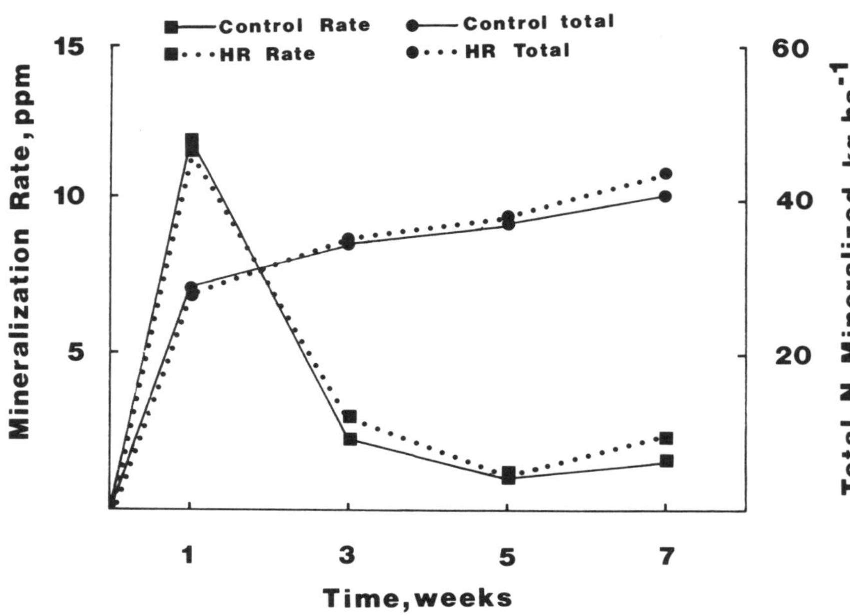

Fig. 4. Household refuse vs. control.

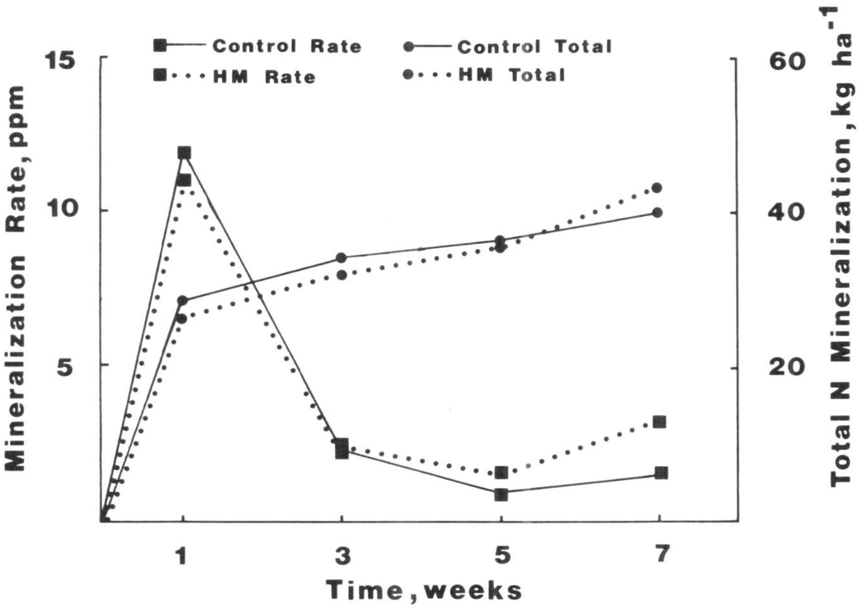

Fig. 5. Horse manure vs. control.

based materials had the greatest positive effect on mineralization with differences in mineralization rates and accumulated N being highly significant. Both of the sludge-based earthworm composts yielded immediate increases in NH_4, whereas the horse manure and household waste-based composts did not yield significant increases until later in the experiment.

The treatment with the commercial earthworm compost mix caused increasing immobilization of N during the first three weeks of incubation. Even after seven weeks, the treatment with this material did not have a significant positive effect on mineralization rate. The composition of the commercial mix is probably the major factor for these results. In addition to the sand and gravel fraction (15–20%), the material consisted of approximately 65% ground bark, and only 15–20% horse and rabbit-manure worm casts. Thus, while a determination of organic-C yielded a relatively low value, it did not account for the predominant forms of C (hemicellulose or lignins) nor the diluting effect of the inorganic contaminants. Figures 1–5 compare the mineralization characteristics of the vermicompost treatments with the unamended soil.

As was stated earlier, the quantity and characteristics of compost C and the C:N ratio will influence N mineralization. Phosphorus and micronutrients (Mo, Fe, Zn) content may also have an indirect effect on mineralization by limiting the activity of soil microorganisms. The comparisons shown in Figs. 1–5 indicate that carbon and nitrogen may be important in affecting the availability of N in amended soils. Kumada *et al.* (1976) ascribed a measure of compost quality based on C:N ratio, % resistant carbon, and N mineralization over a five-week incubation. Our results strongly suggest that such a set of evaluation parameters will be important, if not essential in assessing vermicompost quality.

With this in mind, linear correlation analysis was made for mineralized-N versus total-N content and versus C:N ratio. Even with the inclusion of the commercial mix results, correlation was significant (r = .90) for total-N versus mineralized-N. If the data pair for the commercial mix is excluded, correlation is highly significant (r = .98). There was no significant correlation (r = .57) between mineralized-N and C:N ratio. Again, this may be an indication of the importance of differing carbon compound composition of the composts.

Although a correlation between mineralized-N and total-N content of vermicomposts was anticipated, Castellanos and Pratt (1981), found no relationship between total-N and mineralization in manures and composted manures. Thus, the correlation found in this study may not be as significant as the statistical analysis showed. It is probable, that not enough data pairs (total-N and mineralization) were generated to state confidently a possible 'rule of thumb' concerning any relationship effect on N mineralization.

Expanded experimentation, monitoring, and analyses should be performed in developing evaluation parameters for earthworm composts. Comparisons of differing earthworm composts and earthworm composts with traditional composts will further elucidate factors affecting nutrient availability and value in agri-horticultural use. This work has simply outlined an approach for such evaluations. More data in replicated experiments are needed to validate any results obtained here. Correlation with plant growth-nutrient uptake and quality in greenhouse and field will greatly improve the development of criteria for evaluating the quality and application of earthworm composts as derived from a spectrum of organic wastes.

Acknowledgements

The authors would like to thank Dr. Stephen Gliessman and the UC-Santa Cruz Agroecology Program for support and facilities. A special thanks to Rob Franks of the UC-Santa Cruz Marine Analytical Laboratory for use of equipment and materials. Materials were kindly furnished by Jack and Pat Collier of Collier's Compost Systems, Santa Cruz. Finally, the senior author wishes to thank Dr. Paul Gersper of the Department of Plant and Soil Biology at UC-Berkeley for his continued assistance and support.

Literature cited

Buchanan, M.A. and Gersper, P.L. 1984. Strengthening a Link Between Municipal Waste Processing and Agri-horticultural Industries, Report on Grant # 82-48-2000 to UCAT Program, Berkeley, CA, 102 pp.

Castellanos, J.R. and Pratt, P.F., 1981. Mineralization of manure N-correlation with laboratory indexes. Soil Sci. Soc. America 45: 354–357.

Cooke, G.W. 1963. Fertilizers and Profitable Farming. Chemical Pub. Co., N.Y., N.Y. 122 pp.

Eagle, D.T. 1969. Determination of the N requirements of crops by analysis. *In* Nitrogen and Soil Organic Matter. Proceedings of Conference of National Agricultural Advisory Service, London. pp. 96–104. Edited by Cooke et al. (eds.). Nat. Agric. Adv. Serv., London.

Griffin, G.F. and Laine, A.F. 1983. N mineralization in soils previously amended with organic wastes. Agron. J. 75: 124–129.

Iratini, W.M. and Sommers, L.E. 1982. Total carbon, organic carbon and organic matter. *In* Methods of Soil Analysis. Edited by A.L. Page. Amer. Soc. of Agron., Madison, WI.

Kumada, K., Yoshida, K., Takei, A. and Shimada, N. 1977. Some characteristics of various compost with special reference to nitrogen. *In* Proceedings International Seminar on Soil Environment and Fertilization Management in Intensive Agriculture., Tokyo, Japan. pp. 469–477.

Marumoto, T., Shindo, S.H. and Higashi, T. 1980. Effects of carbonaceous materials on the accumulation of readily mineralizable organic N in soil. Soil Sci. Plant Nutr. 26(2): 185–190.

Nelsen, D.W. and Sommers, L.F. 1982. Total carbon, organic carbon and organic matter. *In* Methods of Soil Analysis. Vol. 2. Edited by Page et al. (eds.). American Society of Agronomy, Madison, WI.

Smith, S.J. and Stanford, G. 1968. Extractable organic *N* and *N* mineralization in soils. Soil Sci. 106(5): 345–351.

Smith, S.J. and Stanford, G. 1971. Evaluation of a chemical index of soil N availability. Soil Sci. 111(4): 228–232.

Stanford, G. and Smith, S.J. 1972. N mineralization potential of soils. Proc. Soil Sci. Soc. America 36: 465–472.

Stanford, G. and Legg, J.O. 1968. Correlation of soil N availability indexes with N uptake by plants. Soil Sci. 105(5): 320–326.

Stevenson, F.J. 1982. Human Chemistry. Wiley-Interscience. 443 pp.

Tinsley, J. 1969. N releasing properties of various types of organic matter. *In* Nitrogen and Soil Organic Matter. National Agricultural Advisory Service, London.

Waring, S.A. and Bremner, J.M. 1964. Ammonium production in soil under waterlogged conditions as an index of nitrogen availability. Nature (London) 201: 951–952.

Williams, D.E., Vlamis, J., Pukite, A. and Corey, J. 1980. Trace element accumulation, movement, and distribution in the soil profile from massive applications of sewage sludge. Soil Sci. 129(2): 119–131.

World Health Organization. 1978. Methods of Analysis of Sewage Sludge, Solid Wastes and Compost. WHO International Ref. Ctr. for Wastes Disp., Switzerland. 48 pp.

THE EFFECT OF CASTS OF *PHERETIMA ALEXANDRI* (BEDDARD) ON THE GROWTH OF *VINCA ROSEA*, AND *ORYZA SATIVA* L.

M. VIKRAM REDDY
North Eastern Hill University, College of Agriculture, Medziphema – 797 106, Nagaland, India

Introduction

The effects of earthworms on increasing plant productivity have been investigated either using living worms or their dead bodies (Satchell, 1958; Edwards and Lofty, 1977, 1980; Graff and Makeschin 1980 and McColl *et al.*, 1982). With the introduction of recent earthworm technology which is commonly known as vermicomposting, the casts produced by the earthworms feeding on large wastes, such as municipal waste water sludges, paper wastes and vegetative plant matter have been used for enhancing plant productivity. However, very little information is available on the effect of these worm-casts on the growth patterns of plants (Nijhawan and Kanwar 1952; Springett and Syres 1978), although such information would help in the establishment of the vermicomposting technology. Thus, the present investigation was carried out to determine the effect of casts of *Pheretima alexandri* (Beddard) on the growth of two plants, one ornamental and medicinal, *Vinca rosea* and one cereal, *Oryza sativa* Linn.

Materials and methods

The casts of *P. alexandri* were collected from uncultivated areas near Medsiphema, India. The soil (0–10 cm layer), the casts, and the mixture of casts and soil (50% casts and 50% soil) were packed into pots of 8 cm diameter at the mouth, 17 cm deep containing an equivalent of about 1 kg of soil. A total of 18 replicate pots, six for each treatment were prepared. Seedlings of *V. rosea* measuring 4.5 cm in shoot-length and two cm in root length with two initial leaves were transplanted into the pots on April 3, 1982 and observation on the growth patterns were recorded until January 3, 1983.

In another experiment, earthworm casts collected were mixed with soils in different proportions, at the usual recommended rates for potassium such as 20 kg, 40 kg and 60 kg K_2O ha^{-1} based on the K_2O content of the casts (Reddy, 1983), and the mixtures were packed in pots. A total of 24 pots were prepared and grouped into four sets each containing six replicates. For the first set, about 310 g of casts were mixed with 2 kg of soil (equivalent to 20 kg K_2O $ha^{-1}2$) and packed in each pot. For the second set, about 620 gms of casts were mixed with 2 kg soil (equivalent to 40 kg K_2O ha^{-1}) and packed in each pot. A fourth set of six pots were prepared, each being packed with 2 kg of soil and treated as the control. Five seeds of *O. sativa* L. were sown in each of the pots. After germination, only three seedlings were kept in each pot and maintained for the growth measurements. The pots were kept in the field.

Earthworms in waste and environmental management
edited by C.A. Edwards and E.F. Neuhauser, pp. 241–248

Table 1. Different growth measurements (in cm) of the shoot system of *V. rosea* grown in soil only, mixture of soil and casts, and casts only

Months	Soil only			Mixture of soil and casts			Casts only		
	TSL_1	TSL_2	Gir	TSL_1	TSL_2	Gir	TSL_1	TSL_2	Gir
April, 1982									
July	10.5	7.5	0.7	19.7	12.0	1.5	17.0	10.25	1.5
August	23.0	16.25	1.2	30.0	22.7	1.7	27.5	22.0	1.7
September	29.6	23.0	1.2	34.0	30.5	1.9	30.5	29.5	1.8
October	35.0	30.0	1.3	41.0	33.5	2.0	37.0	31.5	2.1
November	44.0	38.5	1.5	44.0	39.0	2.2	45.0	39.5	2.4
December	49.8	44.5	1.6	51.6	44.6	2.3	54.5	47.0	3.0
January, 1983	51.5	49.5	1.7	54.5	50.8	3.2	57.7	51.2	3.4

TSL_1 = Total shoot length with leaf; TSL_2 = Total shoot length without leaf; Gir = Circumference or the girth.

(The experiment was started in April 1982 with seedlings of 4.5 cm in shoot length and 2 cm root length)

a

b

Fig. 1. V. rosea grown in soil only, mixture of soil and casts (50% soil + 50% casts) and casts only, *a.* three months old and *b.* eight months old.

The plants, particularly the shoot portions, were probably under winter stress, as the experiment was started in late spring and continued during the winter.

Results

The salient features of growth of *V. rosea* grown in soil, soil mixed with earthworm

Table 2. Some important growth patterns of *V. rosea* grown in soil, mixture of soil and casts, and casts only

Events	Soil only	Mixture of cast and soil	Casts only
Appearance of inflorescence and flowers	After five months, 20 days	After third month; first flower after three months, eight days; second flower after three months, 12 days. On branches, after fifth month	After third month
Ripening of the leaves and appearance of the branching	No branching until the end (13th month)	After fourth month (two branches appeared)	After fourth month (one branch appeared)
Appearance of fruits	No fruiting until the end (13th month)	After five months, 20 days; mean length of the fruit = 1.9 cm; number of fruits = 4; number of seeds/fruit = 9	After five months, 20 days mean length of the fruit – 2.7 cm; number of fruits = 4; number of seed/fruit = 12. Fruits matured earlier and seeds dispersed

(The experiment was started on April 3, 1982).

Fig. 2. Root system of *O. sativa* grown in different proportions of casts mixed with soil viz., 940 gm casts/2 kg mix, 620 gm casts/2 kg mix, and 310 gm casts/2 kg mix.

casts, and earthworm casts only during a period of 10 months are summarized in Table 1. After three months, the total length of the shoot, whether including or excluding the top leaf of the plants, was maximum in plants grown in the soil mixed with casts, and intermediate in the plants grown in only casts. The shoot length was least in plants grown in only soil (Fig. 1a). This growth pattern, *i.e.*, being maximum in soil mixed with casts, continued until the end of the seventh month. However, from the eighth month, the maximum in shoot length intermediate was recorded for those plants grown in casts, in those grown in soil mixed with casts, and in plants grown in soil (Fig. 1b). The circumference of the stem of the plants was also greatest in the plants grown in casts only, intermediate for plants grown in soil mixed with casts, and was only half for plants grown in soil.

Other important parameters of *V. rosea* grown in soil, in the mixture and in casts are summarized in Table 2, which shows that the inflorescence appeared after the third month in plants grown in the casts, and in mixture. The first flower appeared after three months, eight days and the second three days later in plants grown in the mixture. However, the inflorescences appeared in the main shoot of the plants grown in soil only after five months, 20 days, and at the same time in the branches of the plants grown in casts, and in the mixture of soil and casts. The ripened leaves and branching appeared in plants grown in casts only, and in the mixture of soil and casts after the fourth month. No branching appeared in the plants grown in soil only even after the thirteenth month. The fruits appeared earlier, *i.e.*, after five months, 20 days, in the plants grown in casts and in the mixture compared to the plants grown in soil. In the later plants no fruit appeared until the end of the thirteenth month. The mean lengths of the fruits and their seeds were greatest in plants grown in casts only, compared to those grown in the soil/cast mixture, although the number of fruits remained the same. Moreover, the fruits matured earlier on plants grown in casts.

The dry weight of the plants and their root systems of *V. rosea* grown in all treatments are summarized in Table 3. The dry weight of the plants and root systems grown in casts was greatest for the plants and root system grown in casts and less for those grown in the mixture and in soil only. The root systems were twice as heavy for plants grown in the soil/cast mixture and three times as heavy for plants grown in casts only as those grown in soil only.

The growth patterns of *O. sativa*, grown in different proportions of casts mixed with soil, and in soil only (control), are summarized in Table 4 which shows that the length of the whole plant, and the number of adventitious roots was greatest for plants grown in pots with 940 gms of casts with 2 kg^{-1} of soil, intermediate for those in the pots with 620 gms and 310 gms of casts with 2 kg^{-1} of soil. The length was least, and there were fewer adventitious roots for plants grown in control soil (Fig. 2). The length of the root systems was greatest for plants grown in pots with maximum amount of casts and least for plants grown in control soil. The number of tillers were also greatest for plants grown in pots with most casts, numerically being double than that for plants grown in soil, intermediate for plants grown in pots with medium, and least for plants grown with the smallest amount of casts.

Discussion

All the data collected from the two experiments lead to the conclusion that the casts of *P. alexandri* had a significant effect in increasing not only the length and weight of the shoots and root systems for *V. rosea* and *O. sativa*, the number of seeds in the former and the number of tillers in the latter plants but also contributed to the earlier appearance of inflorescence, flowers and fruits. These observations are in accordance with the findings of Nijhawan and Kanwar (1952) who compared the growth of wheat plants grown in casts of different earthworms and the surrounding soil. They reported that the growth in plants grown in casts was double that of plants grown in soil in the first two months. The average height, the number of tillers and the average ear length were more in plants grown in casts than those grown in soil. Edwards and Lofty (1980) recorded total root weights of wheat as significantly greater in plots inoculated with earthworms. By contrast, Bal and Curry (1977) reported that soil containing earthworm castings had no effect in increasing the flower size or number of blooms of annual plants compared to those grown in soil without worm casts.

V. rosea plants grew better in the mixture of casts and soil than in casts only until the end of the seventh month which was probably due to the physical conditions of the mixture in the pots. However, the plants grown in casts only surpassed that of those grown in the mixture after the seventh month. The second experiment used the

Table 3. Dry weight of the root-system of *V. rosea* grown in soil, mixture of soil and casts, and only in casts and uprooted after 13th month

Weight	Soil	Mixture	Casts
Weight of the plant (gms)	5.9	7.2	8.2
Weight of the root system (gms)	0.5	0.9	1.7

Table 4. Different growth measurements of *O. sativa* grown in pots with different proportions of casts mixed with underlying soil, and soil only

Growth measurements	Control pots with soil only	Pots with 940 gms casts 2 kg^{-1} of soil	Pots wth 620 gms casts 2 kg^{-1} of soil	Pots with 310 gms casts 2 kg^{-1} of soil
Whole length of the plants (cm)	62.79 ± 4.70^{a}	81.53 ± 5.81^{a}	64.81 ± 3.76	64.31 ± 5.35^{a}
Whole length of the root system (cm)	27.55 ± 4.67^{a}	47.40 ± 7.21	30.64 ± 3.42	35.83 ± 3.83^{a}
Adventitiousness of the root system	minimum (very thin)	maximum	medium	minimum

[a] Significant at $p < 0.01$.

cereal, *O. sativa* which could be harvested within four months, so the treatment of pots with casts only was excluded. The paddy plants grown in the greatest proportion of casts mixed with soil (*i.e.*, a mixture of about 70% soil and 30% casts) increased significantly in plant-growth. It may therefore be concluded that perennial plants such as *V. rosea* can be grown in casts only or in a mixture containing more casts for better yields. However, seasonal plants such as *O. sativa* should be grown in a mixture of casts and soil.

The effect of casts on the growth of *V. rosea* and *O. sativa* may be due to the presence of 'plant growth substances' identified as indole compounds which could be secreted into the casts, and, in turn, increase the plant growth (Springett and Syres, 1978). They have reported better root and shoot growth in the plants grown in casts than those grown only in soils. Also, Graff and Makeschin (1980) reported that 'yield-influencing substances' were released by the earthworms into the soil which increased the yield of plants. Collier and Livingston (1980) and Appelhof (1981) also reported that the plants grown in casts grew better than those grown in top-soil. In addition, increased plant growth in casts may be attributable to the fact that the castings consisted of digested soil and particles of organic matter and was more chemically neutral than the underlying soils (Bridgens, 1981). Chemical analysis of casts showed that they were richer in various plant nutrients compared to the underlying soil (Edwards and Lofty, 1977; Bridgens, 1981 and Reddy, 1983), and could act as complex-fertilizer granules.

Acknowledgements

This investigation was supported by a grant (Minor research project grant No. 12547) from U.G.C., New Delhi. The author is also grateful to his wife for maintaining the *V. rosea* plants. Mr. Chandrama Ram maintained *O. sativa* plants.

References

Appelhof, M. 1981. Use of *Eisenia foetida* to process organic kitchen wastes. Flowerfield Enterprises, USA. 110 pp.

Bal, R.C. and Curry, L.L. 1977. Culture and agriculture importance of earthworms. Cooperative Extension Serv., Michigan State Univ., Ext. Bull., E-766, 27 pp.

Bridgens, S. 1981. The importance of the earthworm. Span. 22: 14–15.

Collier, J. and Livingston, D. 1980. Conversion of municipal waste-water treatment plant residual sludges into earthworm castings for use as fertile top-soil. Unpublished report, 28 pp.

Edwards, C.A. and Lofty, J.R. 1977. Biology of earthworms (2nd ed.), Chapman and Hall, Ltd., London, 316 pp.

Edwards, C.A. and Lofty, J.R. 1980. Effect of earthworm inoculation upon the root growth of direct drilled cereals. J. Appl. Ecol. 17: 533–543.

Graff, O. and Makeschin, F. 1980. Beeinflussung des Ertrags von Weidelgras (*Lolium multiflorum*) durch ausscheidungen von Regenwurmern dreier Verschiedener Arten. Pedobiologia 20: 176–180.

McColl, H.P., Hart, P.B.S. and Work, F.J. 1982. Influence of earthworm on some soil chemical and physical properties, and the growth of rye grass on a soil after top-soil stripping – a pot experiment. N.Z.J. Agric. Res. 25: 239–243.

Nijhawan, S.C. and Kanwar, J.S. 1952. Physico-chemical properties of earthworm castings and their effect on the productivity of soil. Indian J. Agric. Sci. 22: 357–373.

Reddy, M.V. 1983. Annual cast production by the megascolecid earthworm, *Pheretima alexandri* (Beddard). Comp. Physiol. Ecol. 8: 84–86.

Satchell, J.E. 1958. Earthworm biology and fertility. Soils and Fertil. 21: 209–219.

Springett, J.A. and Syres, J.K. 1978. Effect of earthworm casts on ryegrass seedlings. *In* Proceedings of the Second Australian Conference on Grassland Invertebrate Ecology. pp. 44–47. Edited by T.K. Crosby and R.P. Pottinger. Government Printers, Wellington.

V. Earthworms in land reclamation, soil amelioration and land improvements

CONTENTS

THE ECOLOGY OF EARTHWORMS IN RECLAIMED SOILS AND THEIR INFLUENCE ON SOIL FERTILITY

J.P. CURRY
Department of Agricultural Zoology and Genetics, University College, Belfield, Dublin 4

Introduction

The role of earthworms in improving soil structure and fertility is now widely recognized. They have been shown to play a major role in incorporating organic residues into the soil (Raw, 1962; Van Rhee, 1963; Barley and Kleinig, 1964; Hoogerkamp *et al.*, 1983), in soil mixing and turnover (Barley, 1959, 1961; Evans and Guild, 1947), in incorporating lime and fertilizers into the soil (Stockdill, 1966; Springett, 1983) and in increasing water stable aggregates (Van Rhee, 1977). Worm worked soils exhibit enhanced water holding capacity (Stockdill, 1966), improved water infiltration (Sharpley *et al.*, 1979; Hoogerkamp *et al.*, 1983), improved soil porosity and aeration (Noble *et al.*, 1970; Van Rhee, 1977), and enhanced microbial activity (Satchell, 1983). Significant mineralization of organic N has been attributed to earthworms (Barley and Jennings, 1959; Keogh, 1979; Lee, 1983), and Sharpley and Syers (1976) and Mansell *et al.*, (1978) demonstrated their role in increasing plant availability of mineral P. Hopp and Slater (1948), Waters (1951) and Van Rhee (1965) attributed increases in grass, clover and cereal growth ranging from 28% to 1000% in small pot or box experiments to earthworm activity, and Edwards and Lofty (1980) reported increased plant density, better root penetration and increased root and shoot biomass in direct drilled cereal plots inoculated with earthworms. Improved grass production has been recorded following earthworm establishment in reclaimed grassland (Stockdill and Cossens, 1966; Hoogerkamp *et al.*, 1983). Enhanced plant growth has usually been attributed to improved soil physical and chemical conditions, but a direct hormone-like effect on plants has also been demonstrated (Nielsen, 1951, 1965; Springett and Syers, 1979).

The success of land reclamation by conventional techniques is often limited by poor soil structure and low inherent soil fertility, and even in productive soils, a marked deterioration in the botanical composition of the sward can occur within a number of years (Hoogerkamp *et al.*, 1983). A number of studies indicate that earthworms play an important part in improving reclaimed soils. However, natural colonization and establishment occur fairly slowly, and there is increasing interest in accelerating these processes by deliberate introductions.

This paper briefly reviews some studies on the ecology and effects of earthworms on soil fertility in reclaimed mineral soils and industrial wastes, and data are presented from current work on earthworms in reclaimed peat soils in Ireland.

Earthworms in waste and environmental management
edited by C.A. Edwards and E.F. Neuhauser, pp. 251–261

Earthworms in reclaimed mineral soils

Sown pastures in New Zealand and Australia

Earthworms of European origin have become widely established in sown New Zealand pastures (Martin, 1977), notably the peregrene species *Aporrectodea caliginosa* (Sav.). However, there are still considerable areas of recently improved upland pastures from which lumbricids are absent (Stockdill, 1966). Improved production and an increase in the proportion of ryegrass and clover in such pasture was noted following the introduction of *A. caliginosa* in transplanted turfs (Hamblyn and Dingwall, 1945; Nielsen, 1951). The estimated rate of spread of introduced worms was 10–13 m yr^{-1}. Stockdill (1959, 1966) recorded the disappearance of the surface organic mat following earthworm introductions in improved hill pasture. There was a marked improvement in soil structure, aeration and distribution of nutrients. Dry matter production increased initially by 70%, but in older sites where earthworms were well-established the mean-dry matter increase was 29% (Stockdill and Cossens 1966). A turf cutting and distributing machine suitable for large scale introduction projects was developed, and it was established that complete colonization could be expected from turfs placed at 10 m intervals within 6–7 years under New Zealand conditions (Stockdill, 1982). The expected return on investment in earthworm introduction projects was calculated to be more than 300%.

Barley and Kleinig (1964) and Noble *et al.* (1970) describe successful introduction and establishment of populations in excess of 300 m^{-2} in irrigated sown pasture on sandy loam soil in N.S.W., Australia. There was a noticeable improvement in soil structure with a reduction in C:N ratio, a decline in bulk density, and a reduction in the surface organic mat of undecomposed litter and dung which correlated well with earthworm population density.

Polder soils

Van Rhee (1969a, b) reported earthworm population densities exceeding 200 m^{-2} in orchard soils on drained Dutch polders 26 years after reclamation, but concluded that the rate of natural spread was only 4–6 m y^{-1} and that establishment could be accelerated by deliberate introduction. This author (Van Rhee, 1963, 1969a, b, 1977) conducted a series of introduction experiments in grassland and orchards in recently reclaimed polders and recorded population densities of up to 750 m^{-2} in grass areas, with rather lower densities (140–250 m^{-2}) in orchard soils which received heavy applications of pesticides. *A. caliginosa* was a particularly successful early colonizer and played an important role in the early stages of soil formation. Other species which became established included *Lumbricus rubellus* Hoff., *Aporrectodea rosea* (Sav.), *A. longa* (Ude), *Allolobophora chlorotica* (Sav.), and to a limited extent *Lumbricus terrestris* L. Increased litter incorporation, improved aggregation, aeration and water relationships and the development of mull characteristics were observed following earthworm establishment. The improvement in soil structure was confirmed by the soil morphological studies of Rogaar and Boswinkel (1978).

Recent inoculations into polder grassland were described by Hoogerkamp *et al.* (1983), who estimated rates of population build up with the aid of a dispersal model and used infra red line scanning to detect earthworm distribution in the study sites. This technique detected different patterns of heat exchange depending on whether or not a surface organic mat was present. When earthworms were well-established the surface mat had disappeared resulting in better heat exchange between soil and air and diminished daily temperature fluctuations at the surface. The mat was ingested and incorporated into the soil within three years of worm invasion and a dark colored A_1 soil horizon began to develop, increasing gradually to a thickness of 5–8 cm in 8–9 years. This was accompanied by an increase in the C:N ratio of the topsoil. Earthworm activity increased pore volume, aeration and available moisture at field capacity, and increased infiltration capacity. Better soil conditions influenced root growth and distribution positively; this was reflected in better sward attachment to the soil and fewer tufts detached by grazing cattle. After 10 years, worm plots generally had a better botanical composition with a higher proportion of *Lolium perenne*, urine damaged grass patches were less common, and grass yields were on average 10% higher in worm plots. On the debit side, some damage by treading and soiling of grass was observed on worm plots in wet weather.

Poor mineral soils

Some efforts to improve impoverished soils by stimulating earthworm activity have given encouraging results. Large increases in earthworm numbers and improvement in soil structure occurred in poor vineyard soils which had been fertilized with organic manure (Bosse, 1967). Huhta (1979) reported encouraging results from experimental introductions of *A. caliginosa* into limed coniferous forest plots in Finland, while indigenous *L. rubellus* and *Dendrobaena rubida* (Sav.) populations also benefited from liming and the addition of deciduous litter. Langmaid (1964) reported considerable soil improvement in afforested virgin podzols in New Brunswick where earthworms had become established but Satchell (1980a, b) concluded that heather moor in Yorkshire afforested with birch was unlikely ever to support deep burrowing species.

Tréhen and Bouché (1983) studied the role of earthworm in the amelioration of impoverished heathland soils in central Brittany where organic refuse was spread. Surface dwelling species such as *Eisenia fetida* (Sav.) and *Dendrobaena octaedra* (Sav.) played an important part in the initial composting phase, while a complex of species introduced from pasture could contribute to the improvement of soil structure.

Mining wastes

Acidity, metal toxicity, unfavorable moisture conditions and lack of suitable food are among the factors likely to inhibit earthworm establishment on recently reclaimed mining spoil. Even under favorable conditions, natural colonization is slow. Dunger (1969a) reported the presence of earthworms after five years on reclaimed coal mine spoil in East Germany, but it took 10 years for significant populations to develop and on more acid sites earthworms had not become established after 15 years. Once earthworms had become established under *Aluus glutinosa*, litter incorporation was greatly accelerated and the soil structure changed from moder to mull (Dunger, 1969b). Standen *et al.* (1982) reported significant correlations between site age, shade, moisture content and earthworm numbers and biomass in restored open cast coalmining sites in Co. Durham. Earthworm numbers were still low ($< 33\ m^{-2}$) 11 years after restoration for agriculture, but up to 162 *L. rubellus* m^{-2} were found in 14 year-old colliery spoil heaps naturally colonized by scrub woodland. Significant numbers of *L. terrestris* were found in only a few sites which were more than 50 years old. Surface organic amendment of newly reclaimed sites favored earthworm establishment, both by increasing food supply and by stabilizing temperature and moisture conditions in the surface soil (Dunger, 1969a; Luff and Hutson, 1977).

L. terrestris has been introduced successfully into reclaimed coal mine spoil sites in the U.S.A. ranging in pH from 3.5 to over 7 (Vimmerstedt and Finney, 1977; Hamilton and Vimmerstedt, 1980). Worm activity accelerated the incorporation of litter, and increased exchangeable cations and available P. After 12 years, populations had become widely dispersed throughout an area of afforested calcareous spoil, with numbers and biomass being greater under *Alnus glutinosa* than under *Robinia pseudoacacia* where the litter was less palatable.

Earthworms in reclaimed peat

Earthworms in virgin peat are limited by high moisture levels, low pH and poor quality litter, but peat soils drained and reclaimed for agriculture can support moderate population densities. Thus, Guild (1948) found 9–10 species and population densities ranging from 50–100 m^{-2} in improved hill pastures on peaty soils in Scotland, whereas Baker (1983) recorded 15 species with up to 197 individuals m^{-2} in long-established grassland on reclaimed fen peat in central Ireland. A survey of 24 sites showed that earthworms were scarce in newly reclaimed peat, and population densities only exceeded 100 m^{-2} in sites which were more than 25 years old (Curry and Cotton, 1983). Surface dwelling species typical of raw humus habitats such as *D. octaedra, D. rubida* and *L. rubellus* were common in young sites, whereas in the mature sites typical soil dwelling species such as *A. longa, A. rosea, L. terrestris* and *Octolasion cyaneum* (Sav.) occurred. The two most abundant species in mature sites were *A. caliginosa* and *A. chlorotica*; these were early colonizers and were widely distributed in young sites, but at low population densities.

Currently, large areas of cutover peat consisting of a residual peat layer about 50 cm deep, overlying relict soils derived from limestone till are available for reclamation in central Ireland following industrial peat extraction. Preliminary studies indicated that earthworm colonization could be greatly accelerated by deliberate introductions (Curry and Cotton, 1983). Data are presented in this paper on rates of population development of introduced earthworms, site carrying capacity, and the effects of earthworms on grass production.

Population development in grassland on reclaimed peat

The experimental site was formerly a raised bog at Clonsast, Co. Offaly which had been cutover, leaving behind 50 cm of mixed forest fen peat overlying a well-developed relict soil derived from limestone till. An area of 12 ha, which had been reclaimed initially in 1977, was deep ploughed, cultivated, limed to pH7 and seeded

Table 1. Earthworm densities (nos m^{-2}) and biomass (g m^{-2} preserved material) one and three years after introduction

Worms	After one year				After three years			
	Cattle slurry		No cattle slurry		Cattle slurry		No cattle slurry	
	Mean	Max.	Mean	Max.	Mean	Max.	Mean	Max.
Lumbricus terrestris l.	3	4	5	8	6	8	2	4
L. rubellus (Hoff.)	4	8	4	4	4	4	4	4
L. festivus (Sav.)	1	4	3	8			2	4
Lumbricus immatures	43	76	36	48	124	180	134	168
Aporrectodea caliginosa (Sav.)			3	4				
A. longa (Ude)	4	12	4	12	2	4		
A. rosea (Sav.)			1	4				
Allolobophora chlorotica (Sav.)	84	112	45	56	190	340	100	120
A. tuberculata (Eisen)			1	4				
Aporrectodea/Allolobophora immatures	116	204	101	144	118	144	176	256
Octolasion tyrtaeum (Sav.)							2	4
Dendrobaena rubida (Sav.)	68	132	112	160	6	12	2	4
D. octaedra (Sav.)	1	4			2	4		
Dendrobaena immatures	94	120	202	280	18	20	20	24
Total nos	418		517		470		442	
Biomass	46	56	59	72	80	88	75	77
Cocoons								
L. terrestris								
L. rubellus					6	12		
A. chlorotica	} 151	} 176	} 107	} 184	510	624	500	720
A. caliginosa					70	80	58	76
D. rubida	257	344	633	872	6	12	20	36

with perennial ryegrass/white clover in spring 1980. Preliminary sampling revealed the presence of *A. chlorotica, A. caliginosa* and *L. rubellus* at low population densities (about 20 m^{-2}).

Previous experience indicated that unconfined worms dispersed too rapidly to allow short-term study of their colonizing success (Curry and Cotton, 1983). Open-topped nylon cages were found to prevent dispersion for long enough to ensure population establishment without unduly altering environmental conditions. The cages were made of 0.5 mm nylon mesh supported on wire and wood frames; they were cubic in shape with 50 cm sides. Ten cages were buried to a depth of 25 cm and the soil and grass sods were replaced in their original positions. Cattle slurry of 10% dry wt. was mixed into the soil below the grass sods in every second cage at the rate of 4 l per cage (160 t ha^{-1}) in order to test the hypothesis that earthworms are food limited in newly reclaimed sites. Twenty-two worms extracted from local peaty grassland soils were added to each cage on 4 June 1980. The introduced population consisted on average of 40% *A. chlorotica*, 30% *L. rubellus*, 10% *L. terrestris*, 10% *A. caliginosa* and 10% other species. Six of the cages were hand sorted in May 1981, the remaining four in May 1983, and the earthworm populations recorded are given in Table 1.

Table 2. Grass yields from experimental cages, 1981–1983. Totals from five cuts in 1981, four in 1982 and in 1983

Treatments		Yields (g dry wt m^{-2} ± S.E.)		
		1981	1982	1983
160 t ha^{-1} cattle slurry incorporated into the soil initially, plus	worms present	1825 ± 70	1129 ± 105	855 ± 93
335 kg N, 50 kg P, 200 kg K $ha^{-1}yr^{-1}$	worms absent	1798 ± 78	1130 ± 64	1001 ± 105
66 t ha^{-1} cattle slurry applied annually in March	worms present	1924 ± 42	1316 ± 69	1214 ± 23
plus 200 kg N $ha^{-1}yr^{-1}$	worms absent	1821 ± 94	1049 ± 132	814 ± 146
Only inorganic fertilizers 335 kg N, 50 kg P,	worms present	1575 ± 67	1027 ± 33	801 ± 82
200 kg K $ha^{-1}yr^{-1}$	worms absent	1629 ± 87	1089 ± 67	867 ± 61

Substantial development of populations occurred within one year. Natural colonization by *D. rubida* was particularly marked; this is a very mobile coprophagic species which was probably attracted initially by the cattle slurry, invading neighboring cages later when the slurry was depleted. *A. chlorotica* was the most abundant of the introduced species and was the only one which benefited from the initial cattle slurry amendment. After three years, numbers of *D. rubida* had declined to insignificant levels, and *A. chlorotica* contributed by far the greatest number of individual worms and cocoons. Population densities and biomass were considerably greater than reported previously from peat soils, probably reflecting better drainage, management and soil fertility than in previous studied sites. Allowing for a 20% loss in weight during preservation, earthworm biomass after three years was in the order of 100 g m^{-2}. This is within the range which is characteristic of grasslands on mineral soils, although there were fewer deep working species such as *A. longa* and *L. terrestris* than are usually found in more productive sites (Curry, 1976; Cotton and Curry, 1980a, b).

Effects of earthworms on grass growth

The effects of earthworms on grass production were assessed in a randomized block experiment at Clonsast, using cages similar to those already described. The cages were arranged in six blocks, each of which had six cages. Ten *L. terrestris* and ten *Allolobophora/Aporrectodea* spp. (mainly *A. chlorotica*) were added to three cages in each block in November, 1980, and the fertilizer treatments indicated in Table 2 were applied. Apart from the initial slurry incorporation treatment, the fertilizer schedules were similar to those used in intensive grass production systems on the farm. Grass yields were determined at intervals during 1981–1983, and the results are summarized in Table 2.

Earthworms had no effect on grass production during the first year, but yields were 25% greater in the second year and 49% greater in the third year in worm cages

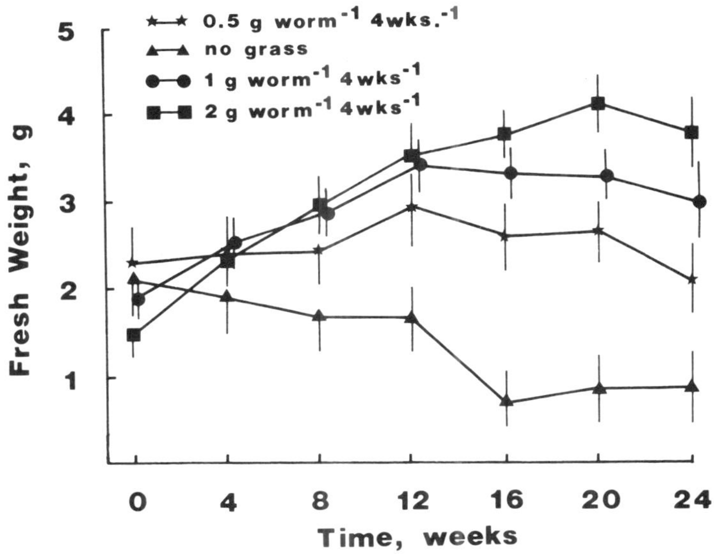

Fig. 1. Growth rates of *L. terrestris* cultured at 15°C in peat with various quantities of chopped grass added. Mean wts. per worm ± S.E.

receiving an annual top dressing of cattle slurry when compared with cages receiving slurry but without added worms. No responses of grass yields to earthworms occurred in the other fertilizer treatments. The cages were only partially successful in retaining earthworms; by the third year, the nylon had deteriorated extensively and by November 1983 some of the 'worm free' cages had population densities approaching 50% of those in the cages which were originally inoculated. Nevertheless, the results indicate clearly a very significant effect of earthworms on grass growth when cattle slurry was a major component of the fertilizer program. This result indicates the importance of earthworm activity in incorporating and mineralizing organic fertilizers.

Studies on L. terrestris

Since deep-working species such as *A. longa* and *L. terrestris* play a major role in soil mixing and litter incorporation in mineral soils, it is important to determine whether the low populations previously reported from reclaimed peat soils (Baker, 1983; Curry and Cotton, 1983) were due to inherent unsuitability of peat as a habitat or merely reflect poor site conditions. Laboratory and field studies were therefore made to assess survival and growth rates of *L. terrestris* in peat, and to determine the earthworm carrying capacity of grassland at Clonsast.

In the laboratory growth rate experiment, worms were cultured at 15°C in 1 litre plastic containers filled with newly reclaimed, root free, soil from Clonsast. There were four levels of feeding: unamended soil and soil with 0.5g, 1.0g and 2.0g of air dried grass chopped in 1 cm lengths added to the surface every four weeks. Two immature, *L. terrestris* were cultured in each container and were weighed at four-week inter-

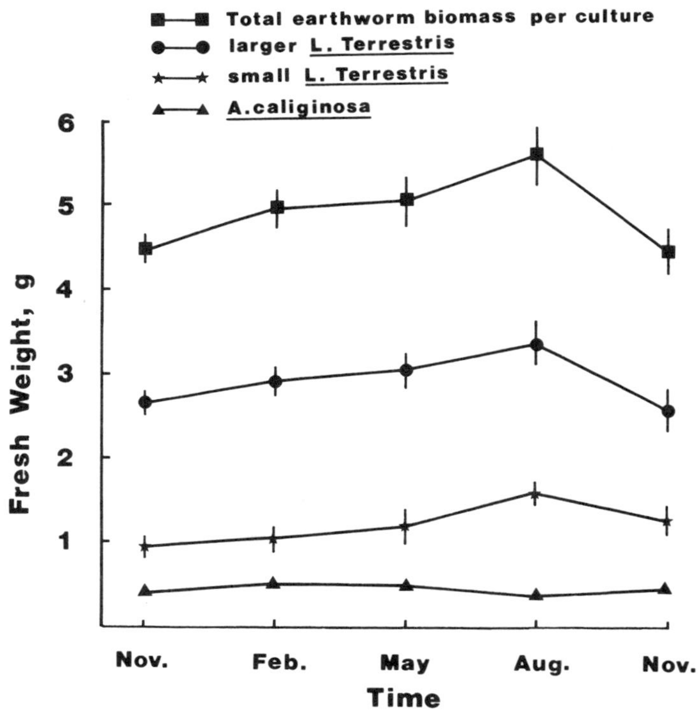

Fig. 2. Growth rates of earthworms in reclaimed peat grassland. Mean wt. per worm ± S.E.

vals. There were 10 replicates per feeding level. The worms lost weight in unamended soil and gained weight at the most sustained rate at the highest level of feeding (Fig. 1). The rate of weight gain between weeks 4 and 20 was 6.6 mg g^{-1} day^{-1} at this level, a faster growth rate than that recorded for worms feeding on *Salix* litter in an earlier experiment (Curry and Bolger, 1984). The decline in weight between weeks 20 and 24 suggests food limitation, indicating that the rate of grass amendment during that period (α 16 mg dry wt. g^{-1} fresh wt. day^{-1}) was inadequate to sustain growth. No such weight loss had occurred in the earlier experiment when worms were provided with similar amounts of *Salix* litter, suggesting that grass litter must be ingested at higher rates than the nutritionally superior *Salix* to maintain growth.

Ten soil cores, 200 cm^2 in surface area and 24 cm deep, were isolated in 0.5 mm aperture nylon mesh bags at Clonsast for determination of carrying capacity in the field. Two *L. terrestris* juveniles weighing about 1 and 2.5 g and two adult/subadult *A. caliginosa* were added to each core; the cores were replaced with fresh soil after six months. All the larger *L. terrestris*, 90% of the smaller specimens and 80% of *A. caliginosa* survived for 12 months. Mean weights of *A. caliginosa* remained fairly constant throughout the 12 months study period, larger *L. terrestris* juveniles gained weight at the rate of 1 mg g^{-1} initial weight day^{-1}, while smaller individuals had an average daily weight gain of 2.5 mg g^{-1} (Fig. 2). The low weight gains suggest that the population was approaching the carrying capacity of the site, with the maximum sustainable biomass being 4.5–5 g per culture or 112.5–125 g m^{-2}.

Previous studies have indicated that *L. terrestris* produces 1–2 cocoons per individual per month in peat (Curry and Bolger, 1984); this is within the range which has been reported for mineral soils (Satchell, 1967).

Conclusions

The Clonsast studies show that well-managed cutover peat can support earthworm populations comparable in density and biomass with those found in fertile mineral soils, and that population development can be accelerated greatly by deliberate introduction of beneficial species. Introduced shallow-working species such as *A. chlorotica* and *A. caliginosa* become established fairly quickly, while deep-working species such as *L. terrestris* with low reproduction and dispersal rates probably require several years to reach significant population densities.

There are many potential benefits from earthworms in reclaimed peat. Curry and Bolger (1984) calculated that a population of *L. terrestris* of mean biomass 100 g m^{-2} would consume 365 g dry wt. of *Salix* litter m^{-2} yr^{-1} in 'energy forest' plantations on cutover peat, thus considerably accelerating nutrient cycling. Deep ploughing results in an admixture of mineral soil and peat, and soil ingesting earthworms could greatly facilitate the intimate mixing of mineral and organic fractions and so enhance soil structural development. Curry and Bolger (1984) estimated that 100 g *L. terrestris* m^{-2} could ingest 1.3 kg of Clonsast soil m^{-2} yr^{-1} and that the entire top 20 cm soil layer could be ingested and turned over in 45 years. Ultimately, beneficial effects on soil fertility should be reflected in enhanced productivity, and the results reported in this paper indicate a strong positive response of grass growth to earthworms in land fertilized with animal manure. Current studies are aimed at exploring further the effects of earthworms on soil properties in cutover peat and the relationships between earthworm activity, fertilizers and grass growth.

Clearly, earthworm introduction can make an important contribution to soil reclamation and improvement in a variety of situations. However, the scale of introduction projects has generally been small because of the large numbers of earthworms required. The sod transplanting technique has given good results in New Zealand hill pasture where the main aim was to introduce the topsoil-dwelling *A. caliginosa* (Stockdill 1982), but its usefulness for deep-burrowing species is restricted. It may be possible to increase its effectiveness by using larger sods and by transplanting in late autumn when maximum numbers of the juvenile stages of deep burrowing species are likely to be present in the upper soil layers. Also, it may be possible to boost the population densities in donor soil prior to transplanting by liberal applications of farmyard manure, municipal sludge, and other organic materials known to favor *L. terrestris*.

Acknowledgements

I am indebted to Bord na Móna for providing facilities for field experiments at Clonsast.

References

Baker, G.H. 1983. The distribution, abundance and species associations of earthworms (Lumbricidae) in a reclaimed peat soil in Ireland. Holarct. Ecol. 6: 74–80.

Barley, K.P. 1959. Earthworms and soil fertility. IV The influence of earthworms on the physical properties of a red-brown earth. Aust. J. Agric. Res. 10: 371–376.

Barley, K.P. 1961. The abundance of earthworms in agricultural land and their possible significance in agriculture. Adv. Agron. 13: 249–268.

Barley, K.P. and Jennings, A.C. 1959. Earthworms and soil fertility. III The influence of earthworms on the availability of nitrogen. Aust. J. Agric. Res. 10: 364–370.

Barley, K.P. and Kleinig, C.R. 1964. The occupation of newly irrigated lands by earthworms. Aust. J. Sci. 26: 290–291.

Bosse, J. 1967. Restoration of biologically impoverished Weinberg soils, an example of earthworm colonization (German, English summary). *In* Progress in Soil Biology. pp. 299–309. Edited by O. Graff and J.E. Satchell. North Holland Publishing Co., Amsterdam.

Cotton, D.C.F. and Curry, J.P. 1980a. The effects of cattle and pig slurry fertilizers on earthworms (Oligochaeta, Lumbricidae) in grassland managed for silage production. Pedobiologia 20: 181–188.

Cotton, D.C.F. and Curry, J.P. 1980b. The response of earthworm populations (Oligochaeta, Lumbricidae) to high applications of pig slurry. Pedobiologia 20: 189–196.

Curry, J.P. 1976. Some effects of animal manures on earthworms in grassland. Pedobiologia 16: 425–438.

Curry, J.P. and Bolger, T. 1984. Growth, reproduction and litter and soil consumption by *Lumbricus terrestris* L. in reclaimed peat. Soil Biol. Biochem. 16: 253–257.

Curry, J.P. and Cotton, D.C.F. 1983. Earthworms and land reclamation. *In* Earthworm Ecology. pp. 215–228. Edited by J.E. Satchell. Chapman and Hall, London.

Dunger, W. 1969a. Fragen der natürlichen und experimentellen Besiedlung kulturfeindlicher Boden durch Lumbriciden. Pedobiologia 9: 146–151.

Dunger, W. 1969b. Über den Anteil der Arthropoden an der Umsetzung des Bestandesabfalles in Anfangs-Bodenbildungen. Pedobiologia 9: 366–371.

Edwards, C.A. and Lofty, J.R. 1980. Effects of earthworm inoculation upon the root growth of direct drilled cereals. J. Appl. Ecol. 17: 533–543.

Evans, A.C. and Guild, W.J.McL. 1947. Studies on the relationships between earthworms and soil fertility. I. Biological studies in the field. Ann. Appl. Biol. 34: 307–330.

Guild, W.J.McL. 1948. Studies on the relationship between earthworms and soil fertility. III. The effect of soil type on the structure of earthworm populations. Ann. Appl. Biol. 35: 181–192.

Hamblyn, C.J. and Dingwall, A.R. 1945. Earthworms. N.Z. J. Agric. 71: 55–58.

Hamilton, W.E. and Vimmerstedt, J.P. 1980. Earthworms on forested spoil banks. *In* Soil Biology as Related to Land Use Practices. pp. 409–417. Edited by D.L. Dindal. E.P.A., Washington.

Hoogerkamp, M., Rogaar, H. and Eijsackers, H.J.P. 1983. Effect of earthworms on grassland on recently reclaimed polder soils in the Netherlands. *In* Earthworm Ecology. pp. 85–105. Edited by J.E. Satchell. Chapman and Hall, London.

Hopp, H. and Slater, C.S. 1948. Influence of earthworms on soil productivity. Soil Sci. 66: 421–428.

Huhta, V. 1979. Effects of liming and deciduous litter on earthworm (Lumbricidae) populations of a spruce forest, with an inoculation experiment on *Allolobophora caliginosa*. Pedobiologia 19: 340–345.

Keogh, R.G. 1979. Lumbricid earthworm activities and nutrient cycling in pasture ecosystems. *In* Proceedings of the second Australian Conference on Grassland Invertebrate Ecology. pp. 49–51. Edited by T.K. Crosby and R.P. Pottinger. Government Printers, Wellington.

Langmaid, K.K. 1964. Some effects of earthworm invasion in virgin podzols. Can. J. Soil Sci. 44: 34–37.

Lee, K.E. 1983. The influence of earthworms and termites on soil nitrogen cycling. *In* New Trends in Soil Biology. Proceedings VIII International Colloquium of Soil Zoology., Louvain-la-Neuve (Belgium). pp. 35–48. Edited by Ph. Lebrun, H.M. André, A. de Meds, C. Grégoire-Wibo and G. Wauthy. Imprimeur Dieu-Brichart, Ottignies-Louvain-la-Neuve.

Luff, M.L. and Hutson, B.R. 1977. Soil faunal populations. *In* Landscape Reclamation Practice. pp. 125–147. Edited by B. Hackett. IPC Science and Technology Press, Guildford.

Mansell, G.P., Syers, J.K. and Gregg, P.E.H. 1978. Plant availability of phosphorus in dead herbage ingested by surface casting earthworms. Soil Biol. Biochem. 13: 163–167.

Martin, N.A. 1977. Guide to the lumbricid earthworms of New Zealand pastures. N.Z. J. Exp. Agric. 5: 301–309.

Nielsen, R.L. 1951. Earthworms and soil fertility. Proc. N.Z. Grassld. Assoc. 13: 158–167.

Nielsen, R.L. 1965. Presence of plant growth substances in earthworms demonstrated by paper chromatography and the Went Pea test. Nature (London) 208: 1113.

Noble, J.C., Gordon, W.T. and Kleinig, C.R. 1970. The influence of earthworms on the development of mats of organic matter under irrigated pastures in Southern Australia. *In* Plant Nutrient and Soil Fertility. Proceedings of the XIth International Grassland Congress., Queensland, Australia. pp. 465–468. Queensland University Press.

Raw, F. 1962. Studies of earthworm populations in orchards. 1. Leaf burial in apple orchards. Ann. Appl. Biol. 50: 389–404.

Rogaar, H. and Boswinkel, J.A. 1978. Some soil morphological effects of earthworm activity; field data and X-ray radiography. Neth. J. Agric. Sci. 26: 145–160.

Satchell, J.E. 1967. Lumbricidae. *In* Soil Biology. pp. 259–322. Edited by A. Burges and F. Raw. Academic Press, London.

Satchell, J.E. 1980a. Potential of the silpho moor experimental birch plots as a habitat for *Lumbricus terrestris*. Soil Biol. Biochem. 12: 317–323.

Satchell, J.E. 1980b. Earthworm populations of experimental birch plots on a *Calluna* podzol. Soil Biol. Biochem. 12: 311–316.

Satchell, J.E. 1983. Earthworm microbiology. *In* Earthworm Ecology. pp. 351–364. Edited by J.E. Satchell. Chapman and Hall, London.

Sharpley, A.N. and Syers, J.K. 1976. Potential role of earthworm casts for the phosphorus enrichment of run-off waters. Soil Biol. Biochem. 8: 341–346.

Sharpley, A.N., Syers, J.K. and Springett, J.A. 1979. Effect of surface-casting earthworms on the transport of phosphorus and nitrogen in surface runoff from pasture. Soil Biol. Biochem. 11: 459–462.

Springett, J.A. 1983. Effects of five species of earthworm on some soil properties. J. App. Ecol. 20: 865–872.

Springett, J.A. and Syers, J.K. 1979. The effect of earthworm casts on ryegrass seedlings. *In* Proceedings of the Second Australian Conference on Grassland Invertebrate Ecology. pp. 44–47. Edited by T.K. Crosby and R.P. Pottinger. Government Printers, Wellington.

Standen, V., Stead, G.B. and Dunning, A. 1982. Lumbricid populations in open cast reclamation sites and colliery spoil heaps in County Durham, U.K. Pedobiologia 24: 57–64.

Stockdill, S.M.J. 1959. Earthworms improve pasture growth. N.Z. J. Agric. 98: 227–233.

Stockdill, S.M.J. 1966. The effect of earthworms on pastures. Proc. N.Z. Ecol. Soc. 13: 68–75.

Stockdill, S.M.J. 1982. Effects of introduced earthworms on the productivity of New Zealand pastures. Pedobiologia 24: 29–35.

Stockdill, S.M.J. and Cossens, G.G., 1966. The role of earthworms in pasture production and moisture conservation. Proc. N.Z. Grassl. Ass. 28: 168–183.

Tréhen, P. and Bouché, M.B. 1983. Place des lumbriciens dans les processus de restauration des sols de lande. *In* New Trends in Soil Biology. Proceedings of the Eighth International Colloquium of Soil Zoology, Louvain-la-Neuve (Belgium). pp. 471–486. Edited by Ph. Lebrun, H.M. André, A. de Meds, C. Grégoire-Wibo and G. Wauthy. Dieu-Brichart, Ottignies, Louvain-la-Neuve.

Van Rhee, J.A. 1963. Earthworm activities and the breakdown of organic matter in agricultural soils. *In* Soil Organisms. pp. 54–59. Edited by G. Doeksen and J. van der Drift. North-Holland Publ. Co., Amsterdam.

Van Rhee, J.A. 1965. Earthworm activity and plant growth in artificial cultures. Plant and Soil 22: 45–48.

Van Rhee, J.A. 1969a. Inoculation of earthworms in a newly drained polder. Pedobiologia 9: 128–132.

Van Rhee, J.A. 1969b. Development of earthworm populations in polder soils. Pedobiologia 9: 133–140.

Van Rhee, J.A. 1977. A study of the effect of earthworms on orchard productivity. Pedobiologia 17: 107–114.

Vimmerstedt, J.P. and Finney, J.H. 1973. Impact of earthworm introduction on litter burial and nutrient distribution in Ohio stripmine spoil banks. Soil Sci. Soc. Amer. Proc. 37: 388–391.

Waters, R.A.S. 1951. Earthworms and the fertility of pasture. Proc. N.Z. Grassl. Assn. 13: 168–175.

EARTHWORMS, SOIL STRUCTURE AND THE REHABILITATION OF FORMER OPENCAST COAL-MINING LAND

V.I. STEWART and J. SCULLION
Bryngwyn Project, Field Laboratory, Penygroes, Dyfed, Wales

A fertile soil is not just a mixture of appropriate ingredients; it is a structure largely created by the organisms living within it. A soil *in situ* under a grass sward is a very different thing from the same material two meters down in a stockpile. If a topsoil has to be removed, stockpiled and then replaced it is very unlikely that it will quickly recover to its former state under grass merely by being spread, drained, limed, fertilized and seeded. Farmers, with the problem of rehabilitating farmland, degraded by the disturbance involved in opencast coal mining, seek to encourage full restoration of biological activity.

The importance of the living component in soils has tended to be more and more down-graded as modern agriculture has turned from the recycling approach of organic systems to an ever greater dependence on agricultural chemicals. The tendency for engineers, architects, agriculturalists and horticulturalists to underestimate the extent to which soil biology is involved in the conditioning of soil for plant growth probably accounts for the frustrations of those who have inherited their handiwork.

Soil texture and soil structure

The properties of soils owe much to the size and the shape of the mineral particles present. Silt and sand particles are roughly spherical and, because of this, have little cohesion, wet or dry. They are liable to dust when dry and compact by slumping when wet. By contrast, clay particles are distinctly platy and cohere strongly to each other or to any smooth surface, much like stacked sheets of greenhouse glass or thin plastic sheets. Hence, the tackiness and stiff feel of a true plastic clay and the slippery, sliminess of silt. The swelling and shrinking that accompanies the entry and withdrawal of water between contiguous clay sheets, is responsible for the intense cracking that can cause a strong clay soil to fragment into an angular, fine blocky tilth when the right sequence of alternating weather conditions allows. Thus, clays can sometimes avoid the threat of waterlogging, that always hangs over a fine particle system, by fracturing into small, strong, blocky fragments that are water retentive within but between, free-draining, aerobic and easily accessible to roots. By contrast, a silt or a sand on drying will fracture only into a widely spaced series of vertical cracks, separating large cloddy blocks, liable to dust on impact when dry and to slump back again into a compact mass at the first sign of rain.

Loams have a mixed particle size composition but, with less than 40% clay, they have insufficient clay to crack themselves into a fine blocky tilth and, with less than 70% sand, have too much fine material to avoid fine particle dominance of the pore space when compact. If a loam is to achieve a satisfactory air/water balance for efficient rooting activity it must somehow be structured into a granular form and the

Earthworms in waste and environmental management
edited by C.A. Edwards and E.F. Neuhauser, pp. 263–272

granules must at least be water stable. A systematic sampling of Soil Survey data for England and Wales suggests that, whereas only one eighth of our mineral soils are strong clays and another eighth sands, three-quarters are some form of loam (Stewart *et al.*, 1980). It is essential, therefore, that we should thoroughly understand the mechanism by which loams can be water-stabely granulated.

The development of water stable, granular soil structure

Any soil with more than 10% clay and an adequate supply of organic matter may acquire a water stable organization providing:

– Somewhere in the system there is an agent capable of aggregating the clay/humus complex, along with the rest of the mineral skeleton, into units large enough to provide for rapid drainage, aeration and easy root penetration between aggregates, and for reserves of nutrients, plus water, within.

– The potentially ephemeral organic agents responsible for water stabilization can be periodically replenished.

It is a common experience to find just such an effective soil structure associated with, and exploited by, grass roots but, under these same circumstances, we also find active populations of burrowing earthworms. The problem is to sort out cause and effect.

Two different hypotheses seem possible:

1. It is the grass itself that is responsible. Therefore, any treatment that promotes the vigor of the grass will simultaneously benefit soil structure.
2. It is earthworms and grass roots working together that are responsible. Therefore, increases in yield of grass should be sought only by means that do not limit the effectiveness of either component and, preferably, benefit both.

The farmer, starting with structurally degraded land, requires to be informed of these alternative hypotheses if he aims to up-grade the condition of his soil whilst at the same time farming profitably.

Evidence of the importance of earthworms from forest and permanent pasture

Two very different soils develop in loam textured soils under natural and semi-natural conditions (Stewart, 1980). These ecologically divergent soils seem to represent the consequences of a fundamental dichotomy in soil development initiated by the presence or absence of burrowing earthworms. A layered form of profile is found where the soil is lime deficient and the litter of the dominant vegetation unpalatable to burrowing earthworms. The organic residues remain to decompose on the surface forming a mat within which the vegetation is rooted. The mineral soil, from which the organic mat is sharply defined, is structurally compact and impedes surface drainage. Little wonder the natural vegetation associated with such soils is shallow rooted, dominated by plants physiologically and morphologically adapted to cope with periodic waterlogging and periodic drought as well as the nutritional consequences of soil acidity. Soils sufficiently less acid to allow an active population of burrowing earthworms, feeding on more palatable vegetation, produce a soil profile that is a uniform blend of mineral and organic matter and there is no surface mat of unincorporated organic debris, no water-holding compaction of the mineral surface and no marked bias towards surface rooting.

Worm-deficient, structurally degraded soils have features in common with soils replaced after opencast mining whereas the worm-worked soil is similar to a healthy soil under a well-managed pasture.

Accepted ideas on soil granulation from agriculture

The possible role of earthworms as initiators of soil granulation has generally not been considered seriously by agriculturalists, possibly because the burrowing earthworm is an inevitable and invariable concomitant of the conditions recognized for other reasons to be essential for productive pasture. The importance of worm channels as an aid to the infiltration of surface water has generally been recognized but not so their role in soil granulation.

In the report of The Agricultural Advisory Council. 'Modern Farming and the Soil' (Strutt, 1970), there is the following statement on the origin of soil crumbs and soil granules:

> They are products of root action and the decomposition of fresh organic matter. They are more common in topsoils than subsoils and are produced more readily by some plant root systems than others. The root systems of grasses vary but usually excel in causing granulation.

This is the view usually accepted by farmers and their advisors. It is the view that explains the standard approach to soil rehabilitation based on grass leys sustained by fertilizing and periodic loosening by subsoiling.

The alternative view implicating the earthworm as a prime causal agent in soil granule formation

By ingesting fresh organic residues, burrowing earthworms incorporate, comminute and begin the process of organic decay. At the same time, by ingesting soil, along with organic matter, they bring clay particles into intimate contact with the early products of organic decay and also with calcium ions introduced from the special gland they possess within their digestive tract. Finally, they mould this mix into a cylindrical ribbon of a diameter allowing for efficient aeration, even when saturated, and coil it into a cast throughout which roots will readily ramify.

However, worm processing of soil is bound to be slow, for they work only of the order of 30–40 tonnes of soil per hectare annually or one fiftieth of the plough layer, but about half of this ends up on the surface where it probably does most good.

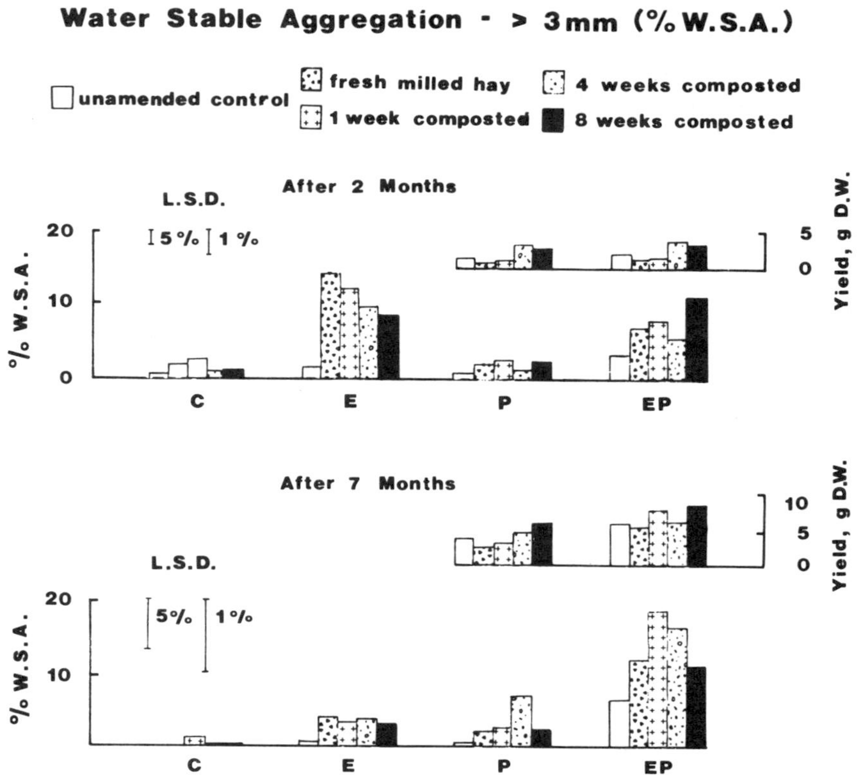

C No earthworms and no plants.
E 2 Lumbricus terrestris earthworms added initially and one each month thereafter.
P Planted to uniform sward of S23 ryegrass. } Yield based on one cut taken at end of experimental period.
EP Earthworms plus plants.

Experimental conditions for Fig. 1 and Fig. 2.
Standard silty clay loam soil of fairly poor structure from 50 cm below surface of permanent pasture. 8% organic matter, originally pH 5.6 but adjusted to pH 7 by addition of lime. Soil screened to pass a 3 mm sieve and placed in 50 cm section of 7 cm diameter, opaque plastic pipe forming a 25 cm column of soil above a 25 cm column of fine sand, soil and sand separated by a thin, fabric, worm barrier. 2% milled, ryegrass hay incorporated after composting under laboratory conditions to achieve various stages of decay. Maintained in greenhouse under conditions favoring continuous growth. (For further details in relation to Fig. 1 see Salih, 1978).

Fig. 1. Effects of earthworms and organic matter on water stable aggregates.

a. Evidence from controlled greenhouse experiments

The following conclusions would appear to be justified from the evidence presented in Fig. 1.

- Addition of organic matter, particularly if fresh, encourages the development of water stable soil granulation but only if earthworms are present to make use of it.
- The granulation achieved by earthworm processing of soil in the presence of a single addition of organic matter is ephemeral. Any initial benefit achieved would

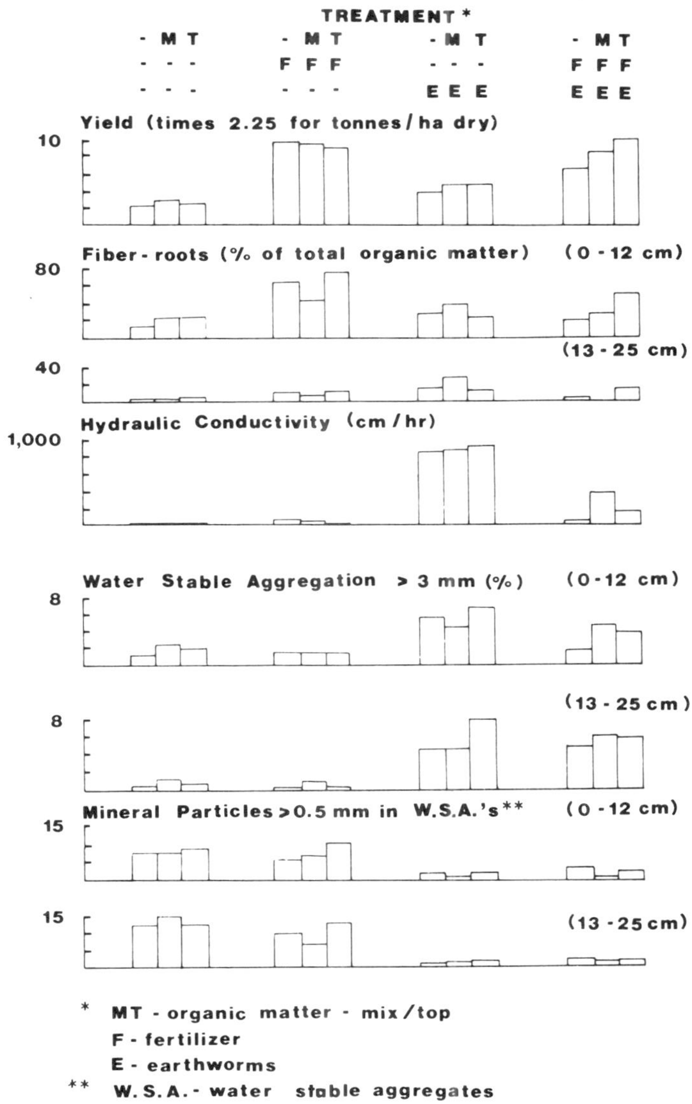

For experimental conditions see Fig. 1 and Al-Bakri, 1984.

Fig. 2. Interactions between fertilizers, organic matter and earthworms under rye grass.

appear to be lost as the organic agents responsible for the water stability are dissipated by further decay.

– When starting with a structurally degraded soil, grass, by itself, even in the presence of added organic matter, can achieve no significant soil granulation but, given time, this can be achieved and sustained by the combination of grass growth in the presence of earthworms.

Further experimental evidence is summarized in Fig. 2. Here there is evidence of a negative interaction between fertilizer and earthworm activity with adverse consequences for soil granulation despite an increase in roots. The evidence from the particle size analysis of water stable granules, points to their having been formed from a selected fraction of the original mineral skeleton available, a feature that is in agreement with the idea that there is an upper limit to the size of solid particle that an earthworm can ingest. Thus, though the received wisdom on roots causing soil granulation suggests that, when farming disturbed land, we should take advantage of the fact that both root activity and yield can be simultaneously increased by the liberal use of readily soluble nitrogen fertilizers, experimental evidence warns that we should now be looking for alternative procedures more wholly in tune with a wider view of the biological basis of soil rehabilitation and the crucial role in this played by the rather sensitive creature, the earthworm.

b. Field evidence from Bryngwyn Farm, Penygroes, South Wales

Two groups of British soils are affected by opencast coal mining. Free-draining sandy loams lying on coal measure sandstone outcrops are 'prized for their ease of working', but are less extensive than the clay and silty clay loams overlying shales and glacial drift (Pringle, 1958). Based on a review of previous research in Britain, Scullion (1984) concluded that up to 85% of restored soils may fall into the heavier loam categories. Soils at Bryngwyn are predominantly clay loam in texture. It is on these fine-textured soils that rehabilitation problems are at their most intractable and where, as indicated by evidence discussed in earlier sections of this report, the role of the earthworm may be crucial.

Evidence relating to the effects of opencast operations on earthworm populations is often contradictory. It ranges from the view that populations recover rapidly (Brook and Bates, 1960) to the statement by Blenkinsop (1957) that 'a complete absence of earthworms is a feature of opencast sites'. Five years after reinstatement populations on the Bryngwyn site were low and their recovery apparently dependent on colonization from undisturbed land (Scullion, 1984).

There is evidence of an association between earthworm activity and improvements in soil structure (Table 1). Whilst various interpretations may be placed on these data, they are reasonably consistent with some of the views expressed earlier:

1. The influence of earthworm activity on soil structure is reduced greatly where organic matter levels are insufficient both to support that activity and to provide a source of aggregate stabilizing agents.
2. Improvements in soil structure, resulting from earthworm activity over a three-year-period, are small in relation to the inherent variation in bulk soil samples from restored land.
3. In the early stages of rehabilitation, earthworm activity appears to improve pore space and microaggregate stabilization, without having any similar effect on macroaggregation.

Table 1. Associations between earthworm casting and soil structure indices

a. Simple correlations

Sampling depth (cms)	Total pore space (v/v)	True crumb (> 0.5 mm)	Dispersion ratio	Dispersion coefficient
0–7.5	0.411[c]	0.087 NS	−0.049 NS	−0.387[c]
7.5–15	−0.020 NS	0.024 NS	−0.280[b]	−0.322[b]

True crumb (> 0.5 mm) = proportion of total soil material < 0.5 mm in water-stable aggregates.
Dispersion ratio = proportion of total silt and clay which is dispersible by shaking in water.
Dispersion coefficient = proportion of total clay which is dispersible by shaking in water.

b. Partial correlations – adjusting for the effects of organic matter level (loss on ignition 300°C)

Sampling depth (cms)	Total pore space (v/v)	True crumb (> 0.5 mm)	Dispersion ratio	Dispersion coefficient
0–7.5	0.134 NS	—	—	−0.068 NS
7.5–15	—	—	−0.233[a]	−0.265[a]

Number of samples = 90.
Note – lower values for dispersion ratio and dispersion coefficient indicate improved conditions.

Earthworms contribute to surface infiltration and this affects drainage performance (Fig. 3). After three years of contrasting management, drain flow patterns on grazed and cut areas, which were initially similar (Scullion, 1984), differed markedly. A similar difference in values for infiltration measured over subsoiling channels, suggests that drainage performance on the cut area is limited by the capacity of these channels to intercept surface water. An alternative explanation is that the capacity of subsoiling channels on the cut area to conduct drainage water is limiting. This seems unlikely, since the rate of water infiltered during field measurements is small compared with the rates of flow recorded as drain discharge.

Evidence that, infiltration over subsoiling channels is the step limiting drainage performance in the above trial, that earthworm biomass and infiltration are both lower on cut areas and, that biomass and infiltration are closely related (Fig. 3b), points to a role for earthworms in maintaining the effectiveness of secondary drainage treatments on restored land.

Further evidence of the part played by earthworms in soil rehabilitation is emerging from trials where randomly selected areas have been treated with carbaryl over the last three years. The effects of carbaryl treatments have not, as yet, been fully assessed, but first impressions indicate:

– Carbaryl reduces surface casting to a level 30–40% that of untreated areas.

– Untreated areas are softer underfoot during periods of wet weather and show clear symptoms of the effects of waterlogging in their swards.

The above observations suggest that whilst earthworms may promote surface infiltration this results in increased waterlogging and vulnerability to poaching during the transition from surface shedding to infiltration and discharge through the soil.

Differing management and manurial regimes affect earthworm populations (Fig. 4). The main conclusions are described below:

Whereas cutting systems of management appear to inhibit the recovery of earth-

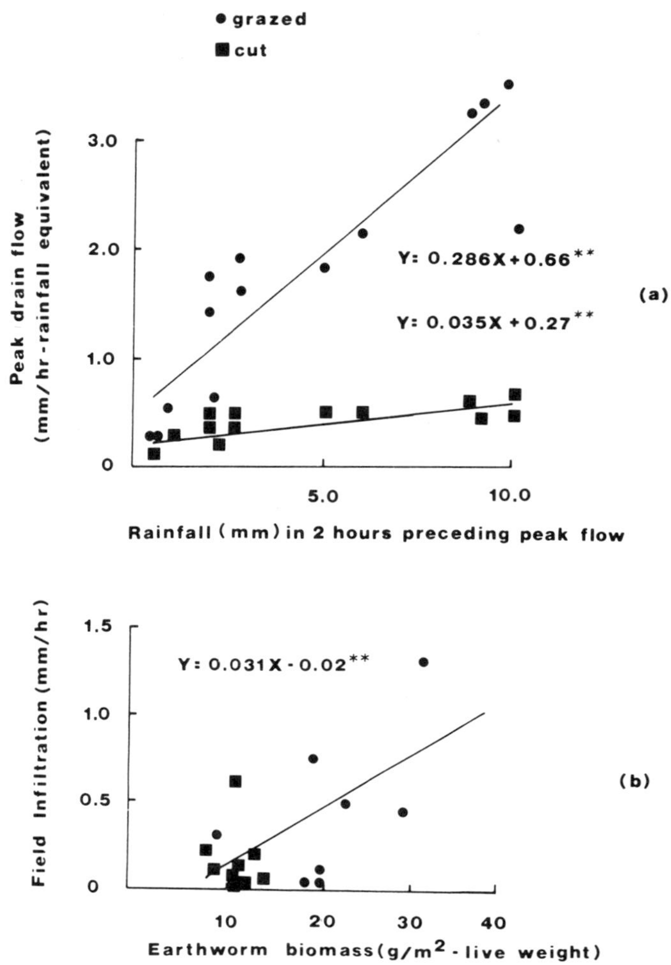

Earthworm populations estimated by formalin extraction.
Infiltration measured by double-ring method (Bertrand, 1965).

Fig. 3. Earthworm activity and the performance of artificial drainage systems; (*a*) drain flow characteristics, (*b*) the relationship between earthworms and infiltration over subsoiling channels.

worm populations on restored land, populations increase markedly under sheep grazing. Data reported elsewhere (Scullion, 1984) indicate similar effects for each of three co-dominant site species; *Aporrectodea caliginosa, Allolobophora chlorotica*, and *Lumbricus rubellus*.

Manurial treatment effects are generally not statistically significant, in part, because of the very high levels of inherent variability on restored land. Nevertheless, results suggest that populations on areas topdressed with fertilizer are smaller compared with those on areas receiving organic manure. More detailed findings

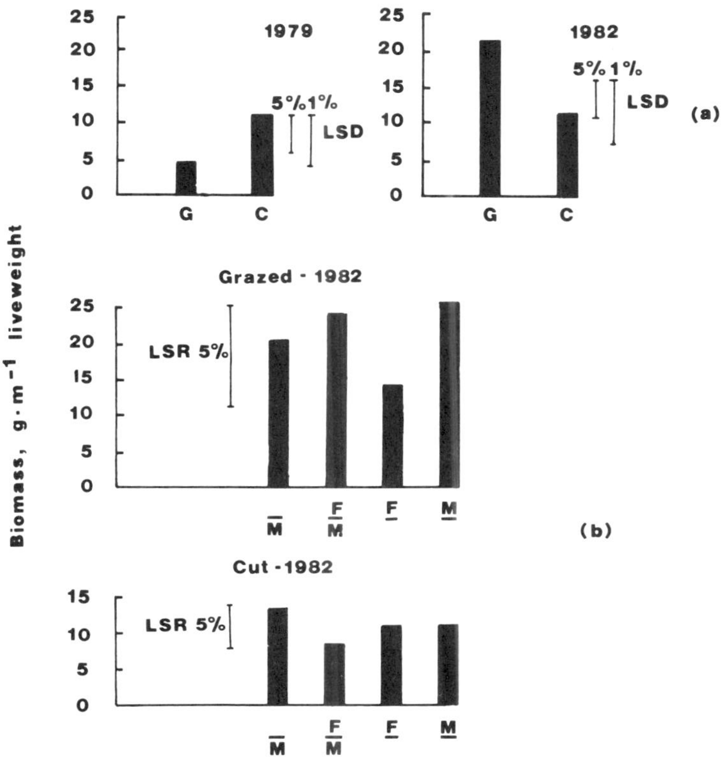

Fig. 4. The effects of various management regimes (*a*) and manurial treatments (*b*) on earthworm biomass.
(*a*) Grazed (*b*) – rotational grazing by sheep only; (*c*) – one or two cuts of hay/silage per year with grazing of aftermaths.
(*b*) $\overline{\mathrm{M}}$ – single dressing (30T/ha) poultry manure worked into plough layer.
$\underline{\mathrm{F}}$ – topdressed compound fertilizer @ 140 Kg N/year in four applications.
$\frac{\mathrm{F}}{\mathrm{M}}$ – combination of $\overline{\mathrm{M}}$ and $\underline{\mathrm{F}}$.
$\underline{\mathrm{M}}$ – topdressed poultry manure @ 15 T/ha per year.

(Scullion, 1984) indicate that levels of surface casting ($p < 0.001$) and numbers of *Aporrectodea caliginosa* ($p < 0.05$) are higher on areas topdressed with manure compared with all other treatments tested.

Where incorporated manure is supplemented with topdressed fertilizer ($\frac{\mathrm{F}}{\mathrm{M}}$) on grazed grass, earthworm biomass is increased. The equivalent response on cut grass is negative. These findings suggest that where fertilizer increases food supplies, as would be the case on grazed but not cut areas, any direct adverse effects on earthworms may be balanced by improved nutrition.

The vast majority of soils in Britain, including those affected by opencast coal mining, are some form of loam. Evidence from greenhouse experiments, which is beginning to be supported by results from field trials, suggests that the earthworm may have a crucial role to play in the structural rehabilitation of these soils. In the light of the above conclusions, an improvement program is being undertaken at Bryngwyn Farm which involves the careful grazing of long-term leys, regularly topdressed with poultry manure to encourage earthworm activity. The provision of effective site drainage is a priority, in particular, during the sensitive transition phase from surface shed to surface infiltration. This necessitates regular subsoiling.

Restored land in Britain is usually returned to normal agriculture after rehabilitative management. It is then frequently said to deteriorate, a feature which is often attributed to poor management on the farmer's part. However, the handover may coincide with the sensitive transition phase when earthworm numbers are building up. Any management regime, which fails to recognize the implications of this and to adapt accordingly, will encounter major problems in the farming of former opencast coal mining land.

References

Al-Bakri, K.H. 1984. Biological Influences on the Development of Soil Structure. Ph.D. thesis. University College of Wales, Aberystwyth. 511 pp.

Bertrand, A.R. 1965. Rate of water uptake in the field. *In* Methods of Soil Analysis. pp. 197–209. Part 1. Edited by C.A. Black. American Society of Agronomy, Wisconsin, U.S.A.

Blenkinsop, A. 1957. Some aspects of the problem of the restoration of opencast sites. Plan. Outl. 4(3): 28–32.

Brook, D.S. and Bates, F. 1960. Grassland in the restoration of opencast coal sites in Yorkshire. J. Brit. Grassl. Soc. 15: 116–123.

Pringle, J. 1958. Opencast coal mining; the restoration problem. *In* International Symposium on Soil Structure, Ghent. pp. 64–72.

Salih, R.O. 1978. The Assessment of Soil Structure and the Influence of Soil Treatment on this Property. Ph.D. thesis, University College of Wales, Aberystwyth.

Scullion, J. 1984. The Assessment of Experimental Techniques Developed to Assist the Rehabilitation of Restored Opencast Coal-mining Land. Ph.D. thesis. University College of Wales, Aberystwyth.

Stewart, V.I. 1980. Soil drainage and soil moisture. *In* Amenity Grassland: An Ecological Perspective. pp. 119–124. Edited by I.H. Rorison and R. Hunt. Wiley, Chichester.

Stewart, V.I., Salih, R.O., Al-Bakri, K.H. and Strong. J. 1980. Earthworms and soil structure. *In* Soil-related pests and diseases. pp. 103–114. Edited by P.A. James. Report Welsh Soils Discussion Group. No. 21.

Stewart V.I. and Salih, R.O. 1981. Priorities for soil use in temperate climates. *In* Biological Husbandry. pp. 19–37. Edited by B. Stonehouse. Butterworths, London.

Strutt, N. 1970. Report of Agricultural Advisory Council on Soil Structure and Soil Fertility. Ministry of Agriculture, Fisheries and Food. H.M.S.O., London.

POTENTIAL USE OF EARTHWORMS IN TROPICAL SOILS

P. LAVELLE and I. BAROIS
E.N.S. Laboratoire de Zoologie, Paris, France

The earthworm fauna of tropical areas is extremely diverse. Thirteen families have been distinguished which include several hundreds and perhaps thousands of species, a great number of which have yet to be described. This diversity is particularly important when compared with the temperate-area fauna of Europe and North America, which have only a few hundred species, most of which are Lumbricidae.

The tropical fauna may be characterized by the generally small size of the species, distribution area and the subsequent existence of a continuum of species with numerous intermediate sub-species intimately adapted to their environment. This is very true of the Mount Nimba area in West Africa, where Omodeo (1958) has described 36 species; 90% of these are endemic. But, beyond this general feeling, our present taxonomical knowledge is for the most part incomplete and scattered, and much remains to be studied. A few species may, however, have large distributions, as, *e.g.*, some *Dichogaster (D. bolaui, D. saliens)*, a few species of *Pheretima*, and the exceptional *Pontoscolex corethrurus* which may be found almost everywhere in the humid tropics where human settlement has modified the environment.

Functional structure of earthworm communities

The earthworm fauna of the tropics is also extremely rich in species from a functional point of view. One finds all the ecological categories described by Bouché (1972) and Lavelle (1981), as well as some particular adaptations which have never been observed in the temperate fauna. Oligohumic endogeics, which feed on the extremely poor organic matter of the deep soil, are found in moist tropical areas, although there are also some mediterranean species (Diaz Cosin and Moreno, 1979). Apart from this category, there are peculiar adaptations such as the carnivorous species of the genus *Agastrodilus* (Lavelle, 1981, 1983), the *Dichogaster* species living at the bottom of Nasutitermitidae termite nests (Lavelle, 1978), and facultative or obligate arboricolous species.

Although tropical earthworm communities include all the ecological categories, they differ from temperate communities in that their populations are predominately geophagous (oligo and above all mesohumic) rather than litter-feeding (epigeic and anecic species). (Fig. 1) Along a thermolatitudinal gradient connecting a set of moist soils from cold temperate, temperate and tropical regions, the sequence of dominance is: epigeic, then anecic species then, in the tropical area, meso- and oligohumic populations (Lavelle, 1983). This is the result of important differences in soil system function (Lavelle, 1984).

Forest and savannah regions have rather different earthworm communities. Litter-feeding populations are relatively more developed in forests and wooded areas than

Earthworms in waste and environmental management
edited by C.A. Edwards and E.F. Neuhauser, pp. 273–279

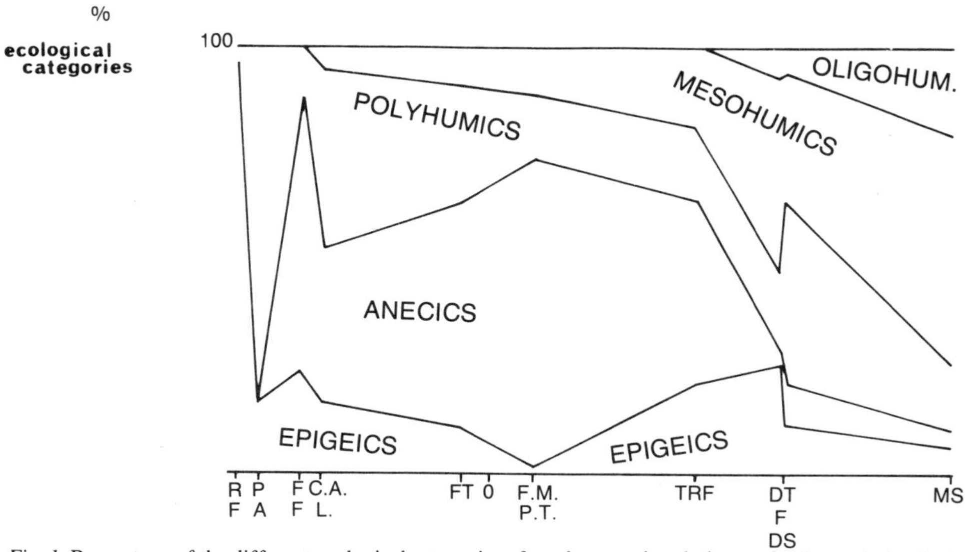

Fig. 1. Percentage of the different ecological categories of earthworms in relation to the thermolatitudinal sequence of vegetation (R.F. = resinous forest; P.A. = alpine grasslands; F.F. = cold deciduous forest; C.A.L. = cold open formations; F.T. = temperate deciduous forests; F.M. = mediterranean forest; P.T. = temperate grassland; T.R.F. = tropical rain forest; D.T.F. = 'dry' tropical forest; D.S. = dry savannas; M.S. = moist savannas; Lavelle, 1983).

in open savannahs, and forest communities are of comparatively low density and biomass. Savannah earthworm populations, on the other hand, are predominantly oligohumic and above all mesohumic. The subordinate role of these functions in forest areas is possibly due to differences in soil moisture regime, since the evapotranspiration of large trees is much greater than that of grasses. In regions of forest-savannah mosaics, the forest and savannah patches differ markedly with respect to species composition and functional community structure. Along the thermolatitudinal gradient (defined above), the humid tropical forests, such as the Amazonian rainforest, have greater anecic and epigeic species populations than savannahs which places them closer to the communities of temperate regions.

For biogeographic reasons, differences in earthworm distributions would also seem to exist at an intercontinental level. Glossoscolecid earthworms include truly anecic species (which live in galleries but feed on soil-surface litter), as in the case of Amazonian rainforest communities. But this is not true of the Megascolecidae, constituting the earthworm communities of the Lacandonian forest in Mexico (Fragoso and Lavelle, 1987) or the Taï forest in the Ivory Coast. We still need much more information about these tropical forest communities, for the function of most species – *i.e.*, their classification according to ecological categories – is by no means clear.

The use of epigeic species in earthworm culture

True epigeic species, those which live in soil-surface litter, are rather uncommon, for litter is generally decomposed so rapidly that it is difficult for them to survive. It would seem that most of the litter-dwelling tropical epigeic species thus enter the soil and have a partly geophagous regime. This is very common in the southern Ivory Coast, and species like *Eudrilus eugeniae* may invade moist organic accumulations up to relatively high latitudes in the southern part of Mali and Upper Volta. *Millsonia lamtoina* is an anecic species, also from Ivory Coast, though the study of its gut content showed that only 10% was litter and the rest was soil (Kanyonyo, 1984).

But real epigeic species do exist. They are generally arboricolous and live in the hanging soils of epiphytes (Orchids, Ferns or Bromeliaceae) or palm trees. These species are generally brightly-colored, fast-moving and active. At Lamto, Wasavo and Omodeo (1966) observed four species, two of which are not found elsewhere, dwelling on the palm *Borassus aethiopum*. In the Lacondonian forest, three different species appear to live exclusively in the hanging soils of Bromeliaceae, very rich in organic matter (Fragoso and Lavelle, 1987). In the flooded forests of the Rio Negro region of southern Venezuela, an undescribed species of Glossoscolecidae climbs along tree-trunks and accumulates in Bromeliaceae mats when the rivers overflow. Several kilos of these earthworms may be found in each mat. When the water subsides, they leave the mats and return to the soil.

There are thus many species, of very diverse size and behavior patterns, which could be used for vermiculture, waste conversion and compost production for the bait market or for feeding small animals, poultry, fish or pigs. But so far only two of these species have been used. *Eudrilus eugeniae* is commonly bred in the United States (Neuhauser *et al.*, 1979), it is reported as transforming toxic cassava peels into fertile compost (Mba, 1983). In Asia, *Perionyx excavatus* is often used for feeding animals, fishes and birds (Guerrero, 1983). But there are many other smaller and larger species which have adapted to forest or savannah climatic conditions and which transform different kinds of organic matter, from decomposing wood to small humified leaves.

The use of oligo and mesohumic endogeic species of earthworms for the restoration of impoverished or eroded soils

Humid savannah soils such as those of Lamto (Ivory Coast), and pastures induced after clear-cutting, like those of Laguna Verde (Mexico), have important earthworm communities essentially dominated by mesohumic populations. These earthworms may ingest up to 10 times their own weight every day, and even 36 times in the case of young *Millsonia anomala*. The whole community may thus ingest up to 850–1150 t dry-soil/ha/year, as is the case in the Lamto savannah. This is equivalent to a 10–15 cm soil layer. (Table 1) This kind of soil processing results in, firstly, the conservation of the soil structure, and secondly, a stable processing of organic reserves.

a. *Conservation of soil structure*

Part of the ingested soil is ejected on the soil surface in the form of casts. Twenty-two to twenty-eight t/ha of these represent 30–40 m^3 of empty space within the soil

Table 1. Earthworm populations from a tropical Mexican pasture and a Coast Ivorian savannah

Population	Density (individuals per ha)	Fresh biomass (kg/ha)	Dry cast production (t/ha/y)
Pontoscolex corethrurus (Laguna Verde, Mexico)	147 000	300	300
Millsonia anomale (Lamto, Ivory Coast)	215 000	250	517
Total geophagous (Lamto savanna, Ivory Coast)	1 800 000 to 4 000 000	359 to 486	850 to 1 150

Lavelle, 1978 (unpublished)

(Lavelle, 1978). It is clear that earthworms thus play an important role in the regulation of soil porosity and the reduction of soil compactness. It has been observed that, in general, the more compact the soil, the higher the cast production. Most of the soil ingested by the worm is ejected into the soil. These casts are well-structured aggregates and assist greatly in structure conservation and in the consequent resistance to soil erosion. Part of this relative structural stability might be due to the addition of polysaccharide to the soil during the transit through the gut.

b. *Processing of organic reserves*

In the Lamto savannahs, the 1200 tonnes of soil ingested annually by geophagous earthworms, represent one third of the total organic matter of the soil, but up to 60% of the organic matter in the top 10 cm. Part of this organic matter is assimilated by the worm during gut transit: about 10% in the case of *Millsonia anomala* (Lavelle, 1978) and 19% in *P. corethrurus* (Barois and Lavelle, 1986). However, the earthworm's most important role is perhaps that of giving a new orientation to the evolution of non-assimilated organic matter rather than assimilating a small part of it. Geophagous earthworms digest complex soil organic matter with the help of their mutualistic relationships with soil microflora. As a matter of fact, they appear not to have enough of the proper enzymatic equipment to effect this digestion by themselves.

After the soil has been ingested, large amounts of water (+ 120%) and water-soluble organic compounds (16% of the soil's total dry weight) are added by the worm. This triggers a great increase in microbial activity which reaches up to six times that in control soil in the posterior part of the gut. The microflora in the ingested soil is highly stimulated by the addition of this water and assimilable organic matter, and a priming effect thus enables it to digest the soil's organic matter for the benefit of the worm. In *P. corethrurus* as we have seen above 19% of the soil's organic matter is assimilated in this way during gut transit. (Fig. 2)

Microbial activity is enhanced further after transit through the gut, since its oxygen absorption is 50% greater than that of the control soil. But some experiments on variation of the soil's water-soluble contents have indicated, that at certain levels, microbial activity may not change, and may sometimes in fact decrease in the casts (Barois, 1982; Lavelle *et al.*, 1980, 1983). As their water-soluble organic contents

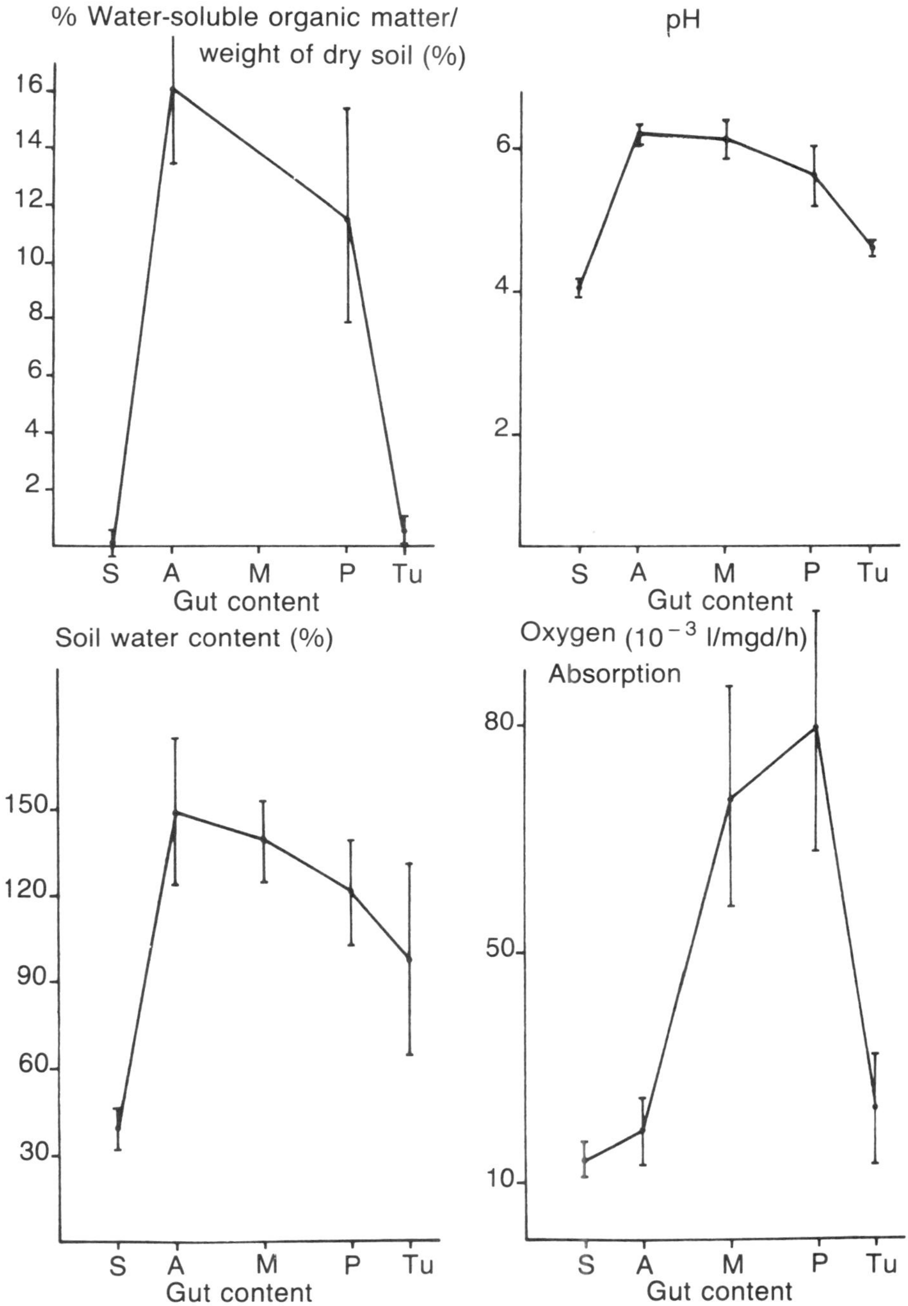

Fig. 2. Variations of diverse parameters in the gut content of *Pontoscolex corethrurus*, anterior part (A), medium (M), posterior (P), soil control (S) and fresh casts (Tu). (Barois and Lavelle, 1986).

Table 2. Ecological categories composition in some earthworm communities

Ecological category	Number of species		
	Lamto[1] (Ivory Coast)	Laguna Verde[1] (Mexico)	Chajul[2] (Mexico)
Epigeic	1	1	2
Anecic	1		
Polyhumic	3	2	1
Mesohumic	1	1	1
Oligohumic	2		1

[1] savannah; [2] forest

increase beyond a certain threshold, the respiratory activity in the casts becomes inhibited. With even greater concentrations, the inhibition decreases and is finally neutralized.

Earthworms would thus appear to effect a very important regulation of the soil's organic-matter metabolism in that they may accelerate or slow down mineralization. Furthermore, it is likely that humification processes are activated at the same time through the effect of residual polysaccharides which are not totally metabolized during gut transit. Although much still remains to be discovered about these complex processes, it may be considered that, in savannah soils, earthworms at once regulate the mineralization rate of reserves and have a stabilizing effect on the non-mineralized parts.

Thus it is clear that soil processes are affected significantly by the absence of adequate endogeic earthworm populations. This would seem to be the case in many tropical grasslands which have the microclimatic conditions and vegetation types of savannahs, but little if any adapted endogeic earthworm populations. For example, the fallow pastures resulting from clear-cutting of tropical forests are more or less comparable to savannahs, but the earthworm populations of the former forest will generally have disappeared through lack of adaptability to the new conditions. A few opportunistic species remain, but these are not always the best suited. This might explain why these soils rapidly become impoverished, losing organic matter and nutrients and becoming subject to erosion.

Composition of earthworm communities

Comparing earthworm communities from the original site before transformation (if it is possible to reconstitute) with the transformed one and with one which has more or less a stabilized community. Thus it would be useful information to know which species is needed for the earthworm community restoration (Table 2).

Lamto is a true savannah region, with an equilibrated savannah community. Laguna Verde is a hundred-year-old induced pasture with climatic conditions similar to that of Lamto. The mesohumic species is *P. corethrurus* an opportunistic species which follows human settlements. This community has no oligohumic species. Of the polyhumic species, one is survivor from the adjacent forest, and the other is *Ocnerodrilus occidentalis*, a peregrine species that colonizes the rhizosphere. Chajul is a tropical forest with an equilibrated forest community, probably comparable to the original forest of the Laguna Verde area. So, where adequate earthworm communities no longer exist, it would seem worthwhile to reconstitute them by introducing savannah species, particularly adaptable like a peregrine species such as *P. corethrurus*. In cases of excessively reduced soil organic reserves, these should also be reconstituted. This could be done with mature compost (from plant biomass or even domestic waste) or perhaps with lombricomposting preparation. With the reconstitution of the soil's organic reserves, and the introduction of earthworms able to regulate the metabolization of these reserves, it is likely that the physical and biological properties of the soil will be improved.

References

Barois, I. and Lavelle, P. 1986. Changes in respiration rate and some physicochemical properties of a tropical soil during transit through *Pontoscolex corethrurus* (Glossoscolecidae, Oligochaeta). Soil Biol. Biochem. 18,5: 539–541.

Bouche, M.B. 1972. Lombriciens de France. Ecologie et Systématique. Ann. Zool., Ecol. Anim. No hors série 1972, 671 pp.

Diaz Cosin, D.J. and Moreno, A.G. 1979. Primera cita en la peninsula ibérica de *Microscolex phosphoreus* (Dugés, 1837) (Oligochaeta, Megascolecidae). Bol. R. Soc. Espanola Hist. Nat. 77: 143–150.

Fragoso, C. and Lavelle, P. 1987. The earthworm community of a Mexican tropical rainforest (Chajul, Chiapas).

Guerrero, R.D. 1983. The culture and use of *Perionyx excavatus* as a protein resource in the Philippines. *In* Earthworm Ecology from Darwin to Vermiculture. pp. 303–313. Edited by J.E. Satchell. Chapman and Hall, London.

Kanyonyo, K.K. 1984. Ecologie Alimentaire du Ver de Terre détritivore *Millsonia Lamtoiana* (Acanthodrilae – Oligochètes) dans la savane de Lamto (Côte d'Ivoire). Thèse Doct. 3° cycle. Univ. Paris VI: 131 pp.

Lavelle, P. 1978. Les Vers de Terres de la Savane de Lamto (Côte d'Ivoire); Peuplements Populations et Fonctions dans l'Écosystème. Thèse Doct. Paris VI. Publ. Labo. Zool. E.N.S. 12. 301 pp.

Lavelle, P., Sow, B. and Schaefer, R. 1980. The geophagous earthworm community in the Lamto savanna (Ivory Coast): Niche partitioning and utilization of soil nutritive resources. *In* Soil Biology as related to Land Use Practices. pp. 653–672. Edited by D. Dindal. Epa, Washington D.C.

Lavelle, P. 1981. Stratégies réproductives chez les Vers de Terre. Acta Oecologica; Oecol. Generalis. 2(2): 117–133.

Lavelle, P. 1983. The structure of earthworm communities. *In* Earthworm Ecology from Darwin to Vermiculture. pp. 449–466. Edited by J. Satchell. Chapman and Hall, London.

Lavelle, P., Zaidi, Z. and Schaefer, R. 1983. Interactions between earthworms, soil organic matter and microflora in an African savanna soil. *In* New Trends in Soil Biology. pp. 253–261. Edited by Ph. Lebrun *et al.* Dieu Brichart, Louvain-la-Neuve.

Lavelle, P. 1984. The soil system in the humid tropics. Biology International 9: 12–18.

Mba, C. 1983. Utilization of *Eudrilus eugeniae* for disposal of cassava peel. *In* Earthworm Ecology from Darwin to Vermiculture. pp. 315–321. Edited by J.E. Satchell. Chapman and Hall, London.

Neuhauser, E.F., Kaplan, D.L. and Hartenstein, R. 1979. Life history of the earthworm *Eudrilus eugeniae*. Rev. Ecol. Biol. Sol 16(4): 525–534.

Omodeo, P. 1958. Oligochètes. *In* La Réserve Naturelle Intégrale du Mont-Nimba. Mém. IFAN 53(1): 9–109.

Wasawo, D.P.S. and Omodeo, P. 1963. Some arboricolous Oligochaeta from Ivory Coast. Mem. Mus. Cir. Storia Naturale; Verena, 11: 211–223.

VI. Earthworms as indicators of environmental contamination

CONTENTS

THE PREDICTION OF FIELD TOXICITY OF CHEMICALS TO EARTHWORMS BY LABORATORY METHODS

G.C. GOATS[1] and C.A. EDWARDS[2]
[1]Entomology Department, Rothamsted Experimental Station, Harpenden, Herts, U.K.; [2]The Ohio State University, Columbus, Ohio

Introduction

Earthworms are important in soil formation and in maintaining soil structure and fertility (Darwin, 1881; Edwards and Lofty, 1977). They are common in a wide range of soils, both in temperate countries and the tropics. They have a number of characteristics which identify them as one of the most suitable soil animals to be used as a key bioindicator organism for testing for pollution by soil chemicals. In addition to their key role in soil fertility, they are found in the great majority of soils, they are large in size and easy to handle, they can be readily collected and identified and are known to be affected by and take up into their tissues a number of organic and inorganic chemicals. Earthworms are easily bred in the laboratory for toxicity testing and their longevity makes it likely that few worms would die during the period of a toxicity test in untreated media.

Because of these characteristics and since the earthworm is such a typical and important member of the soil fauna, it has been selected as a key indicator organism for ecotoxicological testing of the toxicity of industrial chemicals not only by the European Economic Community (E.E.C.), but also by the Organization for Economic Co-operation and Development (O.E.C.D.), the Food and Agriculture Organization of the United Nations (F.A.O.), and by many national pesticide registration authorities and environmental pollution committees. We were asked to review and evaluate all of the currently available methods of testing chemical toxicity to earthworms by E.E.C. and O.E.C.D. If the methods were not satisfactory for predictive purposes, we were asked to develop new methods and test them through a minimum of 20 independent laboratories, worldwide, to assess their reliability and variability of results. This was done, and two new methods, the contact filter paper test and the artificial soil test described later were developed (Edwards, 1983 and 1984).

The effects of chemicals on earthworms and methods of investigating these effects have been reviewed by Satchell (1955), Davey (1963) and Edwards and Thompson (1973). Most reports in the scientific literature refer to field testing of the effects of chemicals, particularly pesticides, on earthworm populations. These are too numerous to list, but range from the effects of lead arsenate, copper sulfate, organochlorines, carbamate compounds, phosphorus compounds, and herbicides (Edwards and Thompson, 1973). When the results are reviewed, it is extremely difficult to assess reliably, the relative toxicity of pesticides due to the considerable variability between test sites, soils, formulations, doses and methods of application used. Field tests are adequate to identify chemicals which are extremely toxic to earthworms but to not identify accurately moderately toxic compounds.

A number of workers have tested the effects of pesticides on earthworms kept in soils in boxes or other containers in the laboratory or in pots. Such tests tend to

Earthworms in waste and environmental management
edited by C.A. Edwards and E.F. Neuhauser, pp. 283–294

produce more consistent and reproducible results because standard numbers of test earthworms of a single species are used and the worms are in intimate contact with the chemical. However, a wide range of soil types have been used of greatly differing adsorptive properties for these laboratory tests, and the L.D. 50 for any particular chemical can differ greatly, depending on the test soil.

Some laboratory toxicity testing methods presented obvious disadvantages and shortcomings and were not investigated further. For instance, large and uniform concentrations of a toxicant were difficult to achieve as a deposit on a glass surface (Busvine, 1957). In the methods involving voluntary ingestion of contaminated food material (Fayolle, 1979; Stringer and Wright, 1973), food repellency affected the valid assessment of mortality. Injection of toxicant into the body tissues of worms (Gilman and Vardanis, 1974; Stenersen *et al.*, 1973) caused direct injury and toxicant placement was uncertain.

Several earthworm toxicity testing techniques were investigated experimentally. These methods included tests in natural soil (Edwards and Lofty, 1972b; Hopkins and Kirk, 1957; Ruppel and Laughlin, 1977), exposure in a silica paste-glass ball matrix (Ferriere *et al.*, 1981), immersion in chemical solution (Ghabbour and Imam, 1967; Lebrun *et al.*, 1981; Martin and Wiggans, 1959) and forced feeding tests (Stringer and Wright, 1976; Wright, 1977). Two new test methods, the contact filter paper tests, and the artificial soil test (Edwards, 1981, 1983, 1984; Goats, 1983) were developed and evaluated. These methods were evaluated using *Eisenia foetida andrei* exposed to four toxic chemicals and a standard toxicant. Two field experiments studied the effects of the same compounds on different soil types and investigated the effect upon numbers of *Lumbricus terrestris* and *Allolobophora longa*.

Materials and methods

a. *Laboratory experiments*

Sexually, mature *E. foetida andrei* weighing 0.6 ± 0.1 g were cultured at 20°C in separated cow manure. Worms were extracted from this, by hand, 24 hours prior to use and stored on damp filter paper to void their gut contents. In each experiment, the technical chemicals thiophanate-methyl, chlordane, carbaryl and analytical grade pentachlorophenol (PCP) were used. Water-soluble chemicals were applied in aqueous solution and volatile organic solvents were used to apply the remaining compounds and evaporated from the test system before the worms were introduced. However, with highly volatile toxicants, or in situations preventing evaporation of the organic solvent, suspensions or emulsions were used and produced in the following manner. The chemical was dissolved in a suitable organic solvent and 200 μl of this was dropped into 199.8 ml of rapidly-swirling distilled water containing 200 μl of 10% Tween 80 in ethanolic solution. The suspension or emulsion so formed was generally stable for 48 hours.

All tests were made in a controlled environment cabinet of 20° ± 1°C at 90% relative humidity. Lighting, where used, was provided by 'daylight' fluorescent lamps delivering 600 ± 200 lux. All tests were accompanied by controls treated with distilled water and solvent blank. Any mortality in the controls resulted in the test being rerun. Mechanical mortality was assessed by prodding the anterior segments of the worm firmly and in the absence of a twitch reaction, assuming and recording death.

i. Natural soil test

A pesticide-free, clay loam soil with 2.5% organic matter content was collected from Geescroft Field, Rothamsted, and air-dried, passed through a 1 cm sieve and rewetted to 20% moisture content. The application of the test compound and subsequent test conditions were as described for the artificial soil method.

ii. Silica paste glass ball test

An emulsion, suspension or solution of the chemical toxicant in 215 ml of distilled water was mixed with 90 g amorphous hydrated silica flour (Levilite®) to produce a thick paste. This mixture was blended with 1425 g of 2 cm diameter glass balls that acted as a supporting matric. The media was held in 2.0 liter glass flasks. Each of four replicates at six concentrations contained 10 worms. The flasks, covered loosely, were stored in the light for 14 days and the worms extracted by washing out onto a fine sieve.

iii. Immersion test

An emulsion, suspension or solution of toxicant was produced in 50 ml of distilled water in a 100 ml glass flask. One worm was immersed in each of four replicates at six concentrations. The flasks, covered loosely, were kept for 18 hours in the light.

iv. Forced feeding tests

The toxicants were dissolved, emulsified or suspended in a 1.5% agar-agar gel and introduced into 1 ml Hamilton syringe (mounted in an Arnold Microapplicator) fitted with a rounded-end glass needle. The worms were anaesthetized by a two-minute immersion in 10% aqueous ethanol solution. The needle was inserted into the foregut of the worm, and 5 μl of gel was injected. The worms were revived by washing in distilled water for 10 min. Ten worms were used in each of four replicates at six concentrations. The worms were kept on damp tissue paper in the light and observed for 72 hours after treatment for assessing mortality.

Each method was tested at least twice for every chemical. The data were analyzed by a computed probit transformation using a maximum likelihood line-fitting technique. The differences between the duplicate results were tested using a χ^2 analysis. Where position and parallelism showed no significant difference at $p < 0.05$, both data sets were combined and a single line calculated. Such data is indicated by an asterisk. Replicates that differed significantly from each other are presented as mean LC_{50} and standard error of the mean LC_{50}.

v. Filter paper contact test

Clear soda-glass vials, 8 cm × 3 cm diameter were lined internally with Whatman No 1

Table 1. Field experiment treatments

Compound	Trade name and % active ingredient	Recommended use	Recommended application rate kg.ha^{-1} a.i	Resultant soil concentration mg.kg^{-1} [a]	Field study application rate kg.ha^{-1} a.i.	Resultant soil concentration mg.kg^{-1} [a]	Application solvent
Chlordane	Syndane 25 25%	worm killer	11.25	5.63	5.0 10.0	2.50 5.00	H_2O
Carbaryl	Sevin 85 85%	worm killer	3.83	1.91	2.5 25.0	1.25 12.50	H_2O
Thiophanate-methyl	Cercobin 50%	Cereal fungcide	0.70	0.35	3.0 6.0	1.50 3.00	H_2O
Pentachlorophenol	Witophen P. 100%	Freshwater snail eradication	—	5.00	12.5 75.0	6.25 37.50	Methanol

[a] Based upon the relationship that an application rate of 2 kg ha^{-1} is equivalent to a resultant soil concentration of 1 mg kg^{-1}, to a depth of 15 cm.

filter paper. The chemical toxicant was applied in 1 ml of a suitable organic solvent or water. Organic solvents were evaporated to dryness under a gentle compressed air-stream and the deposit rewetted with 1 ml of distilled H_2O. Ten vials were treated at each of six concentrations, with one worm contained in each vial. The vials were sealed with an airtight cap and placed on their side in darkness for 48 hours before assessing mortality.

vi. Artificial soil test

An artificial soil, consisting of fine quartz sand, kaolinitic clay, sphagnum peat and $CaCO_3$, in the proportions of 71:20:8:1 (dry weights) with a pH of 6.5 and 48% moisture content was used. The toxicant was applied, dissolved in a minimum volume (not less than 10 ml) of organic solvent or water, and sprayed onto the soil using a fine chromatographic atomiser. Organic solvents were allowed to evaporate before the deposit was incorporated by hand mixing. Four replicates, with 750 g of soil (wet weight) were held in 1.0 liter disposable plastic flasks at each of six concentrations. Each replicate contained 10 worms. The flasks, covered loosely to prevent escape of worms, were stored in the light for 14 days, after which the worms were extracted by hand.

b. Field experiments

The Rothamsted field site (1981 Appletrees Field) was a 19-year-old mixed sward pasture on a pesticide-free, clay loam soil with an organic matter content of 5.0%. The plots were 4.27 m^2 with 0.92 m paths and discard areas.

The Sittingbourne field site (Shell Research Centre Sittingbourne) was a 60-year-old sheep-grazed, mixed sward pasture on a pesticide-free, flinty clay loam with an organic matter content of 6.9%. The plots were 4.5 m^2 with 0.5 m paths and discard areas. Both trials were of a randomized block design with each block containing one plot per treatment and an untreated control. The trials were of factorial design with each chemical applied at two rates (Table 1).

The pasture was tractor-rotovated on the day of treatment. The plots were sprayed using a hand-powered backpack sprayer, then rotovated again the same evening to incorporate the deposit to a depth of 15 cm. The application rates used were based upon manufacturer's recommendations for the most appropriate existing use, and an additional higher rate chosen to produce a partial field mortality. The Rothamsted and Sittingbourne trials were treated on 21 April and 9 April respectively and sampled for earthworms using the formalin extraction technique (Raw, 1959) on two 0.5 m^2 subplots per plot, one and six months after treatment. The numbers of *L. terrestris* and *A. longa* were recorded for each subplot.

Worm numbers per subplot were transformed into logarithms and an analysis of variance used to test for differences between the treatments. No pretreatment sampling was done within the plots, and a pretreatment analysis indicated that the populations varied non-randomly within the original block designs. Therefore, new blocks were defined within which the plots were contiguous spatially and an analysis of covariance used to remove any systematic variation contributed by the pretreatment populations. Residual variation was considerably reduced at the Rothamsted site by

Table 2. Field assessments: effect of treatments on earthworm numbers[a]

Site	Species	Treatment sampling interval	Treatment[b] significance	SED	Chlordane 5 kg.ha^{-1}	Chlordane 10 kg.ha^{-1}	Carbaryl 2.5 kg.ha^{-1}	Carbaryl 25 kg.ha^{-1}	Thiophanate-methyl 3 kg.ha^{-1}	Thiophanate-methyl 6 kg.ha^{-1}	Pentachlorophenol 12.5 kg.ha^{-1}	Pentachlorophenol 75 kg.ha^{-1}
Rothamsted	*L. terrestris*	1 month	**	0.153	96	57	85	4	83	92	69	75
		6 months	**	0.130	92	52	80	105	84	69	88	104
	A. longa	1 month	NS	0.127	35	30	20	0	20	40	115	110
		6 months	***	0.140	82	6	92	103	109	107	104	98
Sittingbourne	*L. terrestris*	1 month	***	0.113	99	61	106	0	96	90	109	117
		6 months	NS	0.089	107	119	106	106	115	98	105	106
	A. longa	1 month	NS	0.095	200	0	188	0	0	63	0	125
		6 months	NS	0.110	89	104	108	103	101	103	105	102

[a] Data expressed as percentage of control values.
[b] Treatment significance determined from variance ratio. ** $p \leq 0.01$; *** $p \leq 0.001$.

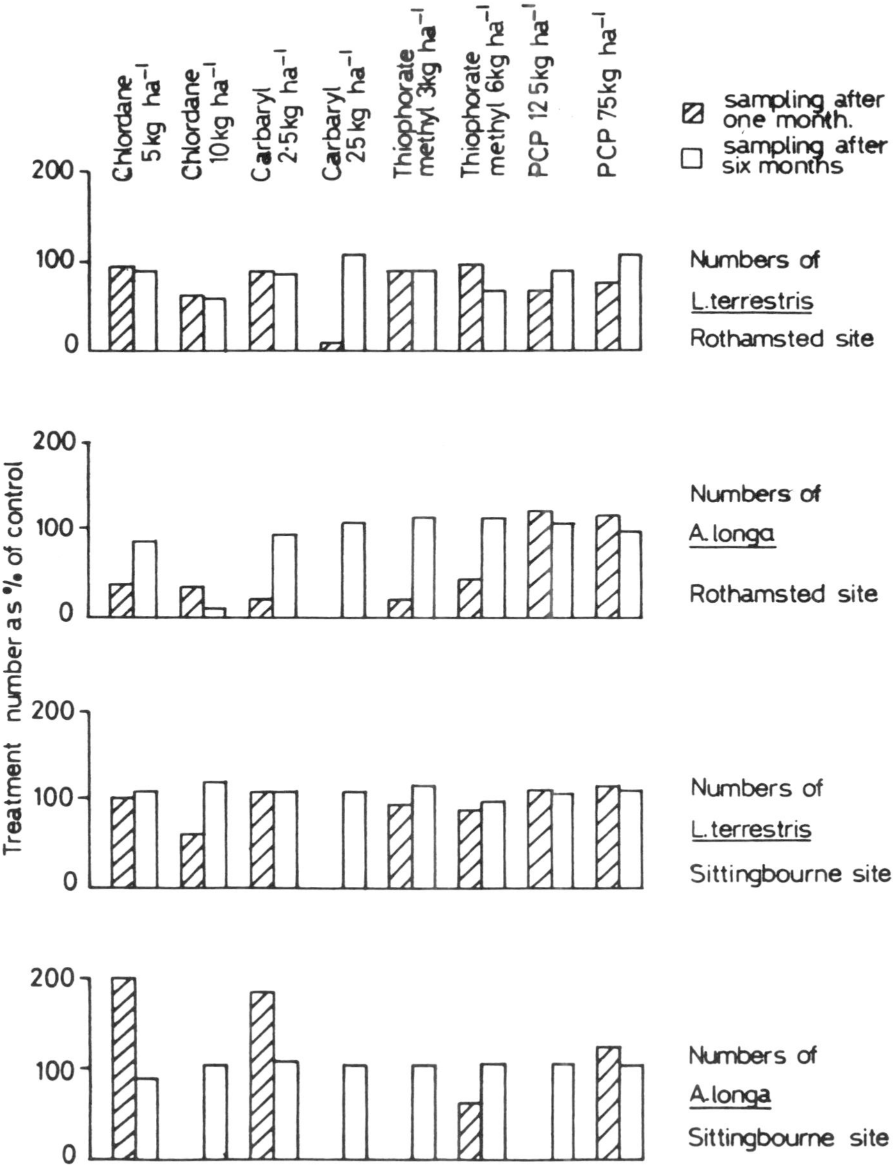

Fig. 1. Results from field experiment assessing the toxicity of chlordane, carbaryl thiophanate methyl and pentachlorophenol to earthworms.

this method. At the Sittingbourne site, a covariance analysis preceeded an analysis of variance only for the six months after treatment sample data, and the covariate used was the mean population numbers from the two subsamples per plot after one month. The data are presented as the numbers of *L. terrestris* and *A. longa* at both

Table 3. LC_{50} toxicity assessments from laboratory tests[a]

Test method	Chlordane	Carbaryl	Thiophanate-methyl	Pentachlorophenol
Filter paper contact gm cm^{-2}	5.63×10^{-7} 0.05650	2.42×10^{-5} 0.08009	5.39×10^{-4} 0.03129	5.28×10^{-6} 0.04358
Artificial soil $mg.kg^{-1}$ (dry weight)	26.960 0.07095	85.740 0.09348	20.060 0.08274	45.53 0.10219
Natural soil $mg.kg^{-1}$ (dry weight)	0.005[b] 0.05302	43.114 0.04373	0.960[b] 0.06086	0.011[b] 0.04950
Silica paste-glass balls $mg.kg^{-1}$ dry weight of silica	23.890 0.02947	112.330 0.04330	44.570 0.09895	266.39[a] 0.02221
Immersion $mg.kg^{-1}$	8.630 0.03978	24.750 0.07587	754.730 0.03753	0.69 0.06691
Forced feeding μg worm	0.069 0.06322	—	1.890 0.011844	—

[a] Data shown at LC_{50} (above), standard error of Log LC_{50} (below).
[b] Replicates combined.

sites, expressed as a percentage of the untreated plot values, calculated from the arithmetic means of the logarithmically transformed data for 0.25 m^2 samples.

Results

i. *Laboratory test results*

The laboratory test probit analysis results (Table 3) and the relative toxicities of the chemicals tested summarize the effects of the chemicals (Table 4). Chlordane was most toxic and carbaryl somewhat less toxic in all the tests, whereas thiophanate-methyl and PCP were much less toxic and could not be ranked. With the toxicant LC_{50} expressed as a ratio of a toxic standard LC_{50} (chlordane) the differences between the ranked toxicities could be quantified (Fig. 2). Carbaryl was considerably less toxic than chlordane in the contact filter paper and natural soil tests. Thiophanate-methyl was less toxic than chlordane in contact filter paper, immersion, natural soil and forced feeding tests. PCP, carbaryl and thiophanate-methyl had similar toxicities in the artificial soil and silica paste-glass ball tests, although PCP appeared to be slightly less toxic in the latter.

ii. *Field experiment results*

The field experiment results (Table 2) (Fig. 1) showed that the numbers of *L. terrestris* were decreased most by chlordane at 10 kg ha^{-1} after one month and by carbaryl at 25 kg ha^{-1} on both sites. A less marked decrease in numbers occur with PCP at 12.5 kg ha^{-1} at the Rothamsted site. After six months, the numbers of *L. terrestris* were decreased by thiophanate-methyl at 6 kg ha^{-1}. Populations of

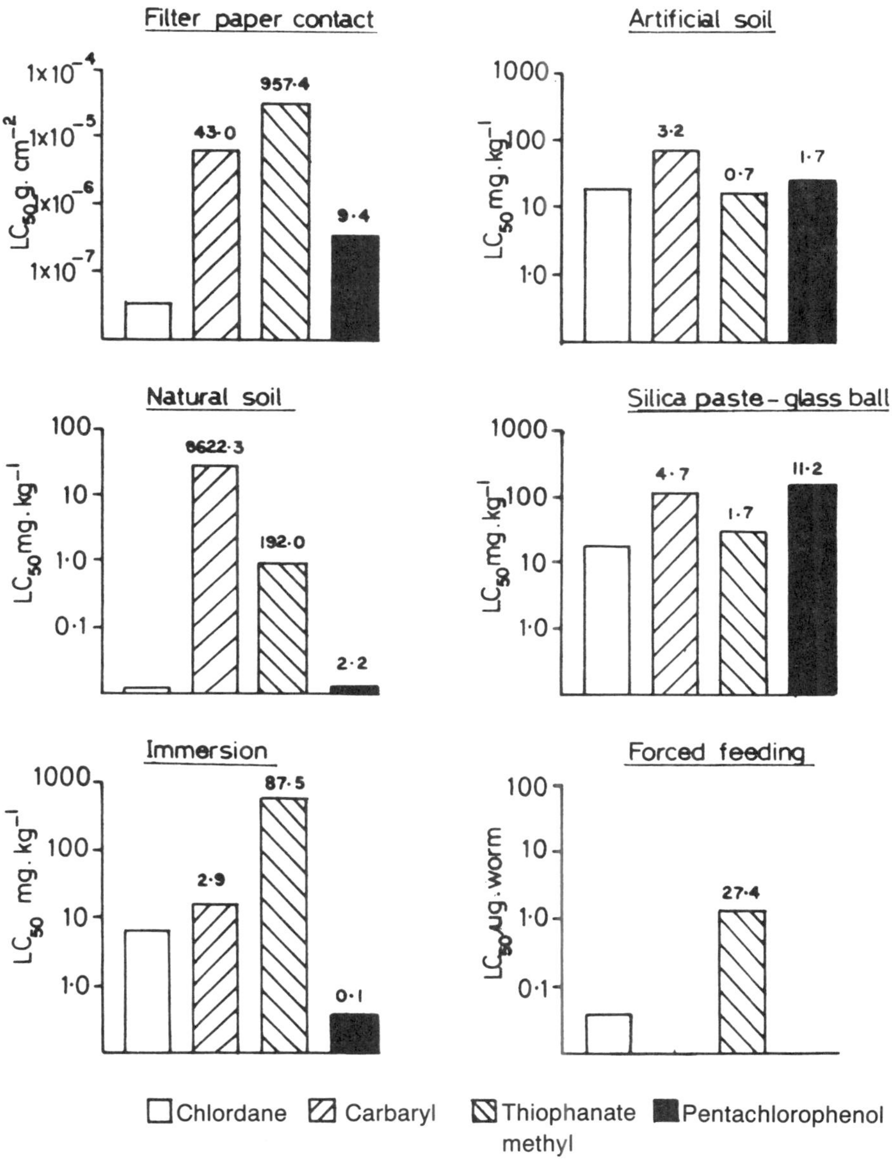

Fig. 2. Toxicity of chlordane (□), carbaryl (▧), thiophanate methyl (▨) and pentachlorophenol (■) to earthworms in laboratory tests.

A. longa after one month were decreased at the Rothamsted site by both rates of thiophanate-methyl, chlordane and carbaryl. However, at the Sittingbourne site, thiophanate-methyl, chlordane, PCP and carbaryl completely eliminated the populations at 3, 10, 12.5 and 25 kg ha^{-1} on the Sittingbourne site, and chlordane at 10 kg ha^{-1} on the Rothamsted site decreased numbers.

Table 4. Laboratory test toxicity ranking

Test method	Toxicity ranking
Filter paper contact	chlordane > PCP > carbaryl > thiophanate-methyl
Artificial soil	Thiophanate-methyl > chlordane > PCP > carbaryl
Natural soil	chlordane > PCP > thiophanate-methyl > carbaryl
Silica paste-glass ball	chlordane > thiophanate-methyl > carbaryl > PCP
Immersion	PCP > chlordane > carbaryl > thiophanate-methyl
Forced feeding	chlordane > thiophanate-methyl

Discussion

The results of field trials indicated that chlordane was toxic to both *L. terrestris* and *A. longa* for as long as six months on the low organic matter soil (Rothamsted), whereas its effect was limited to one month on the higher organic matter soil (Sittingbourne). Carbaryl virtually eliminated both species, on both soil types, after one month, but its effect had disappeared after six months and populations had recovered. Only *A. longa* was affected by thiophanate-methyl such that population decreased after one month on both soil types, although no effect was recorded after six months. PCP caused a limited decrease in numbers of *L. terrestris* after one month on the low organic matter site. Effects of PCP on *A. longa* were still significant after six months on the Rothamsted site.

Thus, *A. longa* seemed more susceptible to the chemical toxicants than *L. terrestris*. In general, chlordane presented a general hazard, and PCP a slight one that was less in a soil with more organic matter. Carbaryl was a severe non-persistent hazard, and thiophanate-methyl a moderate non-persistent hazard, that was independent of soil type. The laboratory methods gave quite different assessments of toxicity depending upon the exposure route, duration and the interaction of test chemicals with the media. The results, which are expressed in different concentration units, depending upon the method used, further complicates a direct comparison between them. Carbaryl and thiophanate-methyl were considerably less toxic than chlordane in the contact filter paper, natural soil, immersion and forced feeding methods, although all three compounds were a hazard in the field. The results of the artificial soil and silica paste-glass ball methods resembled those from the field toxicity experiments most. The results obtained with these two methods were highly correlated (see Heimbach, 1983).

Toxicity testing methods intended for widespread use in international environmental protection programs need to be sensitive and reproducible in predicting field hazards but also to be cheap, simple and rapid. Natural soils, in which the best test species, *E. foetida* survived poorly, are also difficult to standardize. The stable and persistent suspension of non-water soluble materials required in immersion testing involves elaborate formulation techniques. Forced feeding methods are technically demanding and slow.

The results of this work indicates that although the filter paper test was suitable for preliminary screening of chemical toxicity to earthworms, the artificial soil and silica paste-glass ball tests were the most appropriate methods available for laboratory estimation of the field hazards to earthworms and the former method has now been adopted by the E.E.C., O.E.C.D. and the U.K. Pesticide Safety Precautions Scheme. The predictive capability of such laboratory methods could be enhanced by better knowledge of the toxicant-soil component interaction, and if they are extended to also estimate of sublethal toxic effects, such as those on reproduction and feeding.

Acknowledgements

The authors wish to thank Shell (U.K.) Ltd. for supporting this work financially and are grateful to Miss. S.J. Clark and Mr. J.N. Perry of the Statistics Department, Rothamsted Experimental Station, for valuable statistical assistance.

References

Anonymous. 1983. U.K. Pesticides Safety Precautions Scheme. Working Document D6. Ministry of Agriculture, Fisheries and Food, London.

Busvine, J.R. 1957. A critical review of the techniques for testing insecticides Commonwealth Institute of Entomology. London: 132.

Darwin, C. 1881. The Formation of Vegetable Mould through the Action of Worms. Murray, London. 326 pp.

Davey, S.P. 1963. Effects of Chemicals on Earthworms. A Review of the Literature Special Scientific Report. Wildlife No. 74, Washington.

Doane, C.D. 1962. Effects of certain insecticides on earthworms. J. Econ. Entomol. 55(3): 416–418.

Edwards, C.A. 1983. Development of a standardized laboratory method for assessing the toxicity of chemical sustances to earthworms. Report of Commission of the European Communities. EUR 8714 EN, 141 pp.

Edwards, C.A. 1984. Report of the Second Stage in Development of a Standardized Laboratory Method for Assessing the Toxicity of Chemical Substances to Earthworms. Report of Commission of the European Communities. EUR 9360 EN, 99 pp.

Edwards, C.A. and Lofty, J.R. 1972. Pesticides and Earthworms. Report of Rothamsted Experimental Station for 1972. Part 1: 212.

Edwards, C.A. and Thompson, A.R. 1973. Pesticides and the soil fauna. Res. Rev. 45: 1–79.

Edwards C.A. and Lofty, J.R. 1977. Biology of Earthworms. 2nd Edition. Chapman and Hall. 333 pp.

EEC. 1979. Amendment 6 of Council Directive 67/548/EEC. Official Journal of the European Communities No L 259 15.10.79: 10–28.

Fayolle, L. 1979. Consequences of the impact of pollutants on earthworms. III. Laboratory Tests. Documents Pedozoologiques. INRA Lab. Zooecol. Sol 1(1): 34–65.

Ferrière, G., Fayolle, L. and Bouche, M.B. 1981. Un nouvel outil. essentiel pour l'écophysiologie et l'écotoxicologie l'élevage des lombriciens en sol artificial. Pedobiologia 22: 196–201.

Ghabbour, S.I. and Imam, M. 1967. The effect of five herbicides on three Oligochaete species. Rev. d'Ecol. Biol. Sol 4(1): 119–122.

Gilman, A.P. and Vardanis, A. 1974. Carbofuran. Comparative toxicity and metabolism in the worms *Lumbricus terrestris* L. and *Eisenia foetida* S. J. Agric. Food Chem. 22(4): 625–628.

Goats, G.C. 1983. A comparison of field and laboratory methods for testing toxicity to earthworms. Proceedings of the 10th International Congress of Plant Protection. Brighton. 1983. Volume 2: 713.

Goats, G.C. and Edwards, C.A. 1981. Testing the toxicity of industrial chemicals to earthworms. Report of Rothamsted Experimental Station for 1981: Part I: 106.

Goats, G.C. and Edwards, C.A. 1982. Testing the toxicity of industrial chemicals to earthworms. Report of Rothamsted Experimenta Station for 1982: Part I: 104–105.

Heimbach, F. 1983. Private communication.

Hopkins, A.R. and Kirk, V.M. 1957. Effect of several insecticides on the English Red Worm. J. Econ. Entomol. 50(5): 699–700.

Lebrun, P., De Medts, A. and Wauthy, G. 1981. Ecotoxicologie comparée et bioactivité de trois insecticides carbamates sur une population expérimentale de vers de terre, *Lumbricus hurculeus*. Pedobiologia 21: 225–235.

Legg, D.C. 1968. Comparison of various worm-killing chemicals. J. Sports Turf Res. Inst. 44: 47–48.

Lofs-Holmin, A. 1980. Measuring growth of earthworms as a method of testing sublethal toxicity of pesticides. Swed. J. Agric. Res. 10: 25–33.

Martin, L.W. and Wiggans, S.C. 1959. The tolerance of earthworms to certain insecticides, herbicides and fertilizers. Oklahoma State University Experiment Station. Processed Series P-334.

Raw, F. 1959. Estimating earthworm populations by using formalin. Nature (London) 184: 1661.

Ruppel, R.F. and Laughlin, C.W. 1977. Toxicity of some soil pesticides to earthworms. J. Kansas Entomol. Soc. 50(1): 113–118.

Stenersen, J., Gilman, A. and Vardanis, A. 1973. Carbofuran: its toxicity to and metabolism by earthworms, (*Lumbricus terrestris*). J. Agric. Food Chem. 21(2): 166–171.

Stringer, A. and Wright, M.A. 1973. The effect of benomyl and some related compounds on *Lumbricus terrestris* and other earthworms. Pestic. Sci. 4: 165–170.

Stringer, A. and Wright, M.A. 1976. The toxicity of benomyl and some 2-substituted benzimidazoles to the earthworm *Lumbricus terrestris*. Pestic. Sci. 7: 459–464.

Thompson, A.R. 1971. Effects of nine insecticides on the numbers and biomass of earthworms in pasture. Bull. Env. Contam. Toxicol. 5(6): 577–586.

EARTHWORMS AS ECOTOXICOLOGICAL ASSESSMENT TOOLS

CLARENCE A. CALLAHAN
United States Environmental Protection Agency, Corvallis, Oregon 97333, U.S.A.

Introduction

The concerns for the impact of toxic and hazardous materials on the soil environment have continued to increase in recent years and, correspondingly, there have been accelerated efforts in developing methods for evaluation of these effects. Most of these methods have involved assessment of short-term effects primarily of human health (Goyer, 1983). There is, however, an equally important need for assessing long-term toxic impacts on the terrestrial environment, in which humans are considered as one of many components.

This paper reviews (i) the relevant literature concerning earthworms as ecological tools for assessment of the impact of toxic and hazardous wastes; (ii) methods used in past research for measurement of the impact on earthworms at the individual, population and ecological levels; and (iii) present efforts by the Corvallis Environmental Research Laboratory to use earthworms as assessment organisms to measure impact on soil systems. This discussion is based on the premise that to assess the impact of chemicals one must determine the exposure of the chemical toxicant to the organism of interest and then determine the effect (Moriarty, 1983a).

Definitions

A toxic substance or pollutant is defined as a substance introduced to the environment by man that has a deleterious impact on biological organisms (Moriarty, 1983a). Ecotoxicology is the branch of toxicology which studies the ecological effects of toxic substances or pollutants (Truhant, 1975). Ecotoxicological research is concerned specifically with biological problems at the level of populations and communities as they interact with the physical environment. However, the effects of pollutants and toxic materials are often mediated through their impact on individuals and the severity depends upon the dose (Moriarty, 1983a).

Exposure and accumulation

Earthworms are essentially cosmopolitan, given sufficient moisture and adequate temperatures and are vulnerable to most impacts on soil ecosystems, especially those associated with the application of agricultural chemicals (Edwards and Thompson, 1973; Wheatley and Hardman, 1968; Thompson, 1970; Randall *et al.*, 1972). There have been many efforts to assess the impact of toxicants on earthworms with varying success. These studies have shown that earthworms respond to many chemicals and

Earthworms in waste and environmental management
edited by C.A. Edwards and E.F. Neuhauser, pp. 295–301

other factors in many ways, such as increase in body burdens, increased mortality and overall decrease in activities normally associated with viable earthworm populations (Edwards and Thompson, 1973).

Decrease in numbers, as well as effects on their life activities such as feeding, *e.g.*, processing vegetation (Barker, 1958) has been the subject of research in assessment of chemical impacts to earthworms (Edwards and Thompson, 1973). Although there are reports of field evaluations of impacts of chemicals on earthworms these are fewer than laboratory studies. Methods used in field studies appear to be more varied and represent considerable questions as to the value of the results for assessment in an ecological sense (Lofs-Holmin and Boström, this volume, pp. 303 ff). There are problems in assessing density, total numbers, application rates of toxicants, concentrations of toxicants in the field, as well as interpretation of the meaning of the results for many of these studies. Summarization of these research reports is difficult because of the problem of trying to collate the diverse methods since many objectives were used for the studies resulting in difficulties in interpretation of the data.

Effects on individual earthworms

Toxic impacts of chemicals on individual earthworms affects the population in terms of reductions in numbers as well as in removal of a portion of the gene pool. Assessments of exposure to heavy metals is primarily through measurements of body burden (Ireland and Richards, 1981; Carter *et al.*, 1981; Edwards and Lofty, 1977; Gish, 1970).

Growth and reproduction of earthworms can be influenced in the laboratory by heavy metals (Malecki *et al.*, 1982) and these characteristics can also be affected under field conditions. Changes in earthworm densities noted in past studies (Raw, 1962) could be investigated further to determine the dose-response mechanism or other important factors related to changes in numbers. Recent research of the dynamics of contaminants emphasizes the complex nature of the interrelationships encountered in studying effects of heavy metals on earthworms (Ma 1982, 1984; Ma *et al.*, 1983). Cobalt has been shown to increase the maximum weight of *E. fetida* but also causes a lag in its early growth (Neuhauser *et al.*, 1984a). Neuhauser *et al.*, (1984b) exposed *E. fetida* to several heavy metal salts to show that this species is capable of compensatory growth, whereas it did not compensate for loss in reproduction. However, *E. fetida*, was shown to be relatively resilient resuming normal growth and reproduction after the worms were removed from contaminated soil.

Effects of pesticides and herbicides at the individual level have been assessed as qualitative decreases in numbers (Ruppel and Laughlin, 1977). Although some of these studies considered changes in density and species composition, none examined rigorously any ecological variables such as community structure and function (Farnworth and Golley, 1979). There is little information relating soil concentration of chemicals to sublethal effects such as reduced reproduction (Lebrun *et al.*, 1981; DeMedts, 1981), predator-prey relationships, fecundity, changes in egg fertility, growth and differential mortality with age or changes in behavior. There needs to be more evaluation of dose-response relationships under field conditions (Moriarty, 1983a) incorporating a study of those factors important to changes in density, growth, reproduction, and mortality.

Effects on earthworm populations

Ecologically, the impact of toxic materials is at the population level (Moriarty, 1983b). Although the acute impacts of chemicals remove certain individuals from the breeding population, change in numbers (*i.e.*, population size) is the most obvious impact that can be assessed. Fluctuations in numbers are influenced by birth rates, death rates, gains from immigration and losses from emigration. Measurement of these four factors can be used to determine the changes in populations (Moriarty, 1983a). There are few if any field studies with good quantitative data on changes in birth rates and death rates (Lebrun *et al.*, 1981; De Medts, 1981). Studies on population changes due to environmental factors are more plentiful and include measurement of impact of pH on individual growth (Satchell, 1967) and death (Southwell and Majer, 1982), temperature (Standen *et al.*, 1982) and immigration due to low oxygen and high carbon dioxide concentrations. (Schriefer, 1981).

In summary, past research has shown that earthworms accumulate heavy metals and pesticides as well as other organic chemicals, thereby affecting their growth, reproduction and mortality at the individual level. Although impacts of chemicals at the population level are more difficult to document, decreases in numbers, population density, birth rate, and immigration rate occur under field conditions. There has been very little research concerning ecological impact of toxicants especially measuring community structure impact, predator-prey relationships or parasite-host effects. The above discussion serves as a summary of the major problems encountered in efforts to move from the single species, individual organism evaluation through the population level to a more complex multispecies ecological level of earthworm assessments.

Earthworms as an assessment tool for toxic and hazardous wastes

Presently the Hazardous Materials Assessment Team (HMAT) is evaluating earthworm bioassays (48 Hour Contact Test, the 14 Day Soil Test and neurophysiological impact) to assess the toxicity of chemicals to soil systems (Porcella, 1983).

In the 48 Hour Contact Test (CT) (Edwards and Goats, 1982; Edwards, 1983) the chemical being evaluated is dispersed in water or an organic solvent depending upon its solubility characteristics and pipetted into filter paper lined glass vials. A residue of the chemical remains on the filter paper when the solvent is evaporated. Water is used to moisten the paper before placing one worm into the vial which is then capped. The vials are stored on their side in the dark for 48 hours afterwhich they are assessed for mortality. The concentration is expressed as $\mu g/cm^2$ of the chemical on the filter paper lining the vial or as $\mu g/ml$.

The 14 Day Soil Test (ST) (Edwards and Goats, 1982) involves mixing the chemical of interest in a standard mix of sand, clay and peat. Moisture is added to 35% by weight and usually 10 earthworms are placed into 400 grams dry weight of the mixture. Initial and final weights after 14 days allows one to monitor changes in weight as well as mortality with concentration. Chemical concentration is expressed as $\mu g/g$ dry weight of soil.

The neurological assay is an assessment of giant axon conduction velocity (GACV) as related to concentration of chemical and is made with either of the exposure methods described above (Drewes *et al.*, 1983). GACV measurements are made before, during, and after the exposures to monitor the changes as affected by the toxicant.

So far our experience with samples from various industrial waste sites has shown that: the contact test, as expected, is effective in detecting toxicity when the toxicant is water soluble; and the soil test is effective in indicating toxicity of various toxicants (both water-soluble and non-water soluble), including organic chemical pesticides and a mixture of metals, solvents and polynuclear aromatic hydrocarbons. For example, an industrial sample where there was no mortality observed for the contact test, the LC_{50} for the soil test was less than 2% of site soil (w/w). This sample contained high concentrations of dieldrin, endrin, and aldrin. These results are indicative of the difference between the contact test and the soil test. Also, preliminary NTR measurements of this sample leachate (water soluble portion) show significant differences compared to distilled water after exposure of only 15 minutes. Nerve transmission rate measurements have been shown to be very effective in picking up toxicities at lower concentrations and shorter exposures (Drewes and Vining, 1984). For instance the dieldrin LC_{50} by contact test measurements was at 8100 $\mu g/g$ (140.9 $\mu g/cm^2$) whereas there were significant GACV differences at 1.4 $\mu g/g$ (0.0122 $\mu g/cm^2$) within two hours of exposure.

Bioassay results can be used to interpret the impact of toxicants in two ways: toxicity evaluation and ecological impact. The earthworm bioassays described above are sufficient to determine the relative toxicity of chemicals to earthworms (Drewes *et al.*, 1984; Roberts and Dorough, 1984) but as yet have limited ecological significance. The present interpretation of contact test results is limited by the application of the test. For example, in the contact test interpretation of toxicity, results are limited to the relative toxicity of reported standards or data established for that chemical (Edwards and Goats, 1982). That is, the tests are performed in the laboratory and results are compared to known standards also derived from laboratory results. Thus, the reliance on laboratory results is the chief limitation toward field application.

The soil test is limited at this time because the tests results have no ecological correlates as they are presently applied. The endpoints measured in the laboratory have not been correlated conclusively with ecological significance in the field. Of course, the tests could theoretically be extended to a field interpretation, *e.g.*, mortality, reduced growth and reproduction, and changes in physiological functioning with well-designed field experiments. This limitation induces us to better define the use of the endpoints and to recognize the limited relationship between the laboratory results and the field evaluations that may contain ecological relevance.

As an alternative to bioassay assessment, chemical analysis is inadequate when practiced alone and therefore unacceptable under most conditions (Lysyj, 1981). For instance, when unknown chemicals are encountered during an assessment activity, the most widely used method now practiced is to make chemical analyses for identification and concentration levels of EPA priority pollutants. Our experience (and those of many others apparently) with field samples is that frequently we encounter chemicals in mixtures with little known history and no known toxicity levels. Reality is that upon chemical analysis there are many unidentified chemicals in solution for which we have no known toxicological knowledge (Unterberg *et al.*, 1980). There is a distinct advantage in performing bioassay tests under these circumstances because at a minimum one has an estimate of the bioavailability of the complex mixture even when it can't be characterized chemically. The bioassay information can be used to arrive at an acute and hopefully sublethal toxicity measurement which in turn can be used for guidance in solving the waste problem (Schweitzer, 1981).

For assessment of sites with limited exposure, where the toxicants are confined or are not widespread the need is clearly one of toxicity testing. The most immediate need is to determine the actual danger of the chemicals in terms of acute toxicity. The contact test and the soil test (and possibly the GACVT under evaluation) may be used without hesitation in the first example to determine relative toxicity. Test results can be interpreted with respect to known standards and to other studies that have examined similar chemicals and situations. The evaluation of acute and relative toxicity can be made without extrapolations or extentions of the data set beyond the ranges of experience.

When a situation is encountered where the exposure is widespread or uncontrollable, the need is not only one of toxicity determination, but also one of potential ecological impact (Reinecke and Venter, 1984). This situation is clearly more difficult to assess because of the need for more comprehensive information. The above acute toxicity tests appear to be inadequate for ecological assessment as currently used. Evaluations of reduced growth and reproduction (Lofs-Holmin, 1980, 1982), as well as impacts on neurophysiology (Drewes *et al.*, 1984), show promise in assessment of longer term and more complex impacts.

In an attempt to overcome the problems of limited interpretation we are redesigning our experiments for measuring effects in the laboratory that will have more relevance in the field. For instance, sublethal measurements such as growth, reproduction, and neurophysiological changes are planned for the laboratory and field at the same level of exposure. In this way the concentration (*i.e.*, exposure) of the chemical in the organism (*i.e.*, body burden) is the connection to the effect that we would expect to see in the field. If environmentally observed concentrations can be duplicated in the laboratory in terms of body burdens we can make more credible predictions of changes in growth and reproduction and ecological impact. We are using evaluation procedures that should improve our ability to determine the measurable effects in the field. To do this, we are bringing field samples to the laboratory for bioassay evaluation and chemical analysis. These laboratory procedures do not duplicate the field experiments, but do provide data that can be interpretated in terms of risk. These procedures represent a compromise in overcoming the tremendous problems of performing a full field experiment that may be cost prohibitive and can be inconclusive when completed.

The contact test and the soil test appear to be adequate for toxicity assessment of pollutants in hazardous wastes, whereas they have limited interpretation for ecological impact. However, modifications of the ST and well-defined field tests should enable an ecological assessment for sublethal effects of growth, reproduction and neurophysiological functioning. Cooperative research with Cornell and Iowa State Universities is underway to investigate the field concentrations that impact growth, reproduction and density of natural populations.

References cited

Barker, R.J. 1958. Notes on some ecological effects of DDT sprayed on elms. J. Wild. Managem. 22: 269–274.

Carter, A., Hayes, E.A. and Laukulish, L.M. 1981. Earthworms as biological monitors of changes in heavy metal levels in an agricultural soil in British Columbia. *In* Soil Biology as Related to Land Use Practices of the VII International Colloquium of Soil Zoology. pp. 13–24. Edited by D. Dindal. U.S. Environ. Prot. Agcy. Wash., D.C. Office of Pesticides and Toxic Substances. EPA-560/13-80-038 (PB 81-155251).

De Medts, A. 1981. Effects de residus de pesticides sur les Lombricieus en terre de culture. Pediobiologia 21: 439–445.

Drewes, C.D., Landa, K.B. and McFaull, J.L. 1983. Giant nerve fiber activity in intact, freely moving earthworms. J. Exp. Biol. 72: 217–227.

Drewes, C.D. and Vining, E.P. 1984. *In vivo* neurotoxic effects of dieldrin on giant nerve fibers and escape reflex functions in the earthworm, *Eisenia foetida*. Pesticide Biochem. and Physic. 22: 1125–1135.

Drewes, C.D., Vining, E.P. and Callahan, C.A. 1984. Electrophysiological detection of sublethal neurotoxic effects in intact earthworms. This volume, pp. 355 ff.

Edwards, C.A. 1983. Development of a standardized laboratory method for assessing the toxicity of chemical substances to earthworms. Report of the Commission of the European Communities EUR 871A-EN. 141 pp.

Edwards, C.A. and Thompson, A.R. 1973. Pesticides and the soil fauna. Res. Rev. 45: 1–81.

Edwards, C.A. and Goats, G. 1982. Testing the toxicity of industrial chemicals to earthworms. Rothamsted Report for 1982. pp. 104–105.

Edwards, C.A. and Lofty, J.R. 1977. Biology of Earthworms. Chapman and Hall, London. 283 pp.

Farnworth, E.G. and Golley, F.B. (Ed.). 1973. Fragile Ecosystems. Springer-Verlag, New York. 258 pp.

Gish, C.D. 1970. Pesticides in soil. Pestic. Mon. J. 3(4): 241–253.

Goyer, R.A. 1983. Introduction and Overview. *In* Research Needs for Evaluation of Health Effects of Toxic Chemical Waste Dumps. Environm. Health Persp. Vol. 48, 144 pp.

Ireland, M.P. and Richards, K.S. 1981. Metal content, after exposure to cadmium, of two species of earthworms of known differing calcium metabolic activity. Environ. Poll. (A) 26: 69–78.

Lebrun, Ph., Medts, A. and Wauthy, G. 1981. Ecotoxicologie comparée et bioactivité de trois insecticides carbamates sur une population expérimentale de vers de terre, *Lumbricus herculeus*. Pediobiologia 22: 225–235.

Lofs-Holmin, A. 1980. Measuring growth of earthworms as a method of testing sublethal toxicity of pesticides. Swedish J. Agric. Res. 10: 25–33.

Lofs-Holmin, A. 1982. Measuring cocoon production of the earthworm *Allolobophora caliginosa* (Sav.) as a method of testing sublethal toxicity of pesticides. Swedish J. Agric. Res. 12: 117–119.

Lysyj, I. 1981. Novel approaches to screening of aqueous effluenta from hazardous waste sites. *In* Management of Uncontrolled Hazardous Waste Sites. pp. 114–121. Sponsored by U.S. E.P.A. and American Society of Chemical Engineering. Hazardous Materials Control Res. Inst. Silver Springs, MD.

Ma, W. 1982. The influence of soil properties and worm-related factors on the concentration of heavy metals in earthworms. Pediobiologia 24: 109–119.

Ma, W. 1984. Sublethal toxic effects of copper on growth, reproduction and litter breakdown activity in the earthworm *Lumbricus rubellus*, with observations on the influence of temperature and soil pH. Environm. Poll. (Series A) 33: 207–219.

Ma, W., Edleman, Th., van Beersum, I. and Jans, Th. 1983. Uptake of cadmium, zinc, lead, and copper by earthworms near a zinc-smelting complex: Influence of soil pH and organic matter. Bull. Environ. Contam. Toxicol. 30: 424–427.

Malecki, M.R., Neuhauser, E.F. and Loehr, R.C. 1982. The effects of metals on the growth and reproduction of *Eisenia foetida* (Oligochaeta, Lumbricidae). Pediobiologia 24: 129–137.

Moriarty, F. 1983a. Ecotoxicology. Academic Press, New York, N.Y.

Moriarty, F. 1983b. Effects of toxicants on terrestrial animals. *In*: Environmental Toxicology. pp. 185–204. Edited by J.H. Duffus and J.I. Waddington. Interin Document 13, World Health Organization, Copenhagen.

Neuhauser, E.F., Meyer, J., Malecki, M.R. and Thomas, J.M. 1984a. Impact of dietary cobalt supplements on growth and reproduction in the earthworm *Eisenia foetida*. Soil. Biol. Biochem: 16: 521–523.

Neuhauser, E.F., Malecki, M.R. and Loehr, R.C. 1984b. Growth and reproduction of the earthworm

Eisenia foetida after exposure to sublethal concentrations of metals. Pediobiologia: 27: 89–97.

Porcella, D.B. 1983. Protocol for bioassessment of hazardous waste sites. EPA-600/2-83-054. USEPA. Corvallis, Oregon, U.S.A.

Randell, R., Butler, J.D. and Hughes, T.D. 1972. The effects of pesticides on thatch accumulation and earthworm populations in Kentucky bluegrass turf. Hort. Sci. 7(1): 64–65.

Raw,F. 1962. Studies of earthworm populations in orchards. I: leaf burial in apple orchards. Ann. Appl. Biol. 50: 389–404.

Roberts, B.L. and Dorough, H.W. 1984. Hazards of chemicals to earthworms. Env. Toxicol. Chem. 3: 67–78.

Ruppel, R.F. and Laughlin, C.W. 1977. Toxicity of some soil pesticides to earthworms. Jour. Kan. Entom. Soc. 50(1): 113–118.

Satchell, J.E. 1967. Lumbricidae. *In*: Soil Biology. pp. 259–322. Edited by A. Burgess and F. Raw. Academic Press, N.Y.

Schriefer, T. 1981. Regenwurmer (Lumbricidae) auf unterschiedlich abgedeckten Mulldeponien Bestandsaufnahme und Besiedlungsmechanismen. Pedobiologia 22: 153–166.

Schweitzer, G.E. 1981. Risk assessment near uncontrolled hazardous waste sites: role of monitoring data. *In* Management of Uncontrolled Hazardous Waste Sites. pp. 238–247. Sponsored by U.S. E.P.A. and Amer. Soc. of Chem. Eng. Hazardous Materials Control Res. Inst. Silver Springs, MD.

Southwell, L.T. and Majer, J.D. 1982. The survival and growth of the earthworm *Eisenia foetida* (Lumbricidae: Oligochaeta) in alkaline residues associated with the bauxite refining process. Pediobiologia 23: 42–45.

Standen, V., Stead, G.B. and Dunning, A. 1982. Lumbricid populations in open cast reclamation sites and colliery heaps in County Durham, U.K. Pedobiologia 24: 57–64.

Thompson, A.R. 1970. Effects of nine insecticides on the numbers and biomass of earthworms in pasture. Bull. Environ. Contem. Toxicol. 5: 577–586.

Truhaut, R. 1975. Ecotoxicology, a new branch of toxicology: a general asurvey of its aims, methods, and prospects. NATO Science Committee Conference. Mont-Gabriel, Quebec, Canada, 1974. *In* Environmental Sciences Research, Vol. 7: Ecotoxicology Research. pp. 3–23. Edited by A.D. McIntyre and C.F. Mills. Plenum Press, New York.

Unterberg, W., Stone, W.L.and Tafuri, A.N. 1981. Rationale for determining priorities and extent of cleanup of uncontrolled hazardous waste sites. *In* Management of Uncontrolled Hazardous Waste Sites. pp. 188–197. Sponsored by U.S. E.P.A. and Amer. Soc. of Chem.Eng. Hazardous Materials Control Res. Inst. Silver Springs, MD.

Venter, J.M. and Reinecke, A.J. 1988. Sublethal ecotoxicological effects of dieldrin on the earthworm *Eisena foetida* (Oligochaeta). This volume, pp. 337ff.

Wheatley, G.A. and Hardman, J.A. 1968. Organochlorine insecticide residues in earthworms from arable soils. J. Sci. Food Agric. 19(4): 221–225.

THE USE OF EARTHWORMS AND OTHER SOIL ANIMALS IN PESTICIDE TESTING

ASTRID LOFS-HOLMIN and ULLALENA BOSTRÖM
Swedish University of Agricultural Sciences, Uppsala, Sweden

Introduction

In recent years, many literature reviews dealing with the effect of pesticides on the soil fauna have been published. They have all reported the results while very little attention has been paid to the methods used. In forming an opinion on the results, a certain amount of information on methods must be available. The conclusions drawn from the results might otherwise become misleading or even incorrect. There is considerable confusion about experimental designs, which parameters to investigate, which measurements to use and which data are essential for the presentation to enable the experiments to be repeated or at least to enable comparison with similar ones.

This paper provides a brief survey of how experiments have been performed with regard to testing the side-effects of pesticides on soil fauna and on earthworms in particular. We hope that the report will encourage more standardized and adequate testing experiments in the future. As far as possible, only original papers have been used. Summaries or abstracts have been used only if they were detailed. The original languages have been limited to Scandinavian, English, French, German and Dutch. The papers were collected during a limited period of time and thus the review makes no claims to be a complete report of all papers published on the subject. Most references are older than 1981. They have been found through Biosis, Isi, Agris and Agricola or reference lists in the papers.

A total of 168 papers were examined. They include 99 laboratory experiments, 60 with earthworms and 39 with other soil animals, and 121 field experiments, 63 with earthworms and 58 with other soil animals. The complete reference list is given in Boström and Lofs-Holmin (1982).

Summary of experimental designs

i. *Controls*

Seven percent of the field experiments and 14% of the laboratory experiments lacked controls (Fig. 1). In some cases it was impossible to decide whether controls had been used or not.

The group with 'no controls' includes a few laboratory experiments where pesticide degradation or metabolism in soil fauna have been investigated. In such experiments, controls are not relevant. This group also includes experiments for determination of LD-50 values (LD-50 = dose required to kill 50% of the test animals). In such cases it is desirable that the solvent is used as control. Controls situated in other fields than the experiment have not been approved as the abundance and species composition of soil fauna differ widely between different sites. Sampling several days before,

Earthworms in waste and environmental management
edited by C.A. Edwards and E.F. Neuhauser, pp. 303–313

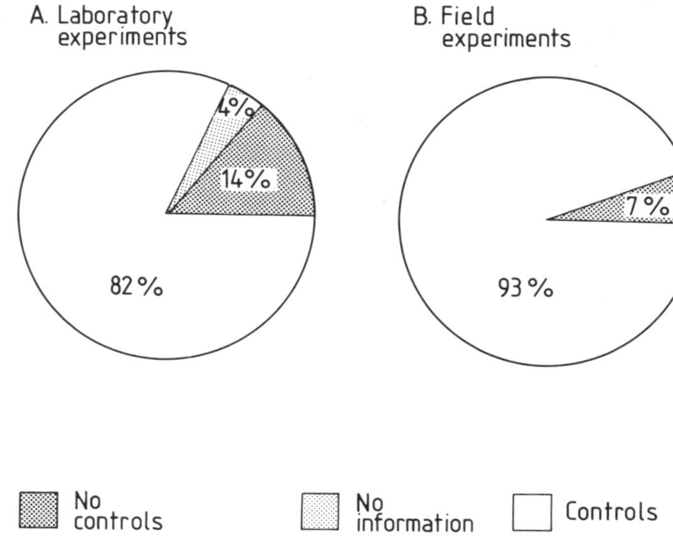

Fig. 1. Use of controls in pesticide experiments on soil fauna.

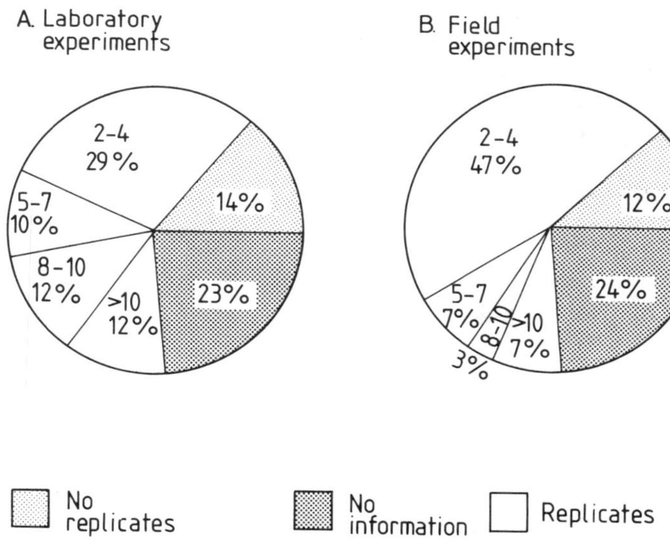

Fig. 2. Use of replicates in pesticide experiments on soil fauna.

compared to sampling after pesticide treatment, has been judged as if no control was used. Faunal composition, number and sampling efficiency vary greatly with time, and are influenced by humidity, temperature, availability of food, etc. Adequate controls should therefore be taken at the same time as the experimental samples and in areas close to treated plots. Unfortunately, in some of the investigations the controls were treated with pesticides.

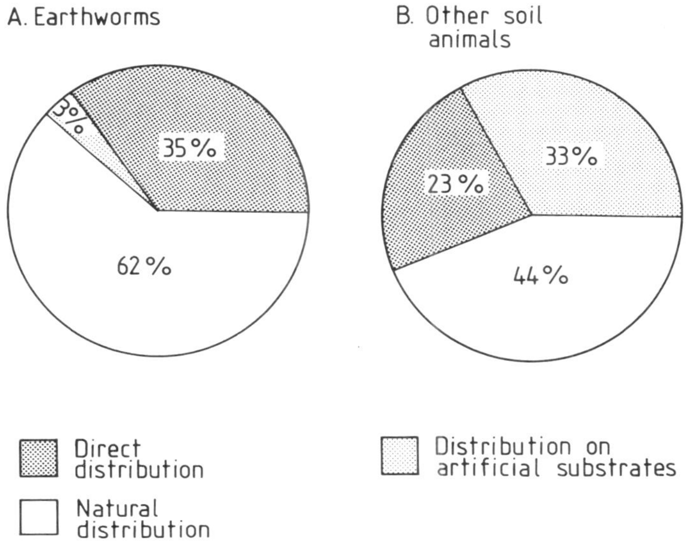

Fig. 3. Distribution of pesticides in laboratory experiments on soil fauna.

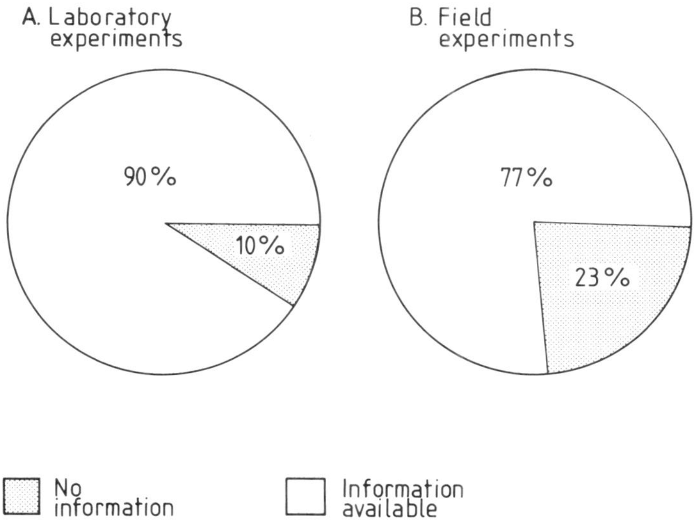

Fig. 4. Dose information on pesticide used in experiments on soil fauna.

ii. *Replicates*

More than one third of laboratory and field investigations either lacked replicates or lacked information on the existence of replicates (Fig. 2). More than half of the field investigations and one third of the laboratory investigations had 2–7 replicates. More than seven replicates are more common in the laboratory than in the field. Individual specimens in a group have not been considered to be replicates, unless individual

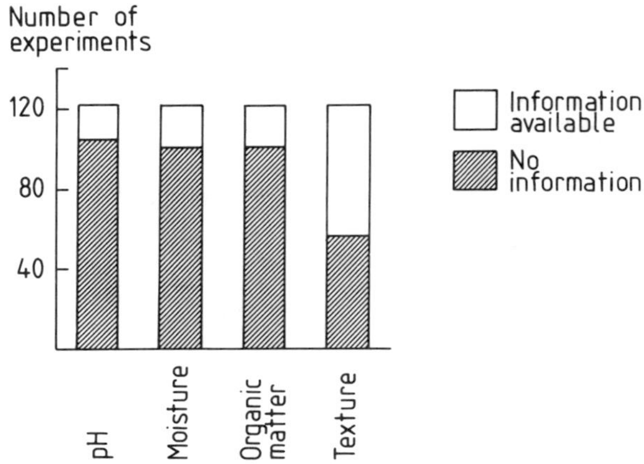

Fig. 5. Soil data information in field experiments with pesticides on soil fauna.

measures of some factor were made. Several groups of separately studied individuals have been indicated as replicates. Different fields or gardens have not been considered relevant replicates for the same reason as above.

iii. *Distribution of pesticides*

The application of pesticides in laboratory experiments differed. When the pesticide was applied to the animals in a similar way as in the field, *i.e.*, in soil, natural or mixed in the laboratory, or on litter, food, etc., this has been termed 'natural distribution'. 'Direct distribution' means dipping or spraying the pesticide directly on or injecting into the experimental animals. Application of the pesticide in petri dishes, on filter paper or in agar was considered as 'distribution on artificial substrates'. This is an unusual way to apply the pesticide to earthworms compared to that used with other soil animals (Fig. 3).

iv. *Information on dose*

If the text, tables or figures enabled us to calculate as least an approximate dose or dose interval used, the experiment was judged as having dose information. Expressions like 'normal', 'routine', 'recommended' and 'common' doses were generally not considered as dose information, if there were no figures to support these vague desriptions. This approach was chosen as all chemicals are not known in all countries and 'normal' can vary widely between different times, purposes of spraying and different countries. Although the judgement was very generous, it was found that 23% of all field investigations lacked acceptable information about dose used, and 10% of the laboratory experiments lacked dose information (Fig. 4).

v. *Substrate and field descriptions*

Information on factors such as temperature was considered to be relevant in all

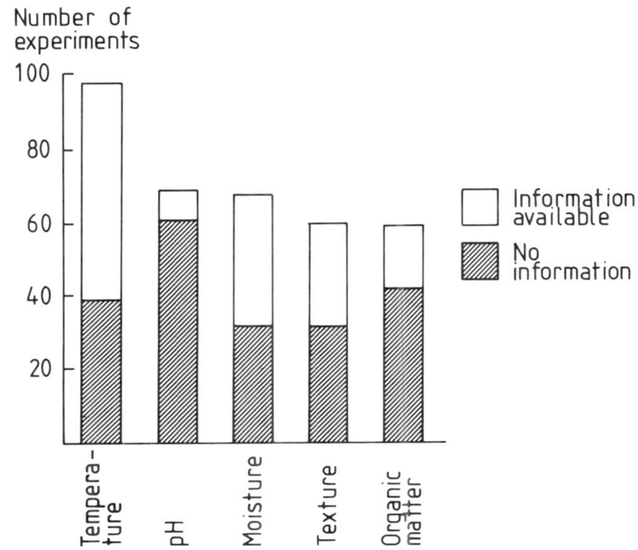

Fig. 6. Relevant information on environmental conditions in laboratory pesticide experiments on soil fauna.

A. Earthworms

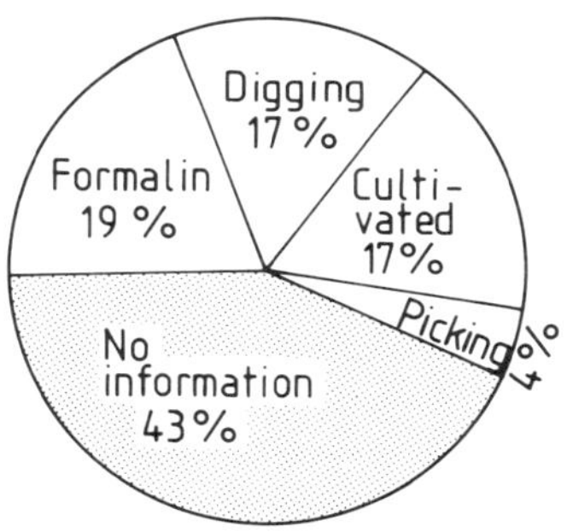

Fig. 7. Methods employed to obtain earthworms for laboratory pesticide experiments.

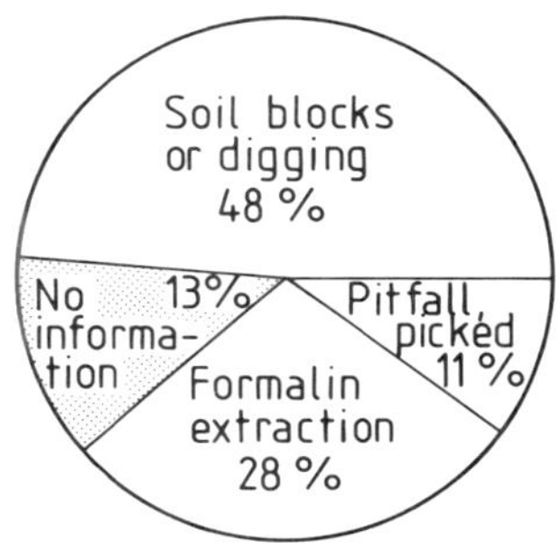

Fig. 8. Methods employed to sample earthworms in field pesticide experiments.

laboratory investigations. Moisture, pH, texture and organic matter content was regarded as important when the pesticides were applied in soil. Texture and organic matter content are not relevant for petri dishes, filter papers, etc. The laboratory trials were judged generously, in the same way as the ones performed in the field. Commercial soil mixtures mentioned only by their trade names have not been considered, as these products can be assumed to be known only to very local readers.

Soil data have always been considered to be relevant in field experiments. Even very sparse information as, *e.g.*, 'acid soil' has been approved as providing information on, in this case, pH.

Despite the very generous judging, it can be said that most field investigations had very sparse information on soil data. Slightly more than half gave information about

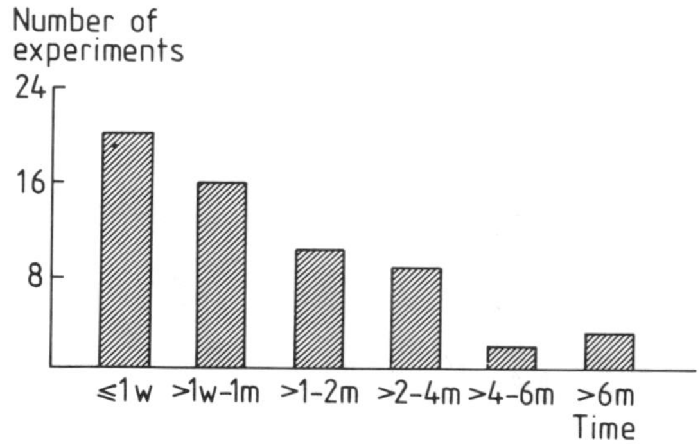

Fig. 9. Duration of laboratory pesticide experiments on earthworms.

soil texture, 20% or less presented information on pH, organic matter content or moisture (Fig. 5). In the laboratory the descriptions were only slightly better; 60% gave temperature data and 10–50% had some kind of relevant substrate description (Fig. 6).

vi. *Sources of soil animals for experiments*

Cultivated earthworms are often bought from earthworm farms which means that the exact age of the animals is not known (Fig. 7). No information was given on how the earthworms were obtained in 43% of the laboratory experiments. Experiments when the earthworms were 'collected' but no information about the method of collection was given are included in this figure. About 20% of the earthworms was captured by watering formalin. It is doubtful whether earthworms influenced by a poison can be recommended as test animals. Soil animals, other than earthworms, used in laboratory experiments were usually (44%) collected from the field through handsorting or extraction from soil blocks. The age, previous pesticide influence, etc., on the animals were thus not known, which makes it difficult to interpret the results and to reproduce the experiments.

All methods of sampling earthworms (Fig. 8) are selective for different species, which is important to consider when conclusions are drawn from the results. The most frequently used method of sampling other soil animals was extraction from soil blocks or cores.

vii. *Length of experiments with earthworms*

The duration of laboratory experiments with earthworms was, as expected, shorter than field experiments. Of the field experiments 68% lasted more than six months, compared to only 5% of the laboratory experiments (Figs 9 and 10).

viii. *Measurements of toxicity in experiments with earthworms*

The measurement of toxicity most often used in laboratory experiments with earthworms is mortality. This includes LD-50 tests. 'Mortality', which means counting

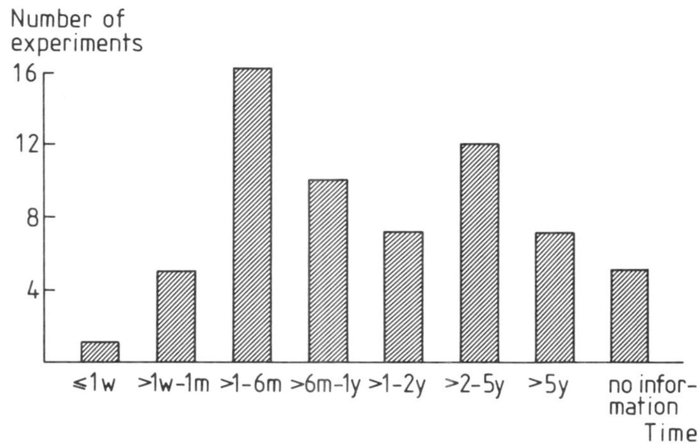

Fig. 10. Duration of field pesticide experiments on earthworms.

survivors, corresponds to 'numbers' in the field experiments. Only when dead earthworms were counted on the soil surface in field experiments was this considered 'mortality'. Measuring the concentration of a pesticide or degradation product in the animals was also common. The group 'physiological, chemical and morphological factors' includes experiments where production of enzymes, tumors, etc., has been used as a measurement of toxicity. 'Behavior' means, *e.g.*, mobility, disappearance rate or burrowing by earthworms (Figs 11 and 12).

ix. *Biological information on earthworms*

In many experiments, earthworms had not been identified. This applied to nearly half of all field experiments and 12% of the laboratory experiments (Fig. 13). The term 'night crawlers' was considered as a species determination although several species in reality can be named as such. There is some evidence that different species of earthworms are infuenced by pesticides in different ways depending on genetical background and behavior. A change in species composition of a faunal group caused by pesticides may remain undiscovered in the field and a laboratory experiment cannot be reproduced if determination to species is not made.

When the earthworms were divided into juveniles and adults this was considered as information on age, although juveniles and adults can be of different age. Nevertheless, some information on the age structure was given. Absolute knowledge of the age of experimental animals can be obtained only when they are raised in laboratory cultures. Most experiments lacked all information on age (Fig. 14).

Discussion

a. *Laboratory tests*

The need for standardized laboratory tests of the effects of chemicals on non-

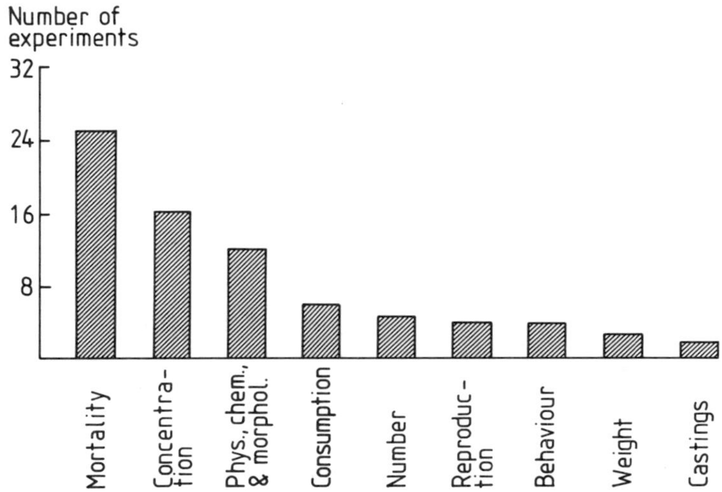

Fig. 11. Measurements of toxicity in laboratory pesticide experiments on earthworms.

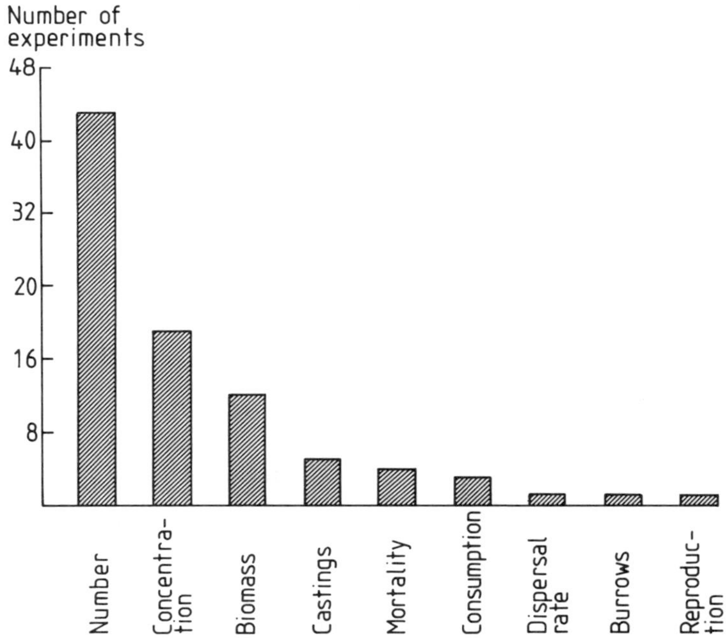

Fig. 12. Measurements of toxicity in field pesticide experiments on earthworms.

harmful soil organisms is obvious. The number of methods used almost equals the number of papers presented on the subject. Lately some effort has been made to achieve international agreement on standard laboratory tests. The test species are intended to be a kind of 'indicator' species. Two criteria which these species should ful-

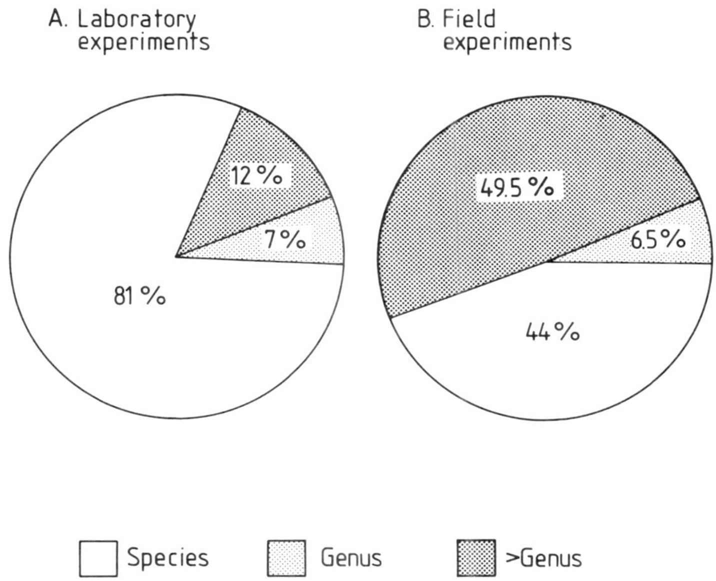

Fig. 13. Taxonomic information on earthworms in pesticide experiments.

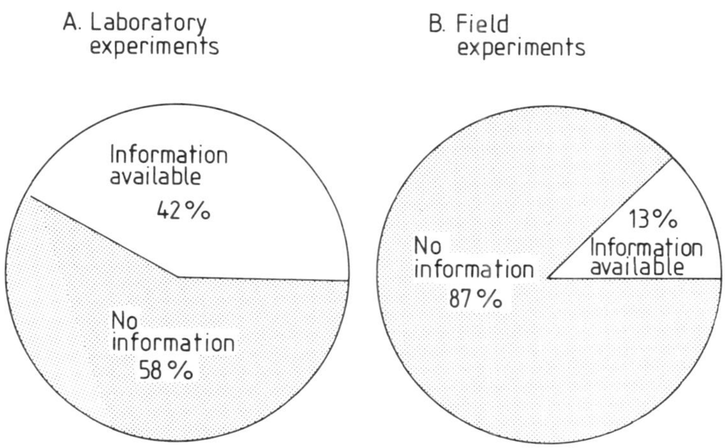

Fig. 14. Age information on earthworms in pesticide experiments.

fil are: They should be representative of the fauna being influenced by the normal use of pesticides. It is necessary that the test species can be cultivated in the laboratory, thus enabling control over age and environmental rearing conditions of the animals. The use of test organisms collected from the field introduces uncertainty in the same factors and there might also be a risk that species are incorrectly determined, particularly as regards immature earthworms. However, there is a risk of changing the genetical composition when a test species is cultured continuously. The cultured population should therefore be refreshed periodically with genetic material from field populations. The animals chosen as test organisms can be exposed to the pesticides in more or less 'natural' ways. If the laboratory tests are intended to simulate field conditions the most natural ways of application are in soil or on plant material, or possibly in the prey for a predatory test species. The drawback is that the composition, *i.e.*, texture, organic matter content, and pH, differ considerably between the sites from which the soil is taken. A compromise is a manufactured soil where the factors mentioned are known, thereby enabling the experiment to be repeated. Application through food can in some cases be well-justified. For example, the earthworm *L. terrestris* often feeds directly on plant material. The difficulties are to determine if and how much of the plant material is being consumed.

An artificial medium on the other hand, has few resemblances to soil. It usually has a certain structure and can hold a suitable amount of water. An example of such a substrate is vermiculite. The advantage of artificial media is that they can easily be standardized, which thus makes replications very easy. The drawback is the difficulty of comparing this distribution to the situation in natural soils. The nature of the substrate is often omitted and the animals are treated directly with the chemicals. This can be done by injections, by dipping them into solutions or by spraying. These methods of application often have the advantage that the dose will be known more exactly than when the pesticide is applied to a substrate. The comparison with field conditions is, however, even more difficult than in the case with artificial substrates.

Laboratory tests can assess either lethal or sublethal effects. Toxicity tests are often done as L.D.-50 or LC-50 tests, *i.e.*, the dose or the concentration when 50% of the experimental animals die. These tests can only give rough information on acute toxicity.

Sublethal tests are designed to detect indirect effects on the organisms. Such effects may be behavior disturbances, retarded development, lowered fertility, etc., and which may cause population changes in the field, although the animals do not suffer from high acute mortality. Such tests are long-term experiments, but many of them are easy to perform.

b. Field tests

Direct side-effects appear when a non-target fauna is hit by the pesticide or when it feeds on prey or on organic matter which has been hit. A change in flora or fauna composition caused by a pesticide can result in indirect effects on non-target groups. As the ecology of most soil animals is poorly known, such effects can be measured easier in field experiments as it is difficult to create natural multi-component habitats in the laboratory. It is more difficult to design reproducible field tests. Field experiments where the amounts and spraying routines correspond to those used in practical management are nevertheless very important. To be able to draw conclusions from a field test, information must be available on the soil type, which influences both the persistence of the pesticide and its action on the organisms. Since the soil fauna tend to be distributed unevenly several replicates are necessary.

Recommendations

The experiments should carefully be planned and detailed information should be given. Some important statements to be born in mind for those who plan pesticide experiments on earthworms.

- A sufficient number of controls and replicates for statistical treatment should be used. In the field the controls and replicates have to be situated close to each other. Pesticide-treated controls are useless.
- The test earthworms must be determined to species and age data given if possible.
- Only species representative of earthworms living in habitats influenced by the pesticide are relevant as test animals in laboratory experiments if conclusions are to be transferred from laboratory results to the field situation.
- The duration of tests should be adapted to the persistence of the pesticide and to the generation time of the test earthworms.
- Attention should be paid to previous influence of pesticides or other chemicals upon the test earthworms if they are not cultured.
- Doses similar to those recommended by the manufacturer should be used.
- Information on the common name and percentage of the active ingredient must be available in the report together with the doses used. Standard units, possible together with local units, are desirable.
- The method of pesticide application should be that commonly used.
- Experimental conditions must be reported. For field experiments this includes at least information on texture, organic matter content, pH and moisture. For laboratory experiments information on temperature is always relevant.

References

Boström, U. and Lofs-Holmin, A. 1982. Testing side effects of pesticides on soil fauna. Swedish University of Agricultural Sciences. Department of Ecology and Environmental Research. Report 12. Uppsala. 70 pp.

EARTHWORM TOXICOLOGICAL TESTS, HAZARD ASSESSMENT AND BIOMONITORING. A METHODOLOGICAL APPROACH

MARCEL B. BOUCHÉ
Laboratoire de zooécologie du sol (INRA) CEPE/CNRS, Montpellier, France

1. Introduction

The aim of this paper is not to cover the wide research field indicated by the title, but to discuss how to improve the design of our research by a better connection between toxicological tests (laboratory), hazard assessment (field) and biomonitoring. There is confusion on the degree by which laboratory experiments can be compared with field trials. To try to obtain information by an inadequate method limits the improvement of the method and obscures the interpretation of results. In this brief account, I present schematically, remarks encouraging the reader to look at papers where the arguments are given more in details (*e.g.*, Bouché, 1984a, b).

2. Toxicological tests

Tests often labelled 'ecotoxicological tests' have basically to answer the question: what is the relation between chemical dose(s) and effect(s) on earthworms.

- Such a test cannot be designed with references to 'natural conditions', because these conditions at the earthworm level are poorly known and probably not reproducible.
- Such a test cannot be made with one species 'representative' of all earthworms. We have no criteria to check the representativeness of a single species.
- Such a test has to be *reproducible*, *i.e.*, must be made using standard products (repetition in the same laboratory), and by synthetic products (readily available between laboratories), with an unique strain of a precisely known species of earthworm which can be used as a standard (*i.e.*, easy to rear and to exchange between laboratories).
- Such a test must give comparable results between chemicals, *i.e.*, must be usable to assess consequences of chemical use on earthworm communities, and consequently on soil properties. Results of tests are needed to interpret observed field hazards or to predict potential risks from new chemicals. Comparability between chemicals made possible a classification of chemical on one scale of toxicity (Fig. 1). To obtain comparable results, the test must use a neutral medium where earthworm crawl and eat, must have a long enough duration for spontaneous chemical change to occur (*e.g.*, hydrolyse) and a biological process leading to the toxic effect (*e.g.*, cholinesterase inhibition leading to death).

A comparison of various technical tests has been made (*e.g.*, Edwards, 1983; Goats and Edwards, 1988; Heimbach, 1985). Among the convenient tests only one medium is synthetic, and permits reproducible and comparable results: the *Artisol* (described

Earthworms in waste and environmental management
edited by C.A. Edwards and E.F. Neuhauser, pp. 315–320

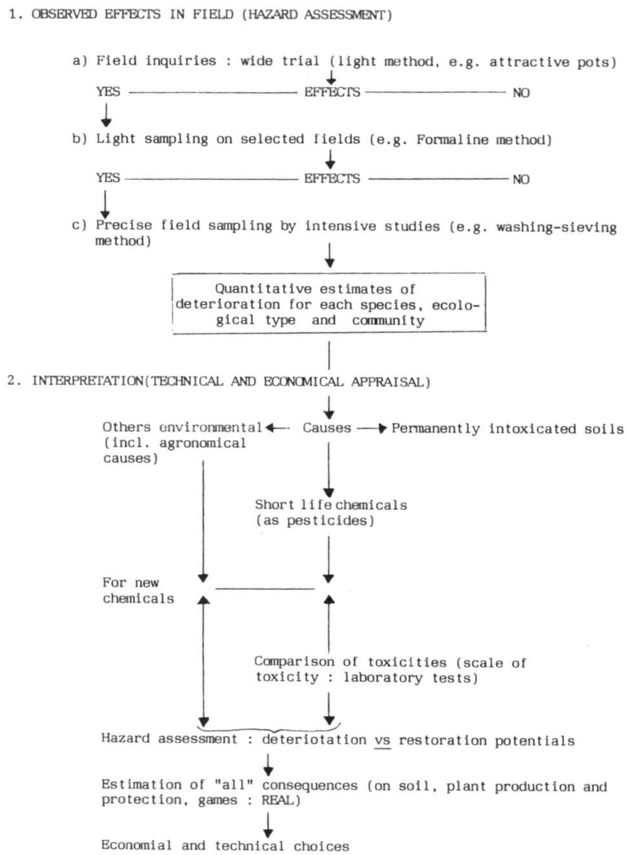

Fig. 1.

by Ferrière *et al.* 1981) following the AFNOR procedure (AFNOR, 1984). Among potentially standard earthworm species the *Eisenia fetida* species-complex was adopted, because it is easy to rear and to transport. On the basis of the work of André (1963) and others, I defined *Eisenia fetida andrei* Bouché, 1972, at the subspecies level because of predictable resistance against this change, while I wrote 'il s'agit très vraisemblablement d'une bonne espèce'. In fact, classical taxonomy has been forgetting this species and Jaenike (1982) 'rediscovered' this species before Sheppard (1988). This important species is now used in both vermiculture and toxicological tests so we have to characterize the strain (*Eisenia andrei*, Bouché, 1972, P2342) by electrophoresis (Robotti, 1982) to be sure we work on the same population.

3. Hazard assessment

A toxicological test gives an answer only in artificial and controlled laboratory conditions on the effect on mortality of one species. In no circumstances can such a test predict directly the hazard to an earthworm community of chemicals; field hazard must be observed in the field.

Conversely, changes in community structure and numbers in field are not dependent directly on toxicity of any one chemical. The behavior of any chemical interferes with its toxicity in soil and plants (adsorption, infiltration, retention by above ground parts), depends on climatic factors (rain, temperature), depends on biological activity (*e.g.*, biodegradation), and on its chemical properties (stability) or others factors such as agricultural practices (sowing, plowing). In turn, earthworm risks vary very much with the behavior (*e.g.*, surface, activity, lethargy) of each stage of each species, this behavior depending on climate, food availability, photoperiod, and agricultural practices. Changes in structure and level of communities also reflects other ecological or agronomical parameters (*e.g.*, cultivation, use of other pesticides). Some plot experiments could give precise results, and more information on processes (Goats and Edwards, 1988). However, the use of plot experiments is restricted to only critical compounds because experimental costs are high and the results are local (*e.g.*, soil, earthworm community, culture and climate are local).

One suspects that so-called harmless chemicals act dramatically against earthworms in the field through sublethal effects such as those described by Reinecke *et al.* (1985). On the other hand, a chemical is often associated with other agronomic stresses (cultivation, other chemical) in a crop rotation system. The state of one community results of the combined effects of such stresses and its ability to recover from the stress.

To improve our ability to predict hazards of new chemicals before approval of use, we need to interpret the effects of present use of classical chemicals in normal field conditions. From this background, we can hope to make by comparison some hazard assessment of new chemicals.

To reduce cost of such studies we have to devize a research strategy:

i. First tests could be very simple, by easy-to-make inquiries relevant to many agronomical and/or environmental conditions. The Lofs-Holmin's pot method seems to be, with some improvement for standardization, a good candidate test.

ii. If some community deterioration is suspected from the first stage test, a semi-quantitative sampling, using the formalin method, could be used in selected treated fields.

iii. Very carefully selected situations could be studied more carefully by destructive quantitative procedures such as digging, sieving, washing procedures (Bouché and Beugnot, 1972).

iv. The interpretation of earthworm community deterioration needs a close knowledge of earthworm species, of their ability to recover, on other climatic, edaphic, human factors acting on this community. Indefinite toxicity of chemicals could be observed by a global soil toxicity test as described by Callahan (1988) or by the electrophysiological detection of Drewes *et al.* (1988) both used as sensor of pollutants. Nevertheless, in many agricultural circumstances, chemicals potentially responsible for some community deteriorations are degraded. The knowledge of climatic events, techniques of field spraying, physical and chemical properties and the position in the *scale of comparative toxicity* give the possibility to interpret the causal agent of deterioration. Conversely, this general knowledge about the agents of deterioration can help to predict the effects of new chemicals before their legal approval.

4. Estimation of consequences of chemical use

In agricultural sites, a pure conservationist position for survival of earthworms is not possible. Some use of chemicals (pesticides, fertilizers) is often a critical need to obtain high yields. Conversely, usually classical pesticide field trials demonstrate the use of a chemical, but take mostly as parameters the yield and effects on a few pests. Destruction of earthworms is not taken in account, in the short term (*e.g.*, release of nitrogen from earthworm bodies increasing yield), mid-term (*e.g.*, no burying of dead leaves, increasing fungal inoculum in subsequent years), or, long term (destruction of soil structure, infiltration decrease, erosion increase) (Bouché, 1974, 1984c).

In fact, we have to take in account all consequences of chemical use, not only the positive ones (such as pest reduction, short time yield increase) or the negative ones (as biological perturbation, long term effects). For earthworms, this can be made thanks to an integration of the role of earthworms in a simulation model, which has been developed for eight years (REAL, Bouché, 1980). At first, it was a conceptual model to select available techniques (or to improve them) to obtain field data to feed later an operational model in term of energy, matter, soil material movements, due to each of the main ecological types of earthworm. All data has been gathered from fields, to avoid assumptions and non-confirmable results. This model is now reaching slowly the operational stage and gives the possibility to estimate from field data the pedological, ecological and plant physiological consequences of any changes for 'all' earthworm species. The model requires humidity, temperature and latitude (photoperiodism) data, plus the influence on structure and level of population of the agricultural technique being studied.

5. Biomonitoring

The study of environmental risks could be improved by the use of earthworms. It is possible to detect the bio-availability of chemicals in soils in the way it was done for PCBs (Tarradellas *et al.*, 1982) or heavy metals (Ma, 1980) from earthworms in natural communities. The introduction of a 'standard' species as *Eisenia fetida* has been done by Marquenie *et al.* (1985). The observations are dependent on:

1. The level of chemical in the soil.
2. The food choices of a particular earthworm stage-species (Ferrière, 1980).
3. The availability of the chemicals in earthworm food and skin contact.

A standard bioaccumulation system, using *Artisol*, and *Eisenia andrei* Bouché, 1972, was used to demonstrate a constant ratio of bioaccumulation after equilibrium for heavy metals. Detoxication of soil by earthworms could be also potentially detected by the same standard process (Bouché, 1984b).

6. Conclusions

In ecological studies, earthworm toxicity must be taken as a question at its right level. They are the largest invertebrate biomass of land surfaces. Because of their great usefulness in agrobiocenoses and natural systems, they have to be protected by a careful choice of chemicals applied to lands. We have techniques available today to make toxicity tests, hazard assessments and technico-economical appraisals: *i.e.*, it is possible to balance between the need of a pesticide and the requirements for earthworm activities in soils.

Because of their wide natural dispersal, their close relation with other main biomasses (*i.e.*, microorganisms, roots, other soil animals), necromasses (humus compound, litter) and their ability to ingest soil minerals compounds (from sand to clay), those animals give us an excellent tool to detect pollutants, to measure their bioavailability in soil, and/or to compare earthworm uptake with a standard bioaccumulation method.

Closely associated with the 'fast' turn-over (Coleman *et al.*, 1983) of elements, earthworms re-cycle most of them rather quickly. Possible chemical transformations at their level, as detoxification, are poorly understood but could be of significance.

References

AFNOR, 1984. Determination de la toxicité d'une substance vis-à-vis des lombriciens (espèce *Eisenia fetida*). Méthode Artisol. Norme X31-250 AFNOR, Paris.

Bouché, M.B. and Beugnot, M. 1972. Contribution à l'approche méthodologique des biocénoses. II. L'extraction des macro-éléments du sol par lavagetamisage. Ann. Zool. Écol. Anim. 4(4): 537–544.

Bouché, M.B. 1974. Pesticides et lombriciens: problèmes méthodologiques et économiques. Phytiatrie-Phytopharmacie 23: 107–116.

Bouché, M.B. 1980. Objectifs, compartimentation et faisabilité du modèle R.E.A.L. Pedobiologia 20(3): 197–211.

Bouché, M.B. 1984a. Ecotoxicologie des lombriciens. I. Ecotoxicité contrôlée. Oecol. Applic. 5(3): 271–287.

Bouché, M.B. 1984b. Ecotoxicologie des lombriciens. II. Surveillance de la contamination des milieux. Oecol. Applic. 5(4): 291–301.

Bouché, M.B. 1984c. Faune et flore telluriques: leur contribution à la fertilité des sols. In Anonyme (Ed.) Fauna et flore auxiliare en agriculture, ACTA, Paris, pp. 105–117.

Callahan, C.A. 1988. Earthworms as ecotoxicological tools (This volume pp. 295–301).

Coleman, D.C., Reid, C.P.P. and Cole, C.V. 1983. Biological strategies of nutrient cycling in soil systems. Adv. Ecol. Res. 13: 1–55.

Drewes, C.D., Vining, E.P. and Callahan, C.A. 1988. Electrophysiological detection of sublethal neurotoxic effects in intact earthworms. (This volume pp. 355–366).

Edwards, C.A. 1983. Development of a standardized laboratory method for assessing the toxicity of chemical substances to earthworms. Report of Commission of European Communities EUR 8714 EN. 141 pp.

Ferrière, G. 1980. Fonctions des lombriciens. VII. Une méthode d'analyse de la matière organique végétale ingeŕée. Pedobiologia 20(4): 263–273.

Ferrière, G., Fayolle, L. and Bouché, M.B. 1981. Un nouvel outil, essentiel pour l'écophysiologie et l'écotoxicologie, l'élevage des lombriciens en sol artificiel. Pedobiologia 22(3): 196–201.

Goats, G.C. and Edwards, C.A. 1988. The prediction of field toxicity to earthworms by laboratory methods. This volume, pp. 283ff.

Heimbach, F. 1985. A comparison of laboratory methods for toxicity testing with earthworms. Present proceedings.

Jaenike, J. 1982. *Eisenia foetida* is two biological species. Megadrilogica 4(1–2): 6–8.

Ma, W. 1983. Biomonitoring of soil pollution: ecotoxicological studies of the effect of soil-borne heavy metals on lumbricid earthworms. Annual Report Research Institute of Nature Management 1982, pp. 83–97.

Marquenie, J.M. and Simmers, J.W. 1988. A method to assess potential bioavailability of contaminants. This volume, pp. 367ff.

Robotti, C.A. 1982. Biochemical polymorphism of earthowrm. IV. Enzymes of *Eisenia fetida andrei*. Monit. Zool. Ital. 16(4): 341–344.

Sheppard, P.S. 1988. Specific differences in cocoon and hatchling production in *E. fetida* and *E. andrei* (This volume, pp. 83–92).

Stewart, V.I. and Scullion, J. 1985. Earthworms, soil structure and the rehabilitation of former opencast coal mining land. This volume, pp. 263ff.

Tarradellas, J., Diercxsens, P. and Bouché, M.B. 1982. Methods of extraction and analysis of PCBs from earthworms. Int. J. Environ. Anal. Chem. 13: 55–67.

Venter, J.H. and Reinecke, A.J. 1988. Sublethal ecotoxicological effects of dieldrin on the earthworm *Eisenia foetida* (Oligochaeta). This volume, pp. 337ff.

EISENIA FOETIDA USED AS A BIOMONITORING TOOL TO PREDICT THE POTENTIAL BIOACCUMULATION OF CONTAMINANTS FROM CONTAMINATED DREDGED MATERIAL

R.G. RHETT, J.W. SIMMERS and C.R. LEE
Waterways Experiment Station, Vicksburg Mississippi, U.S.A.

Introduction

The Clean Water Act in the United States requires that the environmental evaluation of dredged material prior to discharge, include the effects of this disposal on concentration of contaminants through biological processes. This results in a critical need for Corps of Engineers Districts to be able to predict the contamination of animals that may be associated with each of the potential disposal alternative: open-water disposal, upland disposal, and wetland creation. The following is a summary of the bioassay procedures using *Eisenia foetida* to evaluate the potential contaminant mobility into soil-dwelling animals. These tests have been derived from the proposed OECD- and EEC-test procedures and modified to consider accumulation and sub-lethal effects rather than toxicity. The three examples described here are representative of cases in which prediction of contaminant mobility is essential, since confined disposal sites or other upland deposits of dredged material often become highly prolific wildlife habitats. These test procedures are intended to evaluate the potential movement of toxic heavy metals, PCB's, and PAH's from dredged material placed in an upland (oxidized) disposal alternative into soil-dwelling invertebrates as a first step evaluation of contaminant mobility into animals that may colonize the dredged material.

General test description

Specimens of *Eisenia foetida* used for testing were purchased from a local worm grower and placed in a 1.82 m × 1.52 m × 0.30 m wood frame container with a plywood base. A layer of approximately 10–15 cm of horse manure was allowed to accumulate, and chicken meal mash was used as food when fresh manure was not available. The worm beds were located in a partially-shaded greenhouse with temperature not exceeding 27°C. During summer months, the worm beds were watered daily to prevent dehydration.

When worms were needed for testing purpose, they were hand-sorted, rinsed in distilled water, and allowed to drain off excess water on paper towels. A known fresh weight of worms was added to each container of test material. During the test period (usually 28 days) distilled water was added as necessary to maintain proper moisture. Also, a nylon insect screen material was placed over the drain holes in the base and in the lid to prevent escape or the wandering of earthworms between containers. Upon completion of the testing period, the earthworms were sorted from the material, rinsed in distilled water, dried on paper towels, and weighed before and after 48 hrs purging on wet filter paper. The filter paper was changed after 2 hr and 24 hr purging

Earthworms in waste and environmental management
edited by C.A. Edwards and E.F. Neuhauser, pp. 321–328

Table 1. Comparison of weights of *Eisenia foetida* recovered after seven days

Substrate	Initial weight of worms (g wet weight)	Final weight[a] of worms (g wet weight)	Percent loss in seven days
90% WES reference soil + 10% Black Rock sediment	40	17.4 ± 5.4	56
Black Rock reference control sediment	40	33.7 ± 1.8	16
WES reference soil	40	35.0 ± 1.5	12

[a] Mean ± SD of three replications.

at 10°C. Purged worms were then homogenized using a stainless steel Sorvall omnimixer (Dupont Co., Newton, CT 06470) and placed in acid-washed, hexane rinsed glassware. Samples of 5 g fresh weight for heavy metal analysis were then oven-dried at 80°C for 24 hrs and digested with nitric acid. Metal concentrations were measured using atomic absorption spectrometry. Organic compounds were extracted from 25 g of frozen or fresh tissue samples and measured using gas chromotography and mass spectrometry.

Examples of tests

i. *Highly-contaminated salt water sediment*

In this case the *E. foetida* bioassay procedure was applied to predict the contaminant mobility in an upland (oxidized) disposal alternative prior to a dredging project. A highly contaminated sediment from Black Rock Harbor, Bridgeport, CT, was collected and transported to the laboratory for growth-chamber bioassay tests. Because this was a salt water sediment, it was washed until wash water indicated 0 ppt salt, in order to simulate the effect due to rainfall in a disposal site setting. The sediments were then air dried, pulverized, and rewet with distilled water to field capacity before the animals were added.

Initial screening tests indicated that the Black Rock Harbor sediment was quite toxic to the worms while the similarly prepared reference control sediment collected at the mouth of Black Rock Harbor was not. A series of toxicity tests indicated that survival for seven days could only be obtained if the Black Rock sediment was diluted. A local woodland reference soil was chosen as the dilution medium. Mixtures of 10% Black Rock Harbor sediment and 90% WES reference soil were used for the seven-day contact test. Forty grams fresh weight of *E. foetida* were placed in approximately 1 kg each of the 10% Black Rock sediment + 90% WES reference soil mixture, Black Rock Reference Control sediment, and the WES reference soil. The test was conducted in a controlled-temperature growth chamber at 20°C.

Worms in all three treatment groups lost weight during the test procedure (Table 1). Although the worms were weighed and not counted, it was clear that the 56% decrease in animal weight recorded in the 10% Black Rock sediment + 90% WES

Table 2. Black Rock harbor sediment *Eisenia foetida* terrestrial animal bioassay – seven-day test (concentrations $\mu g/g$ dry weight)

Contaminant	90% WES reference soil + 10% Black Rock sediment		Black Rock reference control sediment		WES reference soil		Background worm tissue[a]
	Substrate	Tissue*	Substrate	Tissue*	Substrate	Tissue*	
Cd	2.9	5.83 ± 0.80[a]	2.9	4.26 ± 0.55[b]	0.2	3.80 ± 0.34[c]	4.55 ± 0.21[d]
Cr	191	4.73 ± 1.55[a]	314	6.96 ± 0.28[b]	8.07	2.53 ± 0.06[c]	1.90 ± 0.20[d]
Cu	330	34.1 ± 1.4[a]	438	33.8 ± 1.3[b]	9.86	12.0 ± 0.60[c]	11.25 ± 0.52[d]
Ni	29.6	6.03 ± 2.97[a]	45	3.07 ± 0.84[b]	15.0	3.33 ± 1.18[c]	0.49 ± 0.46[c]
Pb	63.5	7.6 ± 0.13[a]	135	5.13 ± 0.63[b]	10.3	4.7 ± 0.10[c]	5.23 ± 0.40[d]
Zn	176	110.5 ± 2.3[a]	329	112.3 ± 1.5[b]	29.5	102.6 ± 0.6[b]	118.5 ± 2.9[c]
Hg	0.19	< 0.050	0.475	< 0.050	< 0.049	< 0.050	< 0.050

* Mean ± SD of three replications; means within the same row followed by the same letter are not significantly different at the 5% level.

reference soil mixture was due largely to the die-off of some worms. The worms in the other two treatment groups lost less weight, suggesting that fewer worms died during the test procedure. No supplemental food was provided during the seven-day test period.

Analysis of the earthworm tissue for Cd, Cr, Cu, Ni, Pb, Zn, and Hg and comparison of the worm tissue levels to those of the substrates indicated that Cd, Cr, Cu, Ni, and Pb were bioaccumulated (Table 2). Neither Zn nor Hg bioaccumulated.

Some accumulation of Cr, Cu, Ni, and Pb was expected in substrates with high concentrations of these metals; however, the concentrations of these elements in the worm tissues were quite low in comparison with the substrate. In contrast, Cd was found to be consistently present at higher levels in the worm tissue than in the test, control, and reference substrates. There was a higher than desirable background level of Cd (4.55 μg/g) in the worms prior to the test period which may account for the 3.80 $\pm$ 0.34 μg/g Cd in the worms from the WES reference soil, as well as in worms from the other substrates. This could also be due to the expected decrease of worm-tissue Cd in the low Cd medium, or a potential for the *Eisenia* to accumulate Cd from low levels in the media (Hartenstein *et al.*, 1980).

Values in the literature tend to confirm the findings here. The values reported for various earthworm species (Table 3) indicate that Cd levels in worm tissues are generally greater than those of the soil in which they live. Additionally, this suggests that the Cd is less bioavailable in the Black Rock Harbor sediment than in the various soils reported in the literature. The earthworm bioassay was successful in determining that the Black Rock Harbor sediment was quite toxic to soil-dwelling annelids when allowed to dry out under upland conditions. However, there appears to be no significant indication that Cd or other toxic heavy metals will accumulate to elevated levels in soil-dwelling invertebrates if the material is placed in an upland disposal environment and diluted with uncontaminated soil. As indicated previously in Table 2, the comparison of the levels of metals in the three test substrates shows that the toxicity observed in the initial portion of the test was not due to the presence of toxic heavy metals.

ii. *Restoration of pyritic mine spoil using contaminated dredged material*

Contaminated dredged material excavated from a disposal site was used to cover an area of strip mine spoil at Ottawa Il. Perrier, Llopis, and Spaine (1980) have described in detail the physical design and operation of the four demonstration reclamation plots. The plots, which measure about 24 $\times$ 55 meters, are isolated from each other and from the surrounding mine spoil by 0.6 meter high, beveled containment dikes covered with an inert waterproof shroud. Each plot was laid out to effect a diagonal slope, the lowest portion of which was designed to drain water volumes which exceed the field capacity of the soil in the plot.

The control plot, designated Plot I, was fashioned out of untreated mine spoil. Plot II, located just south of Plot I, was overlaid by dredged material to a depth of just under 1 meter. Plot III was overlaid by about 1 meter of dredged material; but, before this was done, about 1.5 metric tons of crushed agricultural limestone was incorporated into the top 15 cm of the underlying mine spoil. Plot IV was treated similarly to Plot III, except that about 2.5 metric tons of limestone was used.

While the principle of direct interest here is the concept of using dredged material

Table 3. Cadmium level in earthworms and soils reported in literature (all concentrations are in $\mu g/g$*)

Species	Tissue Cd	Soil Cd	Locality	Reference
Allolobophora longa	5.7–11.8	0.14–0.99	Denmark	Andersen (1979)
A. caliginosa	6.9–10.9	0.14–0.99	Denmark	Andersen (1979)
A. rosea	10.9–26.9	0.14–0.99	Denmark	Andersen (1979)
A. chlorotica	10.9–16.2	0.14–0.99	Denmark	Andersen (1979)
Lumbricus terrestris	8.8–16.9	0.14–0.99	Denmark	Andersen (1979)
Unidentified species	3.5–17.0	0.11–1.10	Poland	Czarnowska and Japkiewiez (1978)
Allolobophora species *Allolobophora Lumbricus* species and *L. terrestris* combined	5.9–14.4	0.66–1.59	USA	Gish and Christensen (1973)
Allolobophora species, *Lumbricus* species and *Octolasium* species combined	3.1–9.3	0.23–0.80	USA	Van Hook (1974)
Eiseniella tetraeda	3.0 ± 0.3	4.0 ± 0.2	Wales	Ireland (1979)
Dendrobaena veneta	7.0 ± 1.0	4.0 ± 0.2	Wales	Ireland (1979)
L. rubellus	15.0 ± 5.0	2.0 ± 0.1	Wales	Ireland (1979)
L. rubellus	4.0 ± 0.1	4.0 ± 0.2	Wales	Ireland (1979)
L. rubellus	2.5 ± 3.0	4.0 ± 0.3	Wales	Ireland (1979)
Aporrectodea tuberculata, Ap. turgida, Ap. longa, and *L. terrestris* combined	4.8	0.10	USA	Beyer *et al.* (1982)

* From Simmers *et al.* (1984).

for mine spoil restoration, the contaminant mobility aspect was addressed. The Ottawa site has developed into an active wildlife habitat with extensive usage by birds and small mammals. Also, the concern of use of the site for agronomic crops made full assessment of contaminant mobility a necessity. Here a field bioassay was used.

The upper 20 cm of dredged or mine spoil material was tested along with the WES reference soil. The buckets containing test material were buried 20 cm deep in each of the five regions/plot, and 40 g of earthworms were added. A lid sealed the bucket. After 28 days the worms were removed and analyzed.

The results shown in Table 4 indicate some accumulation of Cd and Cu in Test Plot 4 with the remainder of the levels in the worms near those of the background and/or well below sediment levels. Survival and growth of the test earthworms located in plots 2, 3, and 4 was excellent, but the toxicity (possibly due to the low pH of 4) and weight loss were evident on plot 1.

iii. *Confined disposal site, Buffalo,NY*

Between 1972 and 1976, about 7.2×10^5 m^3 dredged materials from Buffalo Harbor (NY) were dumped in an artificial lagoon on the shores of Lake Erie. This operation resulted in the creation of an area composed of an aquatic, a wetland, and an upland environment. Although the dredged materials were heavily contaminated as a result of the activities of several industries including an oil refinery, steel plants, and an aniline dye chemical plant on the waterfront adjacent to the dredging site, a prolific wildlife habitat developed at the disposal site.

Table 4. Ottawa Il. mine spoil restoration project. *Eisenia foetida* terrestrial animal bioassay ($n = 4$)

Test plot	Contaminant μg/g dry weight			
	Cd	Cu	Pb	Ni
2	2.94 ± 0.70	13.75 ± 1.7	2.18 ± 0.65	5.23 ± 1.35
3	3.02 ± 0.39	11.25 ± 0.5	5.4 ± 1.13	5.52 ± 0.21
4	9.67 ± 0.39	21.25 ± 0.5	3.56 ± 0.42	7.55 ± 0.93
Control	3.1 ± 0.34	7.5 ± 1.1	1.32 ± 0.85	5.6 ± 2.5
Sediment (WES reference soil)	< 0.01	27.8	41.4	45.1

Table 5. Contaminant concentrations in *Eisenia foetida* exposed in an upland bioassay

Transect location	Component	Concentration μg/g dry weigt*					
		Cd	Cu	Hg	As	PCB's (10 isomers)	PAH's (22 compounds)
A8 (upland)	Substrate	2.10	116.0	2.10	25.0	0.462	40.93
	Tissue	8.86	27.7	0.48	21.1	3.950	21.06
A6 (transition)	Substrate	0.76	17.3	0.98	17.5	4.426	1.95
	Tissue						
A2 (wetland)	Substrate	2.73	148.0	4.22	38.5	1.004	45.52
	Tissue	11.70	32.1	1.39	24.0	4.520	11.78
B2 (wetland)	Substrate	9.61	334.0	8.50	72.4	0.961	63.96
	Tissue	10.80	57.6	0.81	23.9	6.720	38.88
B6 (transition)	Substrate	5.33	228.0	4.78	58.8	0.743	32.09
	Tissue	17.60	36.2	1.14	35.3	3.620	7.46
B8 (Upland)	Substrate	7.74	269.0	7.45	53.0	0.480	35.10
	Tissue	16.0	46.7	1.77	53.8	3.125	8.49
Earthworm	Substrate	0.39	16.5	0.74	3.40	<0.128	≤3.49
control medium	Tissue	3.04	10.1	0.06	8.72	<0.410	≤0.77

* Substrates reported as dry weight; tissues reported as ash-free dry weight.

This site was selected for investigation because of the recognized ecological value of a continuous sediment/soil interface gradient from a pond with a maximum depth of about 2.00 m to a woodland at about 2.00 m above groundwater levels. This situation also provided an unique opportunity to study the interactions between, on the one hand, the combination of physical conditions and biotic development and, on the other hand, contaminant mobility.

The results of the upland bioassay analyses are shown in Table 5. It is evident from these results that the disposal site is relatively contaminated in comparison to the control substrate, and that there does appear to be a variation in the concentration of metals and organic contaminants along the transects. The highest soil concentrations of contaminants were found generally in the A2, B2 wetland region but also were high in the A8, B8 upland regions. Exposed earthworms also showed higher values of bioaccumulation in these areas as reflected in tissue concentrations of heavy metals and total PCB's (Table 5) and individual PAH's (Table 6).

Table 6. Significant PAH's in worm tissue and substrates – summary of preliminary data

Transect location	Component	Concentration μg/g* Pyrene	Triphenylene	Benzo(e) pyrene	Benzo(b) fluoranthene	Benzo(k) fluoranthene	Benzo(a) pyrene	Benzo(g,h,i) perylene	Ideno (1,2,3-c,d) pyrene
A8	Substrate	2.9	≤0.15	1.8	3.0	1.7	3.8	3.4	3.0
	Tissue	3.9	≤0.015	1.5	2.1	1.3	2.8	1.5	1.3
A6	Substrate	0.53	≤0.15	0.32	0.43	0.25	0.56	0.29	0.47
	Tissue	0.14	0.074	0.099	0.15	0.11	0.16	0.19	0.21
A2	Substrate	2.7	≤0.15	2.1	3.5	1.9	5.2	4.5	3.1
	Tissue	0.53	1.6	0.75	1.3	0.65	1.8	1.1	1.1
B2	Substrate	2.5	≤0.15	4.0	5.6	2.9	8.6	7.6	6.7
	Tissue	≤0.015	3.6	4.2	5.4	2.6	7.4	3.2	5.0
B6	Substrate	1.5	≤0.15	2.1	2.3	1.2	3.6	3.7	2.9
	Tissue	0.093	0.57	0.65	0.54	0.24	0.93	0.93	1.6
B8	Substrate	2.1	≤0.15	2.1	2.4	1.4	3.6	3.9	3.1
	Tissue	0.15	0.69	1.1	0.67	0.37	1.3	1.5	0.83
Earthworm Control Medium	Substrate	≤0.15	≤0.15	≤0.2	≤0.002	0.0047	≤0.002	0.160	0.029

* Substrates reported as dry weight; tissues reported as ash-free dry weight.

Conclusions

The previously summarized animal bioassay procedure has been found to be useful in the evaluation of the contaminant mobility from dredged material placed in upland disposal facilities and in the prediction of contaminant mobility as a part of the decision-making process prior to selection of a disposal alternative. The earthworm *Eisenia foetida* has shown a high level of sensitivity as a bioassay animal in its ability to bioaccumulate various heavy metals, PCB's and PAH's from a variety of contaminated substances. Therefore, it appears to be a valuable tool for predicting or evaluating the contaminant availability from dredged material of either freshwater or marine origin before or after its disposal.

Acknowledgements

This research was sponsored under the U.S. Army Corps of Engineers, Field Verification and Dredging Operations Technical Support Programs, C.C. Calhoun, Program Manager, WES; Drs. W. Klesch, J. Hall, and R. Pierce, Office, Chief of Engineers, Technical Monitors; and C. Hummer, Water Resources Support Center, Technical Monitor.

References

Andersen, C. 1979. Cadmium, lead and calcium content, number and biomass in earthworms (Lumbricidae) from sewage sludge treated soil. Pedobiologia 19: 309–319.

Beyer, W.N., Chaney, R.L. and Mulhern, B.M. 1982. Heavy metal concentrations in earthworms from soil amended with sewage sludge. J. Environ. Qual. 11: 381–385.

Czarnowska, K. and Japkiewiez, K. 1978. Heavy metals in earthworms as an index of soil contamination. Polish J. Soil Sci. 11: 58–61.

Gish, C.D. and Christensen, R.E. 1973. Cadmium, nickel, lead and zinc in earthworms from roadside soil. Env. Sci. Technol. 7: 1060–1062.

Hartenstein, R., Neuhauser, E.F. and Collier, J. 1980. Accumulation of heavy metals in the earthworm *Eisenia foetida*. J. Environ. Qual. 9: 23–26.

Ireland, M.P. 1979. Metal accumulation by the earthworms *Lumbricus rubellus, Dendrobaena veneta* and *Eiseniella tetraeda* living in heavy metal polluted sites. Environ. Pollut. 19: 201–205.

Marquenie, J.M. and Simmers, J.W. 1984. Bioavailability of Heavy Metals, PCB and PCA Components to the Earthworm (*Eisenia foetida*). International Conference of Environmental Contamination, London. (CEP Consultants Ltd) 920 pp.

Perrier, E.R., Llopis, J.L. and Spaine, P.A. 1980. Area strip mine reclamation using dredged material: A field demonstration. Technical Rept. EL-80-4. Environmental laboratory, U.S. Army Engineer Waterways Experiment Station, Vicksburg, Mississippi, U.S.A.

Simmers, J.W., Rhett, R.G. and Lee, C.R. 1983. Application of a Terrestrial Animal Bioassay for Determining Toxic Metal Uptake from Dredged Material. International Congress Heavy Metals on the Environment, Heidelberg. 1284 pp.

Van Hook, R.I. 1974. Cadmium, lead, and zinc distributions between earthworms and soils: potentials for biological accumulation. Bull. Environ. Contamin. Toxicol. 12: 509–512.

A COMPARISON OF LABORATORY METHODS FOR TOXICITY TESTING WITH EARTHWORMS

FRED HEIMBACH
Bayer AG, Agrochemicals Division, Institute for Environmental Biology, Leverkusen, West Germany

Methods for testing toxicity of chemicals to earthworms

Over the last few years, a number of methods for determining the toxicity of chemicals to earthworms have been described (Table 1). In this paper, several of these methods will be compared. They have been intensively worked with and seem to be pertinent for ecotoxicological testing.

Table 1. Earthworm toxicity tests (laboratory)

test with earthworm-related species (Tubifex)
injection
feeding
immersion
*contact filter paper
natural soil
*artificial soil
*'artisol'

* = methods considered by the European Economic Community (EEC) as possible methods for determining the ecotoxicity of chemicals to earthworms.

A method in which an earthworm-related species, *Tubifex tubifex*, was used as test animal, was rejected. *Tubifex* is an aquatic organism which lives in water at the water-sediment interface. In contrast to this, the earthworm species relevant in this context live in arable soils (Edwards and Lofty, 1977). In addition, *Tubifex* is taxonomically not an earthworm species. As we will discuss later, results gained with one species of earthworm cannot always be directly transferred to another species or to other genera of animals.

A test in which the earthworms were injected with the chemical (Stenerson *et al.*, 1973, Stringer and Wright, 1976, Stenerson, 1979) was considered, but found unsuitable. The main reason for this is that the test situation is difficult to relate to field conditions. The method may be useful in metabolism studies, but seems unsuitable as an ecotoxicological test method. The same is true for the feeding method in which the chemical is introduced into the gut of the test animal by means of a small glass tube. Similarly, a test in which the earthworms were immersed in a solution of the test chemical in water was found unsatisfactory since the data are difficult to relate to field conditions. Earlier experiments with this immersion test showed that this method leads to an incorrect toxicity assessment when the results are compared to those gained in a soil test.

Earthworms in waste and environmental management
edited by C.A. Edwards and E.F. Neuhauser, pp. 329–335

No.	Chemical	LC$_{50}$-values (a.i.) Filter Paper (μg/cm^2)	Artifical Soil (mg/kg substrate)
1	Methidathion	0.81	3.6
2	Endosulfan	5.7	6.7
3	Propoxur	21	10
4	DNOC	1.6	21
5	Benomyl	86	22
6	Chloroacetamide	2.4	24
7	Fentinacetat+Maneb	43	27
8	Chlordan	1.3	42
9	Pentachlorophenol	4.2	87
10	Mercaptodimethur	190	129
11	Dialifos	24	133
12	Carbaryl	3.0	174
13	Ethiofencarb	14	262
14	Methaphenamiphos	17	318
15	Bupirimat	15	338
16	Kaliumbromate	530	393
17	Captafol	47	496
18	Captan	290	625

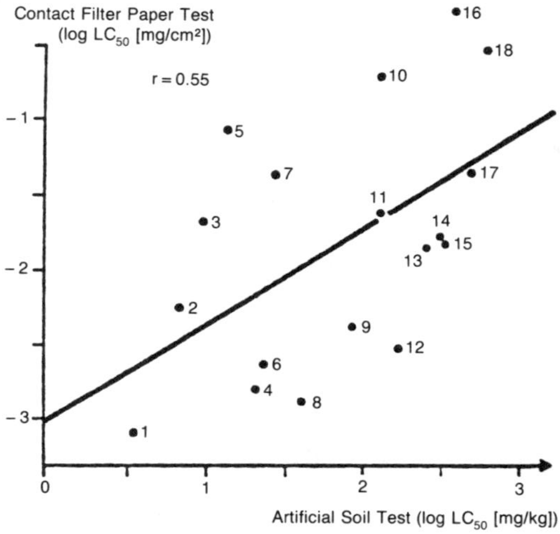

Fig. 1. LC$_{50}$-values using the Contact Filter Paper Test and the Artificial Soil Test and the correlation between these methods (Heimbach, 1984). For pesticides, the common names are shown. For further details, see Heimbach, 1984

Comparison of test methods proposed for EEC legislation

In the Contact Filter Paper Test a piece of wet filter paper is treated with the chemical and then placed in a glass tube. The *Eisenia fetida* are exposed to the filter paper and after 48 hours, the toxicity of the chemical is determined (EEC 1982; Edwards, 1983). The Artificial Soil Test more accurately represents the natural environment of earthworms. An artificial soil composed of sand, a clay mineral, peat and calcium carbonate is mixed with the chemical and water. This mixture is placed in a glass container and 10 earthworms are laid on the surface. After an exposure period of 14 days, during which time the earthworms eat and work their way through the artificial substrate, the number of living earthworms is determined (Edwards, 1983, 1984; Heimbach and Edwards, 1983). The change in weight of the earthworms during the test period gives additional information about sublethal sensitivities of the worms. In the third test, the Artisol Test, the substrate is completely artificial. It consists solely of silica mixed with water and the test chemical. The mixture is kneaded with glass balls. This is placed in a glass container and 10 earthworms are added to the surface. After 14 days, the toxicity of the chemical is determined (Ferrière *et al.*, 1981; Edwards, 1983). The earthworms feed on the silica just as they do in a natural soil. Thus, the substrate allows a high degree of standardization.

Which of these methods is most relevant for an hazard assessment of the toxicity of chemicals to earthworms? While the Contact Filter Paper Test is easy to perform, it gives only information about skin contact toxicity. In contrast to this, both the Artificial Soil Test and the Artisol Test give results on skin and gut toxicity. A comparison of results from these different methods should give us information about their utility. The toxicities of 18 different chemicals using the Contact Filter Paper Test

No.	Chemical	LC_{50}-values (a.i.) Filter paper ($\mu g/cm^2$)	Artisol (mg/kg silica)
2	Endosulfan	5.7	3.0
5	Benomyl	86	18
6	Chloroacetamide	2.4	75
8	Chlordan	1.3	15
9	Pentachlorophenol	4.2	316
10	Mercaptodimethur	190	147
12	Carbaryl	3.0	151
13	Ethiofencarb	14	264
16	Kaliumbromate	530	1000
17	Captafol	47	351
18	Captan	290	691
19	Copper Sulphate	30	1930

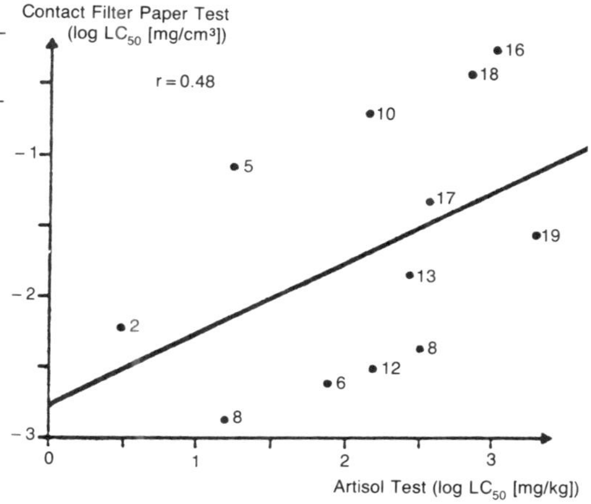

Fig. 2. LC_{50}-values using the Contact Filter Paper Test and the Artisol Test and the correlation between these methods (Heimbach, 1984). For further details, see Fig. 1

and the Artificial Soil Test are summarized in Fig. 1. With the Contact Filter Paper Test, the toxicities ranged from 0.8–530 $\mu g/cm^2$ filter paper, within the Artificial Soil Test from 3.6–625 mg/kg dry weight of substrate. A comparison of these results by a plot based on the logarithmical values of the LC_{50}'s (Heimbach, 1984) demonstrated that there was not a good agreement between the two methods. The deviation of the single points from the regression line is relatively large. The correlation factor of 0.55 shows only a very weak correlation. Whereas, some chemicals tended to be highly toxic in the Contact Filter Paper Test, they were relatively nontoxic in the Artificial Soil Test (see No. 9 and 12). Some were highly toxic in the Artificial Soil Test but of relatively low toxicity in the Contact Filter Paper Test (see No. 3, 5 and 7).

In the Artisol Test, the results gained with the different chemicals (Heimbach, 1984) ranged from 3–1930 mg/kg dry weight silica (Fig. 2). The deviation of the single points from the regression line was also relatively great. The two test methods were poorly correlated. The correlation coefficient of 0.48 was very weak. The relationship of the toxicity of the substances between these methods was about the same as the corresponding relationship between the Contact Filter Paper Test and the Artificial Soil Test. While some chemicals tended to be highly toxic in the Contact Filter Paper Test, they were relatively nontoxic in the Artisol Test (see No. 6, 9 and 12). Some were highly toxic in the Artisol Test but of relatively low toxicity in the Contact Filter Paper Test (see No. 5).

In contrast to this, the comparison of results gained in the Artificial Soil Test and the Artisol Test (Heimbach, 1984) show quite good agreement (Fig. 3). The data demonstrate that the deviation of the single points from the regression line is relatively small. The correlation coefficient was 0.91. Thus, the correlation between the results of these two methods was very strong. This indicates a similar mode of uptake of the chemicals by the earthworms.

For this comparison of the different methods suggested for earthworm toxicity

No.	Chemical	LC$_{50}$-values (a.i.) Artificial Soil (mg/kg substrate)	LC$_{50}$-values (a.i.) Artisol (mg/kg silica)
2	Endosulfan	6.7	3.0
5	Benomyl	22	18
6	Chloroacetamide	24	75
8	Chlordan	42	15
9	Pentachlorophenol	87	316
10	Mercaptodimethur	129	147
12	Carbaryl	174	151
13	Ethiofencarb	262	264
16	Kaliumbromate	393	1000
17	Captafol	496	351
18	Captan	625	691

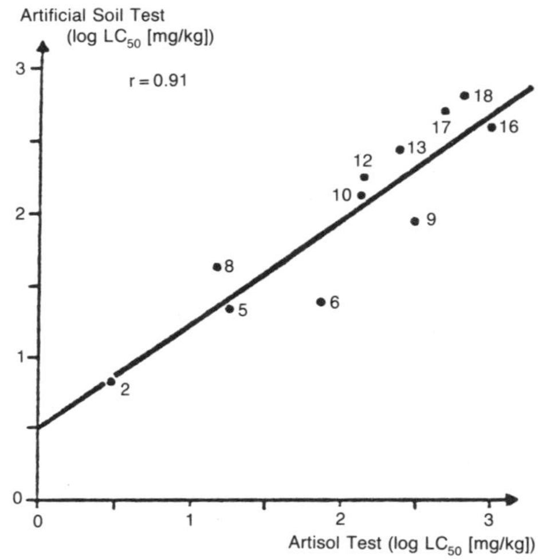

Fig. 3. LC$_{50}$-values using the Artificial Soil Test and the Artisol Test and the correlation between these methods (Heimbach, 1984). For further details, see Fig. 1

testing the following statements can be made: The methods most representative for a hazard assessment are the methods using a natural (Karnak and Hamelink, 1982) or an artificial soil (Lofs-Holmin, 1980; Edwards, 1983, 1984; Haque and Ebing, 1983; Heimbach and Edwards, 1983). Natural soils differ greatly and cannot be standardized to a sufficient degree for legislative purposes. Therefore, substrates have been developed which allow a high degree of standardization and will give reliable data for a hazard assessment. The Artificial Soil Test and the Artisol Test seem to be equally good for determination of the toxicity of chemicals to earthworms. The reasons for the high correlation between these two methods are not fully understood. The mode of uptake of the chemicals, namely through both skin contact and gut uptake is possibly the same. On the other hand, it is also possible that the degree of adsorption of the chemicals to the two substrates is similar and this gives the good correlation. However, uptake by the gut may also depend on the characteristics of the chemicals, and this can be independent of their adsorption to the substrates (Heimbach, 1984). Further experiments on these questions are needed.

Recently the EEC adopted the Artificial Soil Test as the method for investigating the toxicity of chemicals to earthworms (EEC, 1982, 1984). Since our experience has shown that the Artisol Test is easy to handle and to standardize, results obtained in different laboratories using this method might also be of a reliable degree of comparability.

Sensitivity of different test species

Most of the methods described use the earthworm species, *Eisenia fetida*, as the test organism. There are several other species of earthworm that have been used in conjunction with laboratory testing, *e.g.*, *Lumbricus terrestris, Lumbricus rubellus, Allolobophora caliginosa* or *Allolobophora chlorotica*. All of these species present the

No.	Chemical	LC_{50} (mg a.i./kg soil) Lumbricus terrestris	Eisenia foetida
1	Benomyl	3.5	27
2	Terbufos	4.6	6.6
3	Carbofuran	4.7	28
4	Propoxur	5.2	27
5	Methidathion	7.6	4.8
6	Endosulfan	9.0	9.4
7	Aldicarb	26	3.3
8	Fentinacetat+Maneb	44	28
9	Cu-oxichlorid	98	about 900
10	Bupirimat	103	401
11	Methamidophos	110	17
12	Lindane	113	136
13	Calciumcyanamid	124	115
14	Dialifos	174	218
15	Triazophos	210	116
16	Captan	237	612
17	Atrazin	444	131
18	Folpet	459	339
19	Paraquat	>200	>200
20	Triadimefon	>250	>250
21	Chlormequat chloride	>460	>460
22	Sodium chlorate	>750	>750
23	Captafol	>800	516

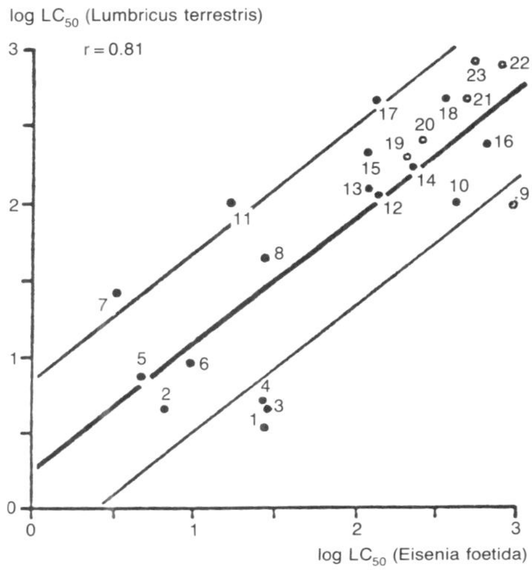

Fig. 4. LC_{50}-values found using *Lumbricus terrestris* and *Eisenia fetida andrei* and the correlation between these species (Heimbach, 1985). The outer lines represent the 95%-confidence limits. Open circles represent chemicals for which only values of '> ... mg/kg' or 'about ... mg/kg' are available. For further details, see Heimbach, 1985.

investigator with difficulties in breeding, maintenance of cultures, and provision of sufficient quantities of organisms for routine testing. In contrast to these species, *Eisenia fetida* can be bred readily in the laboratory and is thus available for experiments throughout the entire year. In regard to *Eisenia fetida*, there has been some question as to whether or not this species is representative or less sensitive to chemicals than the previously mentioned species.

Haque and Ebing (1983) published results from laboratory tests in which *Lumbricus terrestris* was tested with several chemicals in a natural soil. In the same paper they published values with *Eisenia fetida* in the Artificial Soil Test. Adding results from our laboratory in which *Eisenia fetida andrei* (Bouché, 1972) was used in the Artificial Soil Test to those of Haque and Ebing (1983), the sensibility of 23 different chemicals to *Lumbricus terrestris* and *Eisenia fetida andrei* could be compared (Fig. 4).

The LC_{50}-values for the two species lie within the range of variation found in ring tests between different laboratories with one earthworm species (Edwards, 1983). Only a few of the results lie outside of the 95%-confidence limits. Of these, chemicals No. 1, 3, 4 and 7 are highly toxic to both species. The toxicity of chemical No. 11 to *Eisenia fetida andrei* seems to be higher than to *Lumbricus terrestris*; the same is true with No. 17. The toxicity of chemical No. 9 to *E. fetida* is unclear. In general, the toxicity of the chemicals to the two species of earthworms is comparable. Thus, there is a good correlation between the toxicity of the chemicals to the earthworm species tested ($r = 0.81$).

From these experiments the results gained with *Eisenia fetida andrei* as test species are comparable to those obtained with *Lumbricus terrestris*. *Eisenia fetida andrei* seems to be neither more nor less sensitive to chemicals than *Lumbricus terrestris*. It must, however, be mentioned, that some authors have found differences in the sensitivities when comparing *Eisenia fetida* to other earthworm species. Roberts and Dorough (1984) used the Contact Filter Paper Test with *Eisenia fetida* and *Lumbricus rubellus*. They found that the toxicities of the chemicals were quite different between these two species. Thus far, we do not have sufficient results with the chemicals they used in the Artificial Soil Test to determine if real differences in sensitivities of different species exist. Thus, we cannot determine if the species-specific-differences that they found are representative for natural or artificial soils. The results presented in the current discussion demonstrate, that results obtained with the Contact Filter Paper Test cannot be related to those found when the earthworms are tested in a soil. Thus, it is difficult to compare the results of these authors to those obtained in our and other laboratories.

Conclusions

In this paper, different methods for testing the toxicity of chemicals to earthworms in the laboratory have been described. Of these methods, those using a natural soil, namely the Artificial Soil Test, and one using an artificial substrate, namely the Artisol Test, seemed most reliable for examining ecotoxicological testing of the influence of chemicals on earthworms. In these tests the means of exposure of the organisms to the chemicals is similar to the exposure under field conditions. Of advantage with both the Artificial Soil Test and the Artisol Test is that both allow a high degree of standardization. If used routinely by many laboratories, they might provide reproducible and reliable data for evaluating the toxicity of chemicals to earthworms. According to the results of Haque and Ebing (1983), the earthworm *Eisenia fetida* is neither more nor less sensitive to most chemicals than *Lumbricus terrestris*, an earthworm species which is among the most common species in arable soils. Therefore, the Artificial Soil Test, which was recently adopted for use in the EEC guidelines (EEC, 1984), seems to be a practical and reliable method.

References

Bouché, M.B. 1972. Lombriciens de France, Écologie et Systématique. Publ. Institut National de la Recherche Agronomique, Paris. 671 pp.

Edwards, C.A. 1983. Development of a Standardized Laboratory Method for Assessing the Toxicity of Chemical Substances to Earthworms. Commission of the European Communities Doc. EUR 8714 EN. 141 pp.

Edwards, C.A. 1984. Report of the Second Stage in Development of a Standardized Laboratory Method for Assessing the Toxicity of Chemical Substances to Earthworms. Commission of the European Communities. EUR 9360 EN. 98 pp.

Edwards, C.A. and Lofty, J.R. 1977. Biology of Earthworms. 2nd ed. Chapman and Hall, London. 333 pp.

EEC (European Economic Community). 1982. Directive 79/831, Annex V, Part C: Methods for the Determination of Ecotoxicity – level 1. Commission of the European Communitites, DG. XI/127-129/82 Rev. 1.

EEC (European Economic Community). 1984. Directive 79/831, Annex V, Part C: Methods for the Determination of Ecotoxicity – level 1, Earthworms – Artificial Soil. Commission of the European Communities, DG. XI/128/82 Rev. 5.

Ferrière, G., Fayolle, L. and Bouché, M.B. 1981. Un Nouvel Outil, Essentiel pour l'écophysiologie et l'écotoxicologie: l'élevage les lombriciens en sol artificiel. Pedobiologia 22: 196–201.

Haque, A. and Ebing, W. 1983. Toxicity determination of pesticides to earthworms in the soil substrate. PflKrankh. PflSchutz 90(4): 395–408.

Heimbach, F. 1984. Correlations between three different test methods for determining the toxicity of chemicals to earthworms. Pestic. Sci. 15: 605–611.

Heimbach, F. 1985. Comparison of laboratory methods, using *Eisenia foetida* and *Lumbricus terrestris*, for the assessment of the hazard of chemicals to earthworms. PflKrankh. PflSchutz 92(2): 186–193.

Heimbach, F. and Edwards, P. 1983. The toxicity of 2-chloroacetamide and benomyl to earthworms under various test conditions in an artificial soil test. Pestic. Sci. 14: 635–636.

Karnak, R.E. and Hamelink, J.L. 1982. A standardized method for determination the acute toxicity of chemicals to earthworms. Ecotoxicol. Environ. Safety 6(2): 216–222.

Lofs-Holmin, A. 1980. Measuring growth of earthworms as a method of testing sublethal toxicity of pesticides. Swedish J. Agric. Res. 10: 25–33.

Roberts, B.L. and Dorough, H.W. 1984. Relative toxicities of chemicals to the earthworm *Eisenia foetida*. Environ. Toxicol. Chem. 3: 67–78.

Stenerson, J. 1979. Action of pesticides on earthworms. Part 2: Elimination of parathion by the earthworm *Eisenia foetida* (Savigny). Pestic. Sci. 10(2): 104–112.

Stenerson, J., Gilman, A. and Vardanis, A. 1973. Carbofuran: Its toxicity to and metabolism by earthworm (*Lumbricus terrestris*). J. Agr. Food Chem. 21(2): 166–171.

Stringer, A. and Wright, M.A. 1976. The toxicity of benomyl and some related 2-substituted benzimidazoles to the earthworm *Lumbricus terrestris*. Pestic. Sci. 7(5): 459–464.

SUBLETHAL ECOTOXICOLOGICAL EFFECTS OF DIELDRIN ON THE EARTHWORM *EISENIA FOETIDA* (OLIGOCHAETA)

J.M. VENTER and A.J. REINECKE
Department of Zoology, Potchefstroom University for CHE, South Africa

1. Introduction

Setting standards for environmental protection against hazardous chemicals is not an easy task. South Africa is fortunate in that the use of pesticides was never as intense as in Europe or North America and strict control has existed since the early sixties (Van Dyk *et al.*, 1982). Despite this, persistent organochlorine pesticides are used in vast amounts in neighboring countries in southern Africa. The disbandment of the Working Group on Pesticide Residues in 1980 after this WGPR made significant headway in obtaining an overall picture of the level of pesticide residues in the diet of the South African population, is indicative of the fact that there does not seem to be any reason for alarm. Nevertheless, very little was so far accomplished in developing ecotoxicological procedures for clearing pesticides from an environmental contamination point of view.

Van Dyk *et al.* (1982) provides a rather complete and very informative review of the management and determination of pesticide residues in South Africa. The Standing Interdepartmental Advisory Committee Safeguarding Man Against Poisons (INDAC) was established in 1969. Large-scale poisoning of birds and fish in overseas countries led to the establishment of the WGPR in 1972 as part of the National Programme for the Environment. In South Africa, the environmental hazard of pesticides is evaluated on an *ad hoc* basis. Wiese and Bot (1971) have called urgently for more intensive research and pointed out the inadequacy of facilities for clearing pesticides from an environmental contamination point of view. Very little has changed since then. Van Dyk *et al.* (1982) stress the need for more trained and experienced residue chemists as well as the continued close liaison between laboratories still active in the analysis of pesticides. In our opinion, the need for well-trained biologists with a sound background in residue analysis should not be overlooked, since the pesticide problem is in fact rather a biological than a chemical one.

Studies were undertaken by Van Jaarsveld (1969) on the biological effects of dieldrin on freshwater fish. Van Dyk and Greeff (1977) studied the effects of both low and high concentrations of endosulfan on aquatic fauna, Greeff and Van Dyk (1978) on the effects of fenthion pollution on aquatic invertebrates and the effects of dieldrin on guinea fowl (Wiese *et al.*, 1969). Various factors lead to the conclusion that aquatic fauna could not be used as indicators of pollution. No advance was therefore made in developing ecotoxicological evaluation procedures.

Recent EEC legislation on new chemicals makes studies for purposes of environmental protection obligatory. In practice, however, acute toxicity tests are designed typically to determine the concentrations of a chemical which produces harmful responses on the test animal during continuous short-term exposure to the chemical and the relationship between concentration and the intensity or degree of response. Response can be any sort of adverse effect, including death, that may occur within

Earthworms in waste and environmental management
edited by C.A. Edwards and E.F. Neuhauser, pp. 337–353

a short time. Very often, acute toxicity tests are equated with lethal tests aimed at determining the concentration (dose) likely to kill 50% of the test group (LD_{50} or LC_{50}). These tests have little, if any, value for the purpose of setting standards for environmental protection purposes. Ecotoxicological tests are still not adequately developed to fulfil the requirements.

The lastest South African guide to the use of pesticides and fungicides, compiled by Bot *et al.* (1983) of the Plant Protection Research Institute of the Department of Agriculture now also includes the acceptable daily intakes (ADI) of the various pesticides as set by the World Health Organization. Apart from residue data submitted in support of registration, the tolerances laid down by the Food and Drug Administration of the United States Department of Health and Welfare are accepted in South Africa. In other instances, values set in either The Netherlands or in West Germany are accepted (Wiese and Bot, 1971). A standing Advisory Committee on Pesticides 'considers' applications where no tolerances have been established in the abovementioned countries.

The data relied upon, apart from residues, are aimed usually at determining the acute toxicity of a chemical. These tests may be used to determine either lethal or non-lethal effects. Legislators usually require acute lethal toxicity values (LC_{50}) for a particular species. It is general knowledge that within any population of organisms there is a probability distribution of threshold values. As a general assumption this distribution within any population is a lognormal one (a unimodal, positively-skewed distribution typical of most biological variables). Furthermore, it is generally assumed that the relation between stimulus (concentration) and degree of response is a monotonic one (increase in the magnitude of effect as concentration becomes greater). These assumptions are relied upon to construct the well-known dose-mortality rectilinear curve.

The eventual LC_{50} or LD_{50} value obtained from this logarithmic-probability curve provides a sample value with its own limitations. Conventional evaluations recognize this fact and the error of a sample estimate is expressed as the 95% confidence limit. In this way, values are given which define at an acceptable level of probability, the limits within which the LC_{50} for a population might lie. A recent survey of the literature revealed that 95% confidence limits of the 96h LC_{50} values for various substances to various species of fish could be up to 70% of the LC_{50} value determined from the sample. There could be problems with substances of apparent low acute lethal toxicity and with such a substance any large error of estimate must significantly influence conclusions drawn about that substance.

One of the first questions to be answered is whether results can be used to classify chemicals as being of either low or of high toxicity. Secondly, whether further toxicity testing should be done and finally, whether the results are useful in evaluating environmental risks under conditions of chronic exposure to this chemical.

Generally, the answer to the first question is positive if it is accepted that the data only apply to a particular population of the species, tested under the particular test conditions and to its particular life-stage/physiological state. The need for standardization and an internationally reproducible protocol of testing procedure seems logical. But once the information is to be used relevant to environmental protection in various environments, the irrelevance of 'standardized conditions' becomes clear, especially in short-term tests where delayed effects are not taken into account. The detection of dangerous or potentially dangerous substances still remains a useful result of toxicity tests of this nature. The question about the necessity for further testing cannot be answered from data on acute lethality, since the possible effects of chronic exposure cannot be deduced from these tests. Extrapolations of this nature are without a sound objective basis and are usually highly speculative and extremely impractical. Even more unrealistic is the practice of using the LD_{50} test results of one species to predict acceptable concentrations to all species under widely differing conditions and under conditions of chronic exposure.

The question of 'environmentally acceptable' concentrations of a chemical under conditions of chronic exposure cannot be answered from the results of acute lethal tests. The sublethal effects a chemical can have at very low concentrations on behavior, reproduction etc. are not revealed by LD_{50} tests. Using only LD_{50} tests to predict environmental and health impact of a substance is without scientific credence (Sperlingh and McLaughlin, 1976). This is sometimes recognized in the scientific literature but is nevertheless referred to as 'intelligent guesswork'. This is condemned in rather strong language by Brown (1981), probably without appreciation of the very difficult or hitherto almost impossible task of executive officers (especially in developing countries of the Third World) who find themselves in the unenviable position of having to make decisions on the agricultural and other uses of chemicals; often under rather harsh political and economical pressures and with little relevant information at their disposal. This surely underlines the conclusion of Brown (1982) that tests of the effects of long-term exposure to chemicals, 'from which one gets information on how the organism reacts to a chemical in ways *other than* by dying quickly', are essential to obtain data from which exceptable chronic levels could be derived. The necessity of sub-acute testing with a fairly wide variety of species 'representing' various trophic levels and habitats is evident if a satisfactory basis for decision-making is to be established and this is exactly where the zoologist has a major role to play.

Looking at the practical side of these suggestions, and their possible implementation, one is immediately confronted with the lack of workable procedures designed to provide the required information within reasonable time. One must immediately realize that data regarding effects of long-term exposure cannot easily be provided by short-term tests. Ecotoxicological test procedure are time-consuming and probably labor-intensive as well. With environmental and health effects of 2, 4, 5-T and recent industrial accidents as well as the 1976 dioxin incident in Seveso, time and labor (albeit discouraging factors from an economical point of view) should not prevent the development of scientifically sound procedures. The results of such tests might in many cases even provide the information required to counter the emotional and often irrational pressures exerted by uninformed groups which operate under the banners of nature conservation. On the other hand, these results could pinpoint those substances against which the general public and the environment should be protected, results which could in the long run benefit all parties concerned. The purpose of the present study was to contribute towards the development of a more satisfactory basis for correct decision-making.

The search for sublethal effects which might be monitored should continue. If these effects occur at much lower concentrations of the toxic substance they might provide a very sensitive means of detecting its presence. Concentrations at levels preventing homeostatic mechanisms of the organism to cope with the stress could give firm indications of levels to be regarded as dangerous. Various sublethal parameters could be monitored ranging from biochemical to the behavioral. The ideal parameter would be one altered only by specific toxic substances. The inhibition of acetylcholine esterase by organophosphorus insecticides is a good example of such a parameter at the biochemical level. But very little, if any, changes at the organism or population level could be deduced directly from knowledge of effects on the biochemical level (Duffus, 1980). Changes in enzyme activities, electrolyte concentration and blood parameters are convenient biochemical changes to monitor in animals. Eisler (1972) compiled so-called *stress profiles*, trying to assess an overall picture based on as many parameters as possible. Mount and Stephen (1967) defined the *population index* based on the change in biomass of a population measured over at least one generation. It incorporates effects on growth and reproduction and requires considerable time and effort. It is, however, doubtful whether any short cuts exist if sound standards are to be set for environmental protection against hazardous chemicals – especially since phenomena such as biodegradation, bio-concentration, persistence and adsorption are involved and require long-term evaluations.

The present approach was to investigate the effects of toxic substances on growth and reproduction of an earthworm species. Furthermore, the approach recognizes the need for methods and tests with wide ecological significance expressed by researchers of several countries (Lofs-Holmin, 1980). Invertebrates have, until recently, for traditional and other practical reasons, been neglected as experimental animals. Earthworms have been suggested for LD_{50} testing (Edwards, 1978) and for sublethal testing (Lofs-Holmin, 1980). For practical and economical reasons, those parameters indicating the effects of pesticides on the test organisms should be identified for future use in ecotoxicological testing procedures. The aim of this study was therefore also to single out those quantitative effects readily measurable for such purposes.

Pesticides are normally non-specific in their action and the amount used is often far in excess of that required to control the pest. This results in the exposure of wildlife and beneficial soil organisms, such as earthworms, to the negative influences of these chemicals. When studying the effects of pesticides on organisms, the sublethal effects are often overlooked. In the environment where the forces of stress and natural selection are operative, a very slight loss in reproduction capacity or small changes in behavior may have important consequences for the whole population. The sublethal effects of environmental contaminants could in many cases be of great ecological importance.

The side effects resulting from the use of organochlorines have been known since the late fifties and early sixties. Sufficient proof has been provided that these substances do have detrimental effects on ecosystems where bioconcentration in animal tissue has occurred. Researchers such as Edwards (1973), Doane (1962), Davis (1971), Beyer and Gish (1980) and Reinecke and Nash (1984) have provided sufficient evidence that chlorinated substances such as aldrin, dieldrin, DDT, dioxin and others are concentrated up to 10 times in the body tissue of earthworms as compared to the concentration in the surrounding soil. Proof has been provided by Beyer and Gish (1980) that organochlorine pesticides are indeed bioaccumulated in the food chain by predators feeding on earthworms.

The phenomenon of delayed response which is often manifested on different organizational levels on cellular or organismic level, has a direct bearing on survival of a population. The increasing use of pesticides has provided us with an abiotic factor which operates in almost every ecosystem. According to Stenerson (1979), Lofs-Holmin (1980), Oller *et al.* (1980) and Karnak and Hamelink (1982) more information on the effects of pesticides on non-target organisms such as earthworms are required in order to use earthworms as indicators of environmental contamination. The search for indicator species for the terrestrial habitat has only just started according to Lebrun (1980) and Krivolutsky *et al.* (1982). The latter authors consider the earthworm an ideal test organism due to its close contact with the soil.

The present study was undertaken to determine the sublethal effects of the persistent chlorinated pesticide, dieldrin, on the earthworm, *Eisenia foetida* as a typical non-target organism. Dieldrin was banned for domestic and agricultural use in the U.K. in 1964. Most developed countries have followed suite. In the Republic of South Africa, the use of dieldrin for domestic and agricultural purposes was restricted in 1981, shortly after the commencement of this study. Special permission is now required to use it for certain industrial purposes. The restricted use or subsequent banning of dieldrin does not mean that existing stock has not been used since the notice was published in the Government Gazette of 1981-05-01. A conservative estimation of organochlorines sold since 1976, shows a decline from 2,000 tonnes in 1976 to 200 tonnes of active ingredient per annum in 1981. Chlorinated pesticides are being phased out on economic grounds rather than on ecological or environmental grounds (Venter, 1983).

2. Material and methods

Eisenia fetida was reared in the laboratory in environmental control chambers at a temperature of 20°C on an organic substrate (Venter, 1983). According to Lofs-Holmin (1980) sublethal effects of pesticides on the life cycle of earthworms are demonstrated more readily by the growth rate of juveniles, the development of sexual characteristics such as the tuberculae pubertates and the clitellum. Newly-hatched worms were selected and exposed to various concentrations of dieldrin. The pesticide was mixed with the organic culture medium consisting of washed cow manure. Five groups consisting of 25–30 newly-hatched worms in each group were used. One group served as control. The other four groups were exposed to concentrations of 10 mg kg^{-1}, 30 mg kg^{-1}, 50 mg kg^{-1} and 100 mg kg^{-1} (dry weight) respectively. The culture medium was pretreated with dieldrin dissolved in acetone and thoroughly mixed to ensure a uniform distribution of the pesticides. Acetone was allowed to evaporate over a period of 24 hours at 22°C before the medium was moistened to attain a moisture content of approximately 50%. The actual moisture content of the different flasks was confirmed by analysis with an infrared moisture detector.

The newly-hatched worms were weighed at the start of the experiment. The biomass of every individual worm was determined subsequently every 15 days over a period of 90 days. The development of the clitellum was used as one parameter to determine the influence of dieldrin on the sexual development of *Eisenia foetida*. Worms were observed closely every two days starting four weeks after they had hatched. Worms were classified as juvenile, preclitellate and clitellate using the rather

Table 1. Changes in total biomass (g) of five groups consisting of 30 hatchlings each of *Eisenia fetida* over a period of 90 days at various dieldrin concentrations

Period (days)	1	15	30	45	60	75	90
Control	0,1480	1,5846	10,341	18,5238	19,5153	18,5374	17,1764
$\bar{x}$	0,005	0,053	0,345	0,617	0,651	0,618	0,573
SE	—	0,002	0,014	0,022	0,023	0,023	0,022
10 mg kg^{-1}	0,1551	1,2939	8,1407	15,8672	17,9856	18,1852	17,2032
$\bar{x}$	0,005	0,043	0,2714	0,5291	0,599	0,606	0,573
SE	—	0,002	0,012	0,024	0,028	0,028	0,028
30 mg kg^{-1}	0,1838	7,6183	15,2154	17,3715	17,6340	16,4060	–
$\bar{x}$	0,005	0,039	0,254	0,507	0,579	0,588	0,547
SE	—	0,002	0,009	0,024	0,026	0,027	0,027
50 mg kg^{-1}	0,1454	1,0252	6,3498	12,1968	15,3504	15,1224	15,0146
$\bar{x}$	0,005	0,034	0,212	0,407	0,512	0,504	0,501
SE	—	0,002	0,017	0,034	0,042	0,042	0,042
100 mg kg^{-1}	0,1505	0,7503	5,2727	12,5649	15,9661	17,2873	18,5044
$\bar{x}$	0,005	0,025	0,173	0,419	0,532	0,576	0,617
SE	—	0,003	0,014	0,028	0,033	0,036	0,038

(x = mean, SE = standard error)

subjective criterium of clitellum absence, partial development of the clitellum and the presence of a fully developed clitellum. The culture medium of 15 flasks were thoroughly searched for cocoons every second day, starting on day 35. Cocoons were weighed and transferred to multicell containers to be incubated in distilled water. The containers were kept in the dark and the water contents were replaced weekly to prevent bacterial growth.

Dieldrin was extracted from the culture medium by steam distillation. This extraction procedure was used successfully by Nash and is described fully, together with the clean-up procedure by Venter (1983). Analysis of dieldrin was performed gaschromatographically using a Carlo Erba Strumentazione Fractovap model 2150 GC. Various column sizes and packing material were used during the course of the study. Column 1: 1,5 m × 4 mm, 3% OV-17 on 80–100 chromosorb WAW-DMCS; column 2: 2 m × 4 mm, 5% SP2401 on 100–120 Supelcoport; column 3: 2,5 m × 4 mm, 1,5% SP-2250 plus 1,95%, SP-2401 on 100–120 Supelcoport. The columns performed satisfactorily at 254°C. Nitrogen served as carrier gas at a flow rate of 60 cm per minute. A Ni^{63} detector was used at a temperature of 300°C. Nash and Harris (1973) and Suzuki *et al.* (1979) used very similar conditions for GC analysis of organochlorines. Dieldrin contents of worms are expressed in terms of wet weight and that of the various exposure media in terms of dry weight.

3. Results

Influence of dieldrin on growth rate of the earthworm

Growth rate was measured in terms of change in biomass. The growth pattern ob-

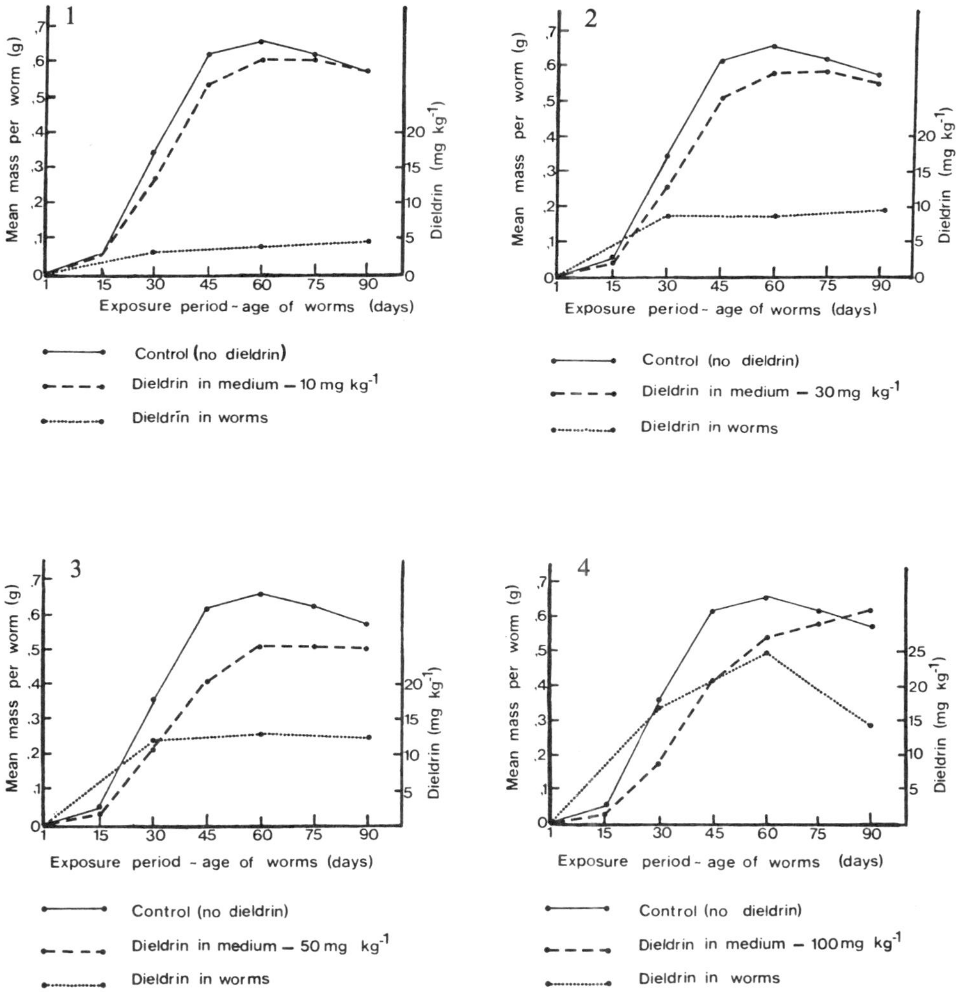

Figs 1–4. Change in biomass of newly hatched specimens of *E. foetida* over a period of 90 days in the controls and in culture mediums containing various concentrations of dieldrin (dry weight) (Dieldrin in worms expressed on a wet weight basis.)

tained during this study, was identical to the classical growth curve of Verhulst. The growth pattern was also similar to the results obtained by Neuhauser *et al.* (1980) for the same species in different culture mediums. Slight changes in soil moisture occurred in the flasks during the course of the study. Considering the wide tolerance range of this species for moisture, it is doubtful whether the decrease in available moisture measured over the period of 90 days could have played an important role in influencing the form of the growth curves. Nevertheless, comparison of the moisture conditions at the various exposure concentrations on days one, thirty, sixty

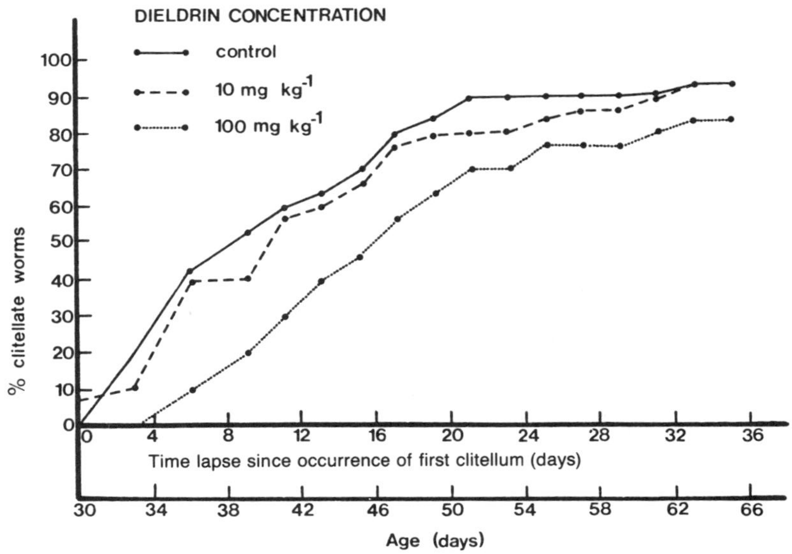

Fig. 5. The rate of development of clitella in *E. foetida* exposed to various dieldrin concentrations (dry weight).

and ninety as well as in the control flasks clearly illustrates the uniformity that was obtained.

The changes observed in the biomass of the various groups are summarized in Table 1 and the growth curves obtained for hatchlings (Figs 1–4). Growth was usually slow over the first few days after which a sharp linear increase in biomass was registered. A mean change in biomass of 570 mg per worm occurred until day 45. Thereafter, the growth rate slowed down considerably and after day 60 the growth curve showed a downward trend until day 90 when the experiment was aborted. The only significant difference between the controls and the various groups was in the case of the groups exposed to 50 and 100 mg kg^{-1}. The highest biomass was registered on day 60. These worms seemed to attain their maximum biomass at a slightly lower rate. Dieldrin only seemed to retard growth slightly at the high concentration of 100 mg kg^{-1}.

Influence of dieldrin on the rate of clitellum development

Lofs-Holmin (1980) reported that the time earthworms need to attain sexual maturity is of great ecological significance, since retardation in clitellum development could affect the population size adversely. Observation showed that clitella were present in worms 30 days of age collected from the control group and the exposure groups of 10 mg kg^{-1} and 30 mg kg^{-1}. In the case of the 50 and 100 mg kg^{-1} groups, the first clitella were present only three days later. The clitellum development of various age groups exposed to different concentrations of dieldrin is summarized in Table 3 and the results for the control group and the groups exposed to 10 and 100 mg kg^{-1} in Figs 5 and 6. The influence of dieldrin on the rate of development of the clitellum is clearly demonstrated.

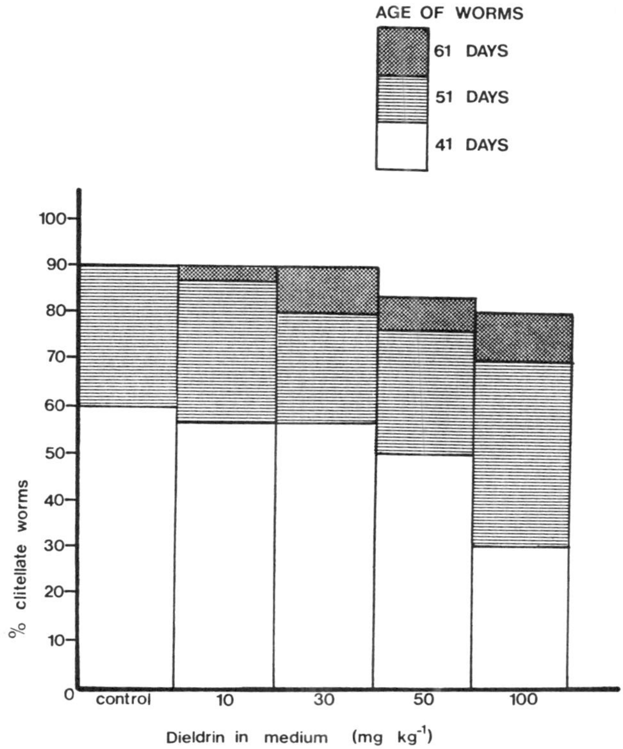

Fig. 6. The percentage clitellate worms (*E. foetida*) after exposure of the groups to various dieldrin concentrations (dry weight).

At low concentration levels, closely related to those under field conditions, the influence of dieldrin was fairly small. Earthworm population numbers in the field should normally recover within one or two seasons provided that the pesticide is degraded fast, has no accumulative characteristics and is not used repeatedly. The worms containing the higher levels of dieldrin contamination (Table 2) were clearly the slowest to develop a clitellum and quite a number of the worms exposed to higher concentrations never developed a clitellum at all, even after 100 days of observation. Nevertheless, the underlying fact is clear. Retardation of clitellum development was an observable sublethal effect of exposure to the organochlorine pesticide.

Influence on cocoon production

The worms of the control group (without dieldrin) produced one cocoon per worm every second day during a production period of 50 days. Worms started producing on the 39th day with the highest rate between days 49 and 69 (Fig. 7). The highest production rate was one cocoon per worm every second day in the control group. Production started slowing down slightly when the worms reach an age of 70 days. This decrease corresponds roughly with the onset of a decrease in biomass. Although the production pattern remained the same the effective production at a concentration

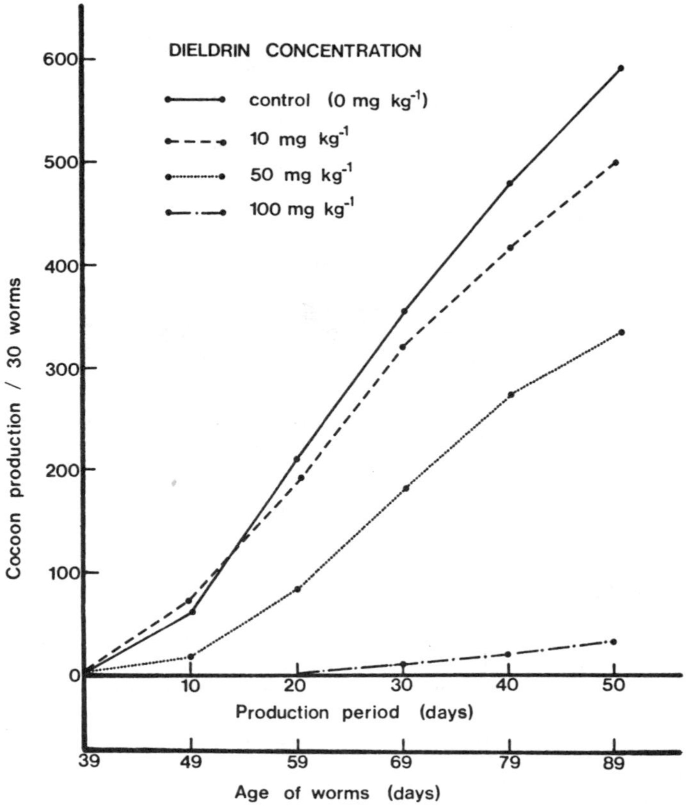

Fig. 7. Accumulated cocoon production of different groups of earthworms (*E. fetida*) in culture media containing various dieldrin concentrations.

Table 2. The mean concentration of dieldrin (expressed in mg kg^{-1}) in specimens of *E. fetida* after exposure for various periods to different concentrations of dieldrin in the culture medium

Dieldrin concentration (mg kg^{-1}) in medium	Exposure period (days)		
	30	60	90
10	3.00 ± 0,12	3,72 ± 0,24	4,62 ± 0,25
30	8,53 ± 0,06	9,87 ± 0,24	9,66 ± 0,22
50	12,08 ± 0,40	12,89 ± 1,32	12,38 ± 0,37
100	16,75 ± 1,05	24,76 ± 0,99	13,92 ± 0,33

of 10 mg^{-1} was lower and each worm produced one cocoon every third day. At 30 and 50 mg kg^{-1}, the production was one cocoon per worm every fifth day. The onset of cocoon production was also retarded. At a concentration of 100 mg kg^{-1} cocoon, production started only after the worms reached an age of 65 days. Considering the actual production period the mean cocoon production was one cocoon per worm every 20 days (see also Fig. 7). The accumulated cocoon production differed at various dieldrin concentrations (Table 4).

Table 3. The rate of development of clitella in *E. foetida* at various dieldrin concentrations

Dieldrin (mg kg^{-1})	No. of worms	% clitellates at age (days)		
		41	51	61
Control (0)	30	60,0	90,0	90,0
10	30	56,6	86,6	90,0
30	30	56,6	80,0	90,0
50	30	50,0	76,0	83,0
100	30	30,0	70,0	80,0

Table 4. Accumulated cocoon production of *E. foetida* at various dieldrin concentrations

Age[a] of worms (days)	No. of worms	Dieldrin (mg kg^{-1})				
		Control	10	30	50	100
49	30	65	69	14	15	0
59	30	209	196	84	85	0
69	30	352	323	183	182	0
79	30	469	418	269	268	23
89	30	586	499	334	334	36

[a] Also represents the exposure period since hatchlings were used.

Table 5. Compiled results of the mean cocoon mass of batches of 75 cocoons of *E. foetida* produced by worms reared in various dieldrin concentrations

	Dieldrin concentration (mg kg^{-1})				
Total mass of 75 cocoons (g)	0	10	30	50	100
	1,031	1,036	0,915	0,923	0,471[a]
$\overline{x}$	0,014	0,014	0,012	0,012	0,013
SD	0,002	0,003	0,003	0,003	0,004

($\overline{x}$ = mean, SD = standard deviation).
[a] Mass of 36 cocoons.

Influence on cocoon mass

Hartenstein *et al.* (1979) showed a definite relation between worm biomass and cocoon mass. Table 5 presents the results obtained in our study. The worms in the control groups clearly produced the cocoons with the highest biomass. The first 75 cocoons produced by each exposure group were used for biomass determinations. The present results confirm the findings of Hartenstein *et al.* (1979) mentioned above. Another peculiar phenomenon was observed. Cocoons from the same exposure groups had similar shapes. A very prominent difference in shape as well as size was observed between cocoons from the control group and the groups exposed at 30 mg kg^{-1} and 100 mg kg^{-1}.

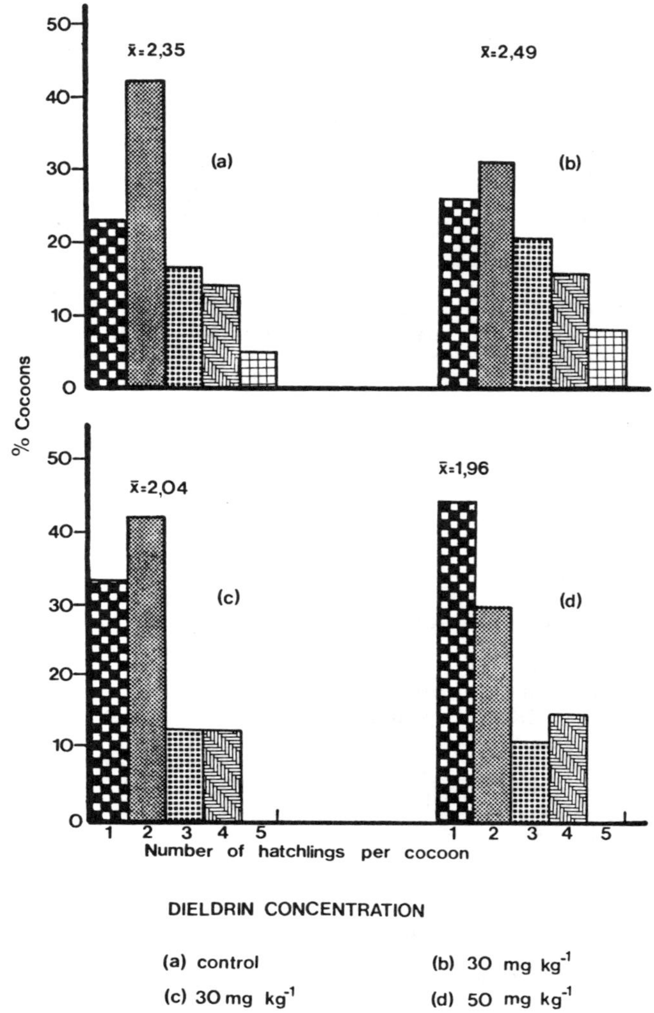

Fig. 8. The number of hatchlings per cocoon of *Eisenia foetida* in (*a*) the control and an organic medium containing; (*b*) 10 mg kg^{-1}; (*c*) 30 mg kg^{-1}; (*d*) 50 mg kg^{-1} dieldrin.

Influence on the number of hatchlings per cocoon

Although many earthworm species produce only one descendent per cocoon, *Eisenia foetida* is known to produce two or more (Evans and Guild, 1948; Vail, 1974; Tsukamoto and Watanabe, 1977). During the present study one instance was noted where 13 hatchlings emerged from a single cocoon. No relation was found between the cocoon mass and the number of hatchlings emerging from the cocoon. The parent worms were exposed to various concentrations of dieldrin (Table 2). Relatively fewer hatchlings were produced by cocoons formed at higher dieldrin concentrations, except for the control group where fewer hatchlings were obtained than in the group

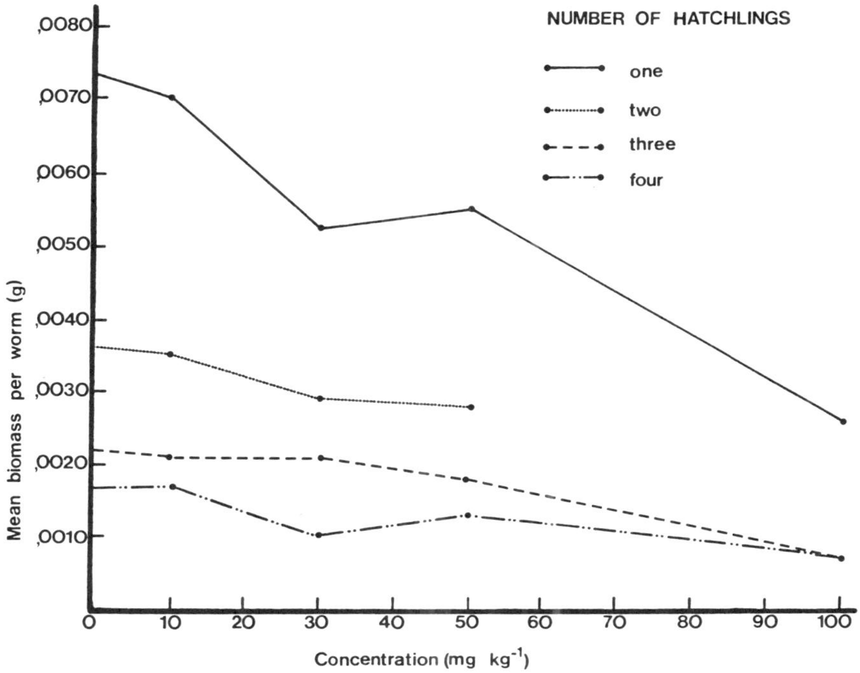

Fig. 9. The mean biomass of hatchlings per *E. foetida* that emerged from cocoons produced in various concentrations of dieldrin – where one (*a*); two (*b*); three (*c*) and four (*d*) descendants emerged from each cocoon.

exposed to 10 mg kg^{-1}. The highest percentage of cocoons producing two hatchlings per cocoon, were found in the control and the 30 mg kg^{-1} groups. In the exposure groups of 50 mg kg^{-1} and 100 mg kg^{-1} the majority of cocoons produced only one hatchling per cocoon. A maximum of five hatchlings emerged from the control group and the group exposed to 10 mg kg^{-1}. The maximum number was slightly lower in the case of the groups exposed to higher concentrations.

Figure 8 illustrates the results obtained for the control and three exposure groups.

Influence on biomass of hatchlings of E. foetida

Reinecke and Visser (1981) found for *E. rosea* and *A. trapezoides* that the bigger of the two hatchlings will normally emerge first from a cocoon. The same basic phenomenon was observed in the present study. The bigger descendant (based on biomass) emerged first in 85% of the observed cases. Figure 9 compares the mean biomass of hatchlings at the various concentrations. It was found that the mean biomass of hatchlings decreased as the dieldrin concentration in the culture medium increased (Fig. 9). The individual biomass of worms decreased as the number of hatchlings increased (Fig. 10).

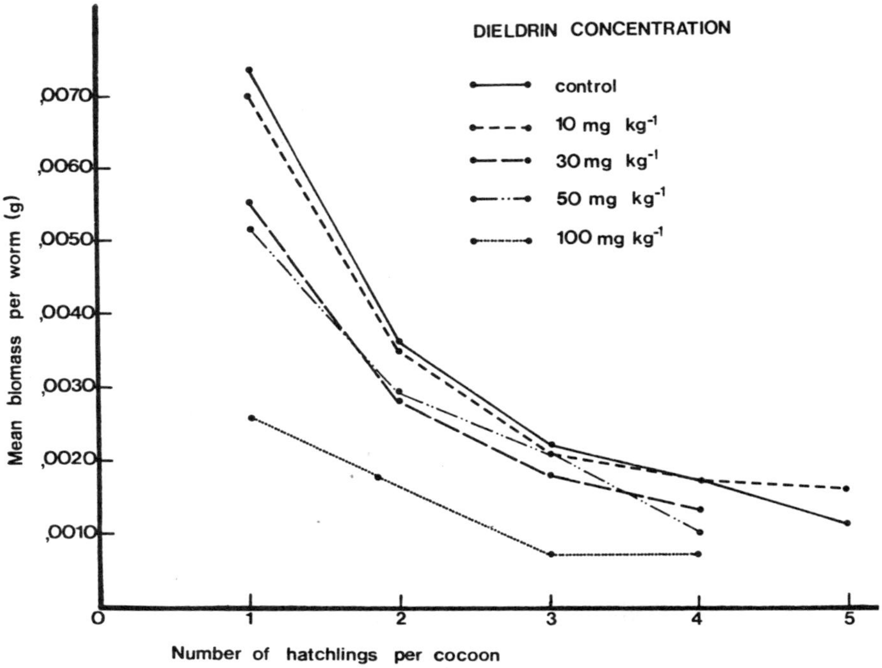

Fig. 10. The mean biomass of newly hatched individuals of *E. foetida* in relation to the number of hatchlings per cocoon (*a*) control; (*b*) 10 mg kg^{-1}; (*c*) 30 mg kg^{-1}; (*d*) 50 mg kg^{-1} and (*e*) 100 mg kg^{-1} dieldrin.

Influence on incubation period of cocoons

Incubation periods of cocoon of *E. foetida* were studied by Tsukamoto and Watanabe (1976) and Kriel (1980). They determined periods of 25,2 and 23,3 days at 20°C respectively. In the present study where the cocoons were kept at a constant temperature of 20°C a much longer mean incubation period was determined. No significant difference was observed between the control and the group exposed to 10 mg kg^{-1} dieldrin although the cocoons produced at 30 and 50 mg kg^{-1} took a week longer than the controls to hatch. The statistical significance of these differences were not assessed.

Influence of dieldrin on hatching success of cocoons

Vail (1974), Tsukamoto and Watanabe (1977), Hartenstein (1979) and Kriel (1980) studied the hatching success of *E. foetida* cocoons. Watanabe and Tsukamoto (1976) concluded that temperature influenced the hatching success of cocoons. Kriel (1980) observed a hatching success above 90% while in the studies of Vail (1974) and Hartenstein *et al.* (1979) hatching success was closer to 80%. The same success rate was obtained during our pilot experiments.

The hatching success was very low compared to the results of other authors

Table 6. The number of hatchlings per cocoon for specimens of *E. foetida* exposed to various dieldrin concentrations

	Dieldrin (mg kg^{-1})				
	Control	10	30	50	100
Number of cocoons	43	39	24	27	8
Total no. of hatchlings	101	97	49	53	13
$\bar{x}$	2,349	2,487	2,042	1,963	1,875
SD	1,131	1,204	0,999	1,091	1,246
SE	0,173	0,201	0,204	0,210	0,441

($\bar{x}$ = mean, SD = standard deviation, SE = standard error)

Table 7. The hatching success of cocoons of *E. foetida* produced in the control group and in culture media containing various dieldrin concentrations

Dieldrin (mg kg^{-1})	Number of cocoons	Number of cocoons hatched	% cocoons hatched
Control	74	43	58,1
10	75	39	52,0
30	75	24	32,0
50	75	27	36,0
100	36	8	22,2

(Table 7). A corresponding decrease in the percentage of cocoons that hatched and the increase in the dieldrin concentration in the medium was, however, evident. The study was repeated with 25 cocoons from each group. In this experiment, 76% of the cocoons hatched although the same pattern was repeated which lead to the conclusion that the hatching success was definitely influenced by the presence of the pesticide.

Discussion

The results indicate clearly that dieldrin will, even at normal agricultural doses of 4 kg ha^{-1} to 15 kg ha^{-1}, retard growth in *E. foetida*. Very little is known about the influence of pesticides on the development of sexual characters in earthworms. A decrease in growth rate will evidently retard the development of the tuberculae pubertates and the clitellum. This is confirmed by Lofs-Holmin (1980). Van Rhee (1977) showed that the presence of DDT caused only 33% of the experimental worms (*A. caliginosa*) to attain sexual maturity. Since a decrease in the reproductive potential would affect population density, this sublethal effect of the pesticide is of prime importance for purposes of ecotoxicological evaluation. The present study showed that all the exposure concentrations of dieldrin had a negative effect on cocoon production. The onset and rate of cocoon production were affected. The quality and quanti ty of available food could be expected to affect cocoon production. Dieldrin could, of course, have an influence on the feeding behavior and thus affect growth and reproduction indirectly. Higher concentrations of dieldrin corresponded with the production of smaller cocoons, which seems logical if the biomass of the worms is affected adversely. The inverse relationship observed between the concentration of dieldrin in the culture medium and the number of hatchlings per cocoon is difficult to explain. Nevertheless, since the growth rate of a population can be determined by the number of offspring, this aspect is of direct ecological significance. An increase in the dieldrin concentration led to a corresponding decrease in biomass of the newly-hatched worms. Apart from the fact that hatching success was affected by dieldrin, the smaller hatchlings emerging from the cocoons would most probably have a much lower survival rate than bigger ones.

The results showed clearly that various parameters could be used to demonstrate sublethal effects of dieldrin on the earthworm species, *Eisenia foetida*. These parameters could in future be used to study chronic ecotoxicological effects in the soil environment. A basic methodology was developed which could be used elsewhere, although many problems concerning the culture medium and its standardization are still to be solved. However, different species, even of the same genus, could vary considerably in their sublethal responses to the same chemical. In our opinion, this emphasizes the need to continue the ecotoxicological evaluation of a greater range of chemicals with a greater variety of species in order to gain a better understanding of the use of earthworms as biomonitoring tools. The predictive value of quantifiable sublethal effects in ecotoxicological studies seems clear. Before and after studies of contaminated environments should serve as timely warning devices of overburdened ecosystems. The lowly earthworm may well play an important role in helping to prevent hazardous build-up of toxic substances in the soil environment. Hopefully the day will soon come to pass when a chemical was described as 'non-toxic' for non-target species merely on the grounds that acute lethality, based on conventional LD_{50}-studies, does not occur.

References

Beyer, W.N. and Gish, C.D. 1980. Persistence in earthworms and potential hazards to birds of soil applied DDT, Dieldrin and Heptachlor. J. Appl. Ecol. 17: 295–307.

Brown, V.M. 1982. The use of toxicity data in environmental toxicology. Environment and Quality of Life, International Symposium, Sophia Antipolis. Van Bonne, 1980, CEC Luxembourg.

Bot, J., Sweet, S. and Hollings, N. 1983. A guide to the use of pesticides and fungicides in the RSA, Government Printer, Pretoria. 367 pp.

Davis, B.N.K. 1971. Laboratory studies on the uptake of dieldrin and DDT by earthworms. Soil Biol. Biochem. 3: 221–233.

Doane, C.C. 1962. Effects of certain insecticides on earthworms. J. Econ. Entomol. 55(3): 416–418.

Duffus, J.H. 1980. Environmental toxicology. Edward Arnold, London. 164 pp.

Edwards, C.A. 1973. Persistent Pesticides in the Environment. Chemical Rubber Co., Cleveland. 171 pp.

Edwards, C.A. 1978. Testing chemicals for toxicity to earthworms. Rothamsted Experimental Station, Harpenden, England, 7 pp.

Eisler, R. 1972. Pesticide-induced stress profiles. *In* Marine Polution and Sea Life. Edited by M. Ruivo. Fishing News, West Byfleet for FAO. pp. 318–322.

Evans, A.C. and Guild, W.J.McL. 1948. Studies on the relationship between earthworms and soil fertility. Ann. Appl. Biol. 34(3): 307–330.

Greeff, C.G. and Van Dyk, L.P. 1978. Preliminary investigation on the effect of fenthion on aquatic invertebrates. Chemosphere 7: 393.

Hartenstein, R., Neuhauser, E.F. and Kaplan, D.C. 1979. Reproductive potential of the earthworm *Eisenia foetida*. Oecologia 43: 329–340.

Karnak, R.E. and Hamelink, J.L. 1982. A standardized method of determining the acute toxicity of chemicals to earthworms. Ecotoxicol. Environ. Safety 6: 216–222.

Kriel, J.R. 1980. Die invloed van temperatuur op die voortplantingskoers van *Eisenia foetida* (Oligochaeta) Unpublished M.Sc.-thesis, Potchefstroom University for CHE, R.S.A. 102 pp.

Krivolutsky, D., Turcaninova, V. and Mikhaltsova, Z. 1982. Earthworms as bioindicators of radioactive soil pollution. Pedobiologia 23: 263–265.

Lebrun, P. 1980. A preliminary study of the use of some soil mites in bioassays for pesticide residue detection. *In* Soil Biology as Related to Land Use Practices. pp. 42–55. Proceedings of the VIIth International Colloquium of Soil Zoology. Edited by D.L. Dindal. EPA, Washington, DC.

Lofs-Holmin, A. 1980. Measuring growth of earthworms as a method of testing sublethal toxicity of pesticides. Swedish J. Agric. Res. 10: 25–33.

Mount, D.I. and Stephan, D.E. 1967. A method for establishing acceptable toxicant levels for fish – malathion and the butoxyethanol ester of 2,4-D. Trans. Am. Fish. Soc. 96: 185–193.

Nash, R.G. and Harris, W.G. 1973. Chlorinated hydrocarbon insecticide residues in crops and soil. J. Environ. Qual. 2(2): 269–273.

Neuhauser, E.F., Kaplan, D.L., Malecki, M.R. and Hartenstein, R. 1980. Materials supporting weight gain by the earthworm *Eisenia foetida* in waste conversion systems. Agric. Wastes 2: 43–60.

Oller, W.L., Cairns, T., Bowman, M.C. and Fishbein, L. 1980. A Toxicological risk assessment procedure: a proposal for a surveillance index for hazardous chemicals. Arch. Environ. Contam. Toxicol. 9: 483–490.

Reinecke, A.J. and Nash, R.G. 1984. Toxicity of 2,3,7,8-TCDD and short term bioaccumulation by earthworms. Soil Biol. Biochem. 16(1): 45–49.

Reinecke, A.J. and Visser, F.A. 1981. Number and size of hatchlings from cocoons of the earthworm species *Eisenia rosea* and *Allolobophora trapezoides* (Oligochaeta). Rev. d'Ecol. Biol. Sol 18(4): 473–483.

Sperlingh, F. and McLaughlin, J.L. 1976. Toxicological tests. Biological parameters and the acute LD_{50}-tests. J. Ass. Off. Anal. Chem. 59(4): 734–736.

Stenerson, J. 1979. Action of pesticides on earthworms. Part I. The toxicity of cholinesterase-inhibiting insecticides to earthworms as evaluated by laboratory tests. Pestic. Sci. 10: 104–112.

Suzuki, T., Ishikawa, K., Sato, N. and Sakai, K. 1979. Determination of chlorinated pesticide residues in foods. I. Rapid screening method for chlorinated pesticides in milk. J. Ass. Off. Anal. Chem. 62(3): 681–684.

Tsukamoto, J. and Watanabe, H. 1977. Influence of temperature on hatching and growth of *Eisenia foetida*. Pedobiologia 17: 338–342.

Vail, V.A. 1974. Observations on the hatchlings of *Eisenia foetida* and *Bimastos tumidus* (Oligochaeta: Lumbricidae). Bull. Tall Timb. Res. Stat., Florida 16: 1–8.

Van Dyk, L.P. and Greeff, C.G. 1977. Endosulfan pollution of rivers and streams in the Loskopdam cotton-growing area. Agrochemphysica 9: 71.

Van Dyk, L.P., Wiese, I.H. and Mullen, J.E.C. 1982. Management and determination of pesticides residues in South Africa. Res. Rev. 82: 37–124.

Van Jaarsveld, J.H. 1969. Die rol van dieldrin in waterbesoedeling met spesiale verwysing na die invloed daarvan op varswaterfauna. Ph.D.-thesis, Rand Afrikaans University, Johannesburg.

Van Rhee, J.A. 1977. Effects of soil pollution on earthworms. Pedobiologia 17: 201–208.

Venter, J.M. 1983. Die invloed van 'n gechloreerde pestisied, dieldrin, op die voortplanting en levensloop van *Eisenia foetida* (Oligochaeta). Unpublished M.Sc.-thesis. Potchefstroom University for CHE. 163 pp.

Wiese, I.H., Basson, N.C.J., Van der Vyver, J.H. and Van der Merwe, J.H. 1969. Toxicology and dynamics of dieldrin in the crowned guinea-fowl *Numida meleagris*. Phytophylactica 1: 161.

Wiese, I.H. and Bot, J. 1971. Pesticide regulation in South Africa. Res. Rev. 35: 49.

ELECTROPHYSIOLOGICAL DETECTION OF SUBLETHAL NEUROTOXIC EFFECTS IN INTACT EARTHWORMS

CHARLES D. DREWES[1], ELIZABETH P. VINING[1] and
CLARENCE A. CALLAHAN[2]
[1]*Department of Zoology, Iowa State University, Ames, Iowa 50011, USA;*
[2]*United States Environmental Protection Agency, Corvallis Environmental Research Laboratory, 200 SW 35th Street, Corvallis, Oregon 97330, USA*

Introduction

Earthworms are continually exposed to and accumulate a wide variety of pesticides and other toxic substances (Edwards, 1966; 1973a, b; Gish, 1970; Edwards and Thompson, 1973; Thompson, 1973; Gish and Hughes, 1982). One of the most common approaches to assessing possible adverse biological effects of these toxicants is determination of mortality (LC_{50}) following treatments with known dosages of chemicals. However, many soil contaminants are not very lethal to earthworms (Davey, 1963; Edwards, 1979; Thompson and Edwards, 1974), and therefore it seems likely that the sublethal effects of the contaminants (*e.g.*, effects on growth, reproduction, or behavior) may be much more critical in diminishing natural populations (Thompson, 1973). The nervous system of earthworms is a logical site for investigating such sublethal effects because (*i*) it controls behavior and many other biological processes, and (*ii*) many contaminants are primarily neurotoxic.

During the past few years we have been involved in developing sensitive and reliable methods for quantifying sublethal neurotoxic effects in earthworms. To do this we selected one biologically important reflex behavior, the worm's rapid escape response, and developed a means of non-invasively assessing alterations in the bioelectrical events underlying the response. The reflex response consists of a rapid longitudinal shortening triggered by various threatening stimuli, such as substrate vibration or touch. The escape reflex is particularly conducive to such assessments because it is mediated by giant nerve fibers which generate large, rapidly conducted impulses that are easily detected with electrodes placed along the ventral surface of the worm (Drewes, 1984). The reflex itself is important from an ecological standpoint because it represents a fundamental means for escaping from potential predators. It is our hope that this reflex may be used for ecotoxicological assessment in much the same way that clinicians use spinal reflexes in animals and humans for electro-diagnosis of neurotoxicity or neuropathology.

The techniques we have developed provide a number of experimental advantages. These include:

1. the sensitivity of the method, which allows quantification of sublethal neurotoxic effects which might otherwise be difficult to quantify in observations of overt behavior;
2. no requirements for anesthesia or immobilization;
3. capabilities for repeated testing of the same worms throughout long-term or chronic exposure;
4. capabilities for assessing neurotoxicity in all ages (including embryonic stages) and

Earthworms in waste and environmental management
edited by C.A. Edwards and E.F. Neuhauser, pp. 355–366

virtually all species of earthworms, and
5. capabilities for simultaneously assessing disruptions in central as well as peripheral nervous system function in different body regions.

Other factors to consider are the relatively low cost of implementing the recording system (less than $10,000) and the relatively short time required for training personnel to effectively utilize the technique. If necessary, the recording set-up could be made portable for rapid assessment of neurotoxicity in freshly collected worms at or near field sites.

Electrophysiological testing procedures

A variety of recording grids can be used for detection of electrical impulses in the giant fibers of earthworms (Drewes *et al.*, 1978; Balter *et al.*, 1980). However, for toxicological purposes the preferred system consists of an array of parallel, metalic grid lines etched onto a printed circuit board (for details, see O'Gara *et al.*, 1982). The overall size of these grids, as well as electrode spacings, can be varied in accordance with the dimension of the worms. Spacing between grid lines for embryonic worms may be as small as 250 μm (O'Gara *et al.*, 1982), or for large adult specimens up to several mm (Drewes *et al.*, 1983). An earthworm is positioned with its ventral surface in contact with four pairs of recording electrodes (Fig. 1). Signals from the electrode pairs are fed into differential recording amplifiers, filtered, and displayed as multiple traces on the storage oscilloscope (Drewes and McFall, 1980; Drewes *et al.*, 1983). For most species of worms impulse amplitudes range from approximately 20–100 μV, with signal-to-noise ratios varying from about 2:1 to 10:1.

A number of devices can be used to stimulate giant nerve fiber impulse activity. For tactile stimulation, a light and focal touch with a small hand-held stimulus probe is delivered to some portion of the body, usually the head or tail. Mechanical contact of the animal with the tip of the probe can be used to make or break a simple electrical circuit, thus providing a transient electrical signal which is used to trigger a single sweep of the oscilloscope (Drewes and McFall, 1980; O'Gara *et al.*, 1982). An alternative is to glue a lightweight extension onto the needle of a phonograph cartridge. When the animal is touched with the tip of the extension, the transient output of the cartridge is sufficient to trigger the oscilloscope sweep (Vining and Drewes, 1985). For the purpose of producing a precisely controlled tactile stimulus, a lightweight rod can be glued to the diaphragm of a small speaker. The speaker is then mounted on a micromanipulator and driven by electrical pulses from a square-pulse stimulator (Drewes and Vining, 1988). The production of precisely controlled stimuli is particularly desirable for assessing changes in touch sensitivity of escape reflexes following toxicological treatment.

Since some toxicants could selectively block touch sensory pathways, it may be desirable to test for giant nerve fiber excitability by circumventing sensory pathways. In this case electrical stimuli, delivered via the grid, can be used for direct electrical excitation of giant fibers (Vining and Drewes, 1985).

Regardless of the specific mode of giant fiber stimulation, giant fiber responses in normal animals can be reliably evoked during many minutes or hours of repeated

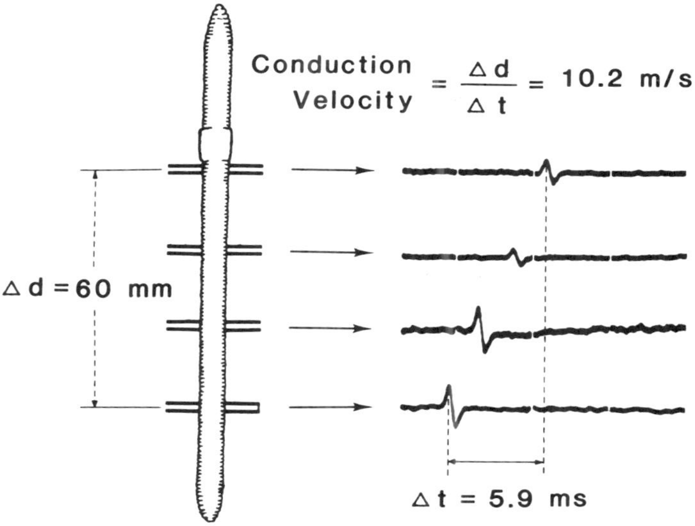

Fig. 1. Calculation of conduction velocity for giant nerve fiber impulses. Four electrode pairs in contact with the ventral surface of a worm (*Lumbricus terrestris*) were used to obtain the four-channel oscilloscope recording shown at the right. A single, anteriorly conducted lateral giant fiber impulse was evoked by very light tactile stimulation to the tail.

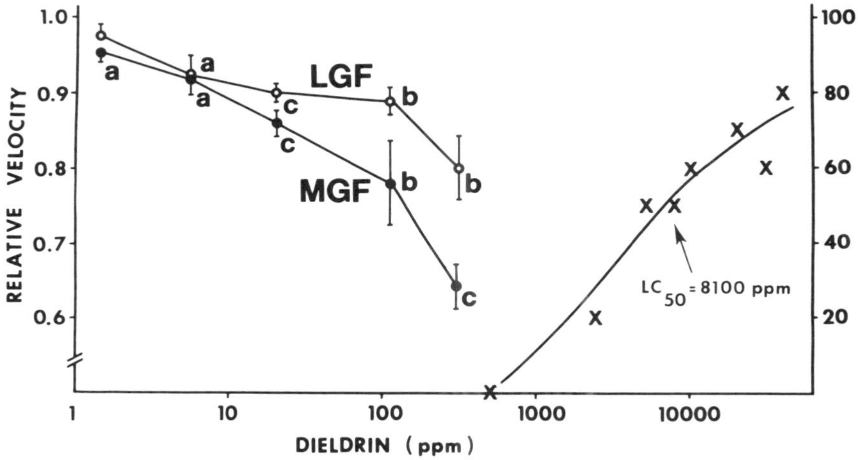

Fig. 2. Toxicity effects in *Eisenia foetida* following 48 h contact exposure to dieldrin. Mortality is shown on the right coordinate, each 'x' indicating % mortality for a group of 10 worms. Relative conduction velocity is shown on the left coordinate, each dot indicating the mean ± SEM ($n = 6$ animals). Note the sublethal reductions in relative velocity of medial (dots) and lateral (circles) giant nerve fibers. Lower case letters indicate statistical significance: (*a*) $p < 0.05$; (*b*) $p < 0.01$; (*c*) $p < 0.001$. (Modified from Drewes *et al.*, 1984).

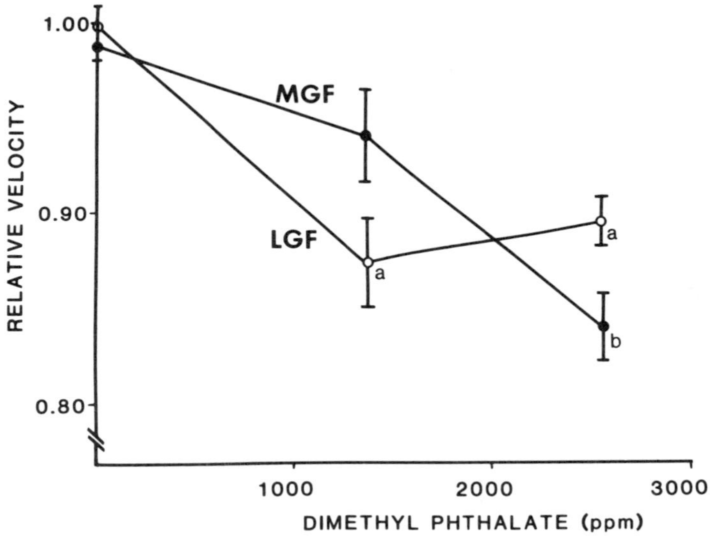

Fig. 3. Conduction velocity effects in *Eisenia foetida* following two-week artificial soil exposure to dimethyl phthalate. Relative conduction velocity values in both medial (dots) and lateral (circles) giant nerve fibers were significantly reduced by treatments. Each point indicates a mean ± SEM ($n = 6$ animals). Lower case letters indicate statistical significance: (*a*) $p < 0.01$; (*b*) $p < 0.001$.

testing. If stimulus strengths are just adequate for evoking one or two giant fiber spikes, muscle responses are small and animals make little or no attempt to crawl away. On the other hand, use of stimulus strengths which evoke high frequency trains of giant fiber spikes will often result in vigorous withdrawal and crawling responses, thus displacing the animal from the recording sites and delaying subsequent testing. Worms which have crawled away from the recording sites can be readily coaxed to crawl back into position by gently stroking head or tail segments with a fire-polished glass rod.

An oscilloscope records one lateral giant nerve fiber impulse recorded from an adult specimen of *Lumbricus terrestris* (Fig. 1). The figure illustrates how conduction velocity values are determined from such records. Several precautions should be taken when obtaining representative conduction velocity values:

1. a reliable mean is obtained from three to six replicate measurements;
2. replicate measurements should be made within the same body region, since velocities systematically vary in different portions of the same animal (O'Gara *et al.*, 1982);
3. measurements should be made using the first impulse in a train of impulses, since conduction velocity is faster for the second and later spikes in a high-frequency train (Drewes *et al.*, 1978);
4. the site of stimulation should always be well anterior or posterior to all recording sites;
5. measurements should be made only when worms are oriented in a straight line and in a reasonably elongated state; and
6. all measurements should be made at a constant temperature.

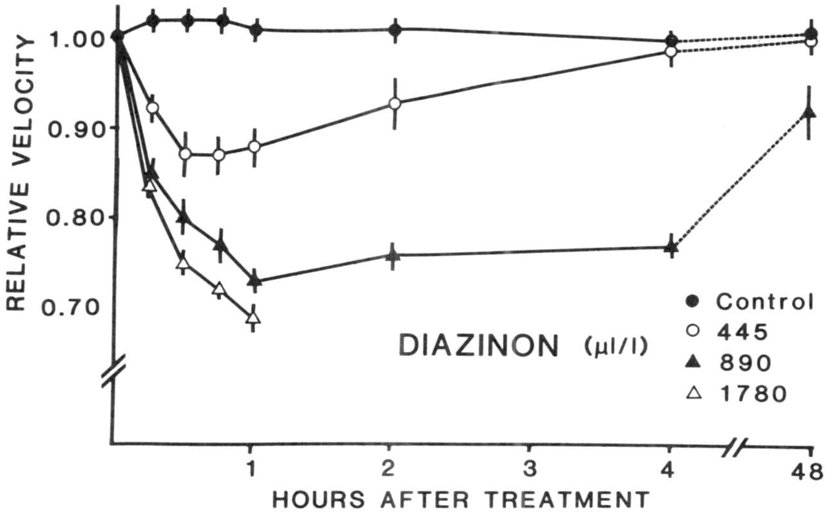

Fig. 4. Conduction velocity effects in *Lumbricus terrestris* following a two-minute immersion exposure to diazinon. Points indicate mean relative velocity ± SEM (n = 6 animals) for the medial giant nerve fiber. Dose-dependent decreases in velocity were evident, with approximately one-half of the total reduction occurring in the first 15 min following treatment. Worms at the highest concentration were dead by 2 h, but worms in the other groups recovered.

If these precautions are met, the intra-animal variability in lateral and medial giant fiber conduction velocities is very low when comparing means from the same animal on successive days. In fact, given a small group of adult worms, the combination of medial and lateral giant fiber velocities from various body regions is sufficiently unique that each animal can be unequivocally distinguished from the other. Thus velocity values together with other characteristics, such as sensory field boundaries for the giant fibers, may be used as a type of neurophysiological 'fingerprint' for each animal (Drewes *et al.*, 1983).

Mean velocities are determined at various times during exposure, or at the termination of exposure, and compared to original values. Pre-treatment and post-treatment values may be represented either as absolute velocities (*e.g.*, see Fig. 6) or as relative velocities (*e.g.*, see Figs 2, 3 and 4). Relative velocity is defined as the ratio of the mean velocity at a particular time after treatment to mean velocity before treatment (Drewes and Vining, 1984). For a group of worms, the mean relative velocity after treatment can then be compared to the initial relative velocity value (*i.e.*, 1.0) using a t-test for related measures.

Contact exposure

This procedure involves exposure of individual worms to an aqueous solution or suspension of the toxicant (Goats and Edwards, 1981; Drewes and Vining, 1988). A single animal is placed within a closed vial which is lined with appropriately shaped pieces of filter paper. Stock solutions and serial dilutions of the chemical are prepared in acetone, and an aliquot of the solution is pipetted into each vial. The acetone is then evaporated leaving a relatively uniform distribution of the chemical on the paper. Next an aliquot of water is pipetted into the vial, thus saturating the paper. After pretesting, the worm is placed in the vial for a prescribed time and retested. Concentrations in the vial can be expressed as parts of chemical per million parts of water (ppm) or μg chemical per cm^2 of paper (Goats and Edwards, 1982). In control groups, the vials are lined with filter paper and treated with acetone. After evaporation of the acetone, an aliquot of water and a worm are placed in each vial. Worms treated in this way survive for many days without statistically significant changes in neurophysiological parameters (Drewes and Vining, 1984).

Examples of results obtained following 48 h contact exposure of adult *Eisenia foetida* to dieldrin are shown in Fig. 2. The results indicate giant fiber conduction velocities were decreased significantly at concentrations as low as 1–10 ppm. The reduction was dose-dependent, with relative velocity declining to approximately 0.65 (medial giant fiber) and 0.85 (lateral giant fiber) at 288 ppm. By comparison the LC_{50} value using identical exposure testing was at least three orders of magnitude greater than the threshold concentration for neurotoxic effects (LC_{50} = 8100 ppm; 95% C1 = 4500 – 14,500 ppm.

These results clearly demonstrate that contact exposure, in conjunction with non-invasive neurophysiological testing, can be used as a sensitive means of detecting sublethal toxic effects. Based on contact exposures to a wide variety of other pesticides and toxic compounds, it appears that comparable sublethal reductions in conduction velocity often occur (Drewes *et al.*, 1984, 1987). Although the sensitivity for detecting dieldrin effects is especially high, effects of other toxicants are usually evident at concentrations ranging from one or two order of magnitude below lethal levels. In unusual cases, such as contact exposure to dimethyl phthalate (a plasticizing agent and insect repellent ingredient), no sublethal effects on conduction velocity were apparent after 48 h.

Artificial soil exposure

Chronic exposure to chemicals, within a context more similar to soil environment, can be achieved using artificial soil exposure (Goats and Edwards, 1981). The dry constituents of artificial soil include fine sand, kaolin, and peat, which are mixed together in a percentage weight ratio of 70:20:10. If insoluble in water, the test chemical can be dissolved at a known concentration in an acetone/water solution, and mixed into the dry artificial soil mixture. The acetone and water are then evaporated off, leaving a relatively uniform distribution of the chemical in the soil. Then water is added to rehydrate the mixture; we routinely use 50 g water per 100 g dry mixture. The hydrated mixture can then be apportioned to completely fill individual containers, such as glass petri dishes. Chemical concentrations in the containers can be expressed as mg of chemical per kg of dry soil (= ppm). Control groups are treated identically, except the test chemical is not included in the mixture.

Some representative results obtained following a two-week exposure to dimethyl phthalate (DMP) are shown in Fig. 3. The results show that DMP caused sublethal and statistically significant reductions in conduction velocity in both medial and lateral giant nerve fibers. No significant reductions in velocity were evident in control

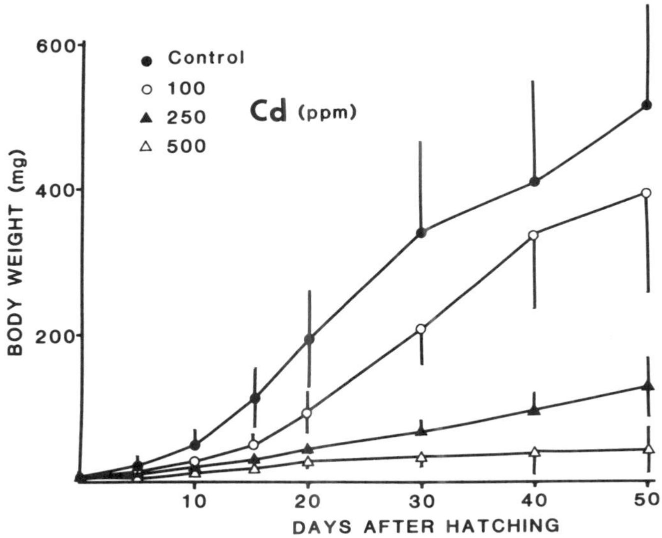

Fig. 5. Effects of cadmium treatment on postembryonic growth in *Eisenia foetida*. Each point represents the mean weight (± SD) of six worms. A dose-dependent stunting of body growth was evident.

groups (0 ppm). Apart from the DMP-induced changes in conduction, there were no other apparent behavior or neurophysiological effects. Presumably, these changes were a consequence of prolonged exposure, since no DMP-induced reductions in conduction velocity were evident in 48 h contact tests.

Immersion exposure

Most earthworms can naturally withstand complete submersion in water for long periods of time, thus providing another possible mode of exposure for toxicity studies. In contrast to contact and artificial exposure, immersion exposure in chemically treated water provides a means of studying toxicity under conditions of nearly maximal surface area for exposure and optimal uptake (Lord *et al.*, 1980). For the purpose of rapidly 'loading' an animal with a toxicant, a carrier solvent, such as dimethylsulfoxide (DMSO), can be mixed with water. In these cases neurotoxic effects of the chemical can occur after exposure for only a few minutes.

Figure 4 shows results obtained with adult earthworms (*Lumbricus terrestris*) immersed for 2 min in an aqueous solution (10% DMSO) of the organophosphate pesticide, diazinon. After immersion, animals were rinsed for 1 min in water and then tested for changes in conduction velocity. The results show a dose-dependent reduction in velocity within 15 min after exposure. Worms exposed at the highest concentration were dead by 2 h after exposure. In addition to decreases in conduction velocity, other neurophysiological symptoms were evident. These included a gradual loss of the motor neuron and longitudinal muscle electrical responses that normally accompany medial giant nerve fiber impulse activity (Drewes *et al.*, 1980). As shown in

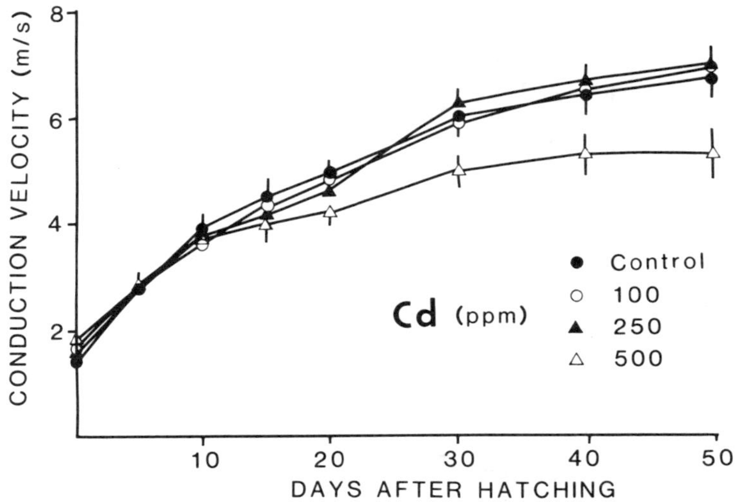

Fig. 6. Effects of Cd treatment on postembryonic development of medial giant fiber conduction velocity in *Eisenia foetida*. At the two lowest concentrations velocity increases were essentially identical to those in the control group. At the highest concentration, velocity was reduced in comparison to the control group. Each point represents the mean ± SEM (n = 6 animals). These data are from the same worms shown in Fig. 5.

Fig. 4, many of the animals achieved full recovery from the treatment, although the speed of recovery appeared inversely related to the chemical concentration.

Growth effects

Although some of the previously described exposure protocols can be used for long-term chemical exposure, none provides a food source. The provision of a food source, in combination with chemical treatment, raises the possibility of studying treatment-related effects on somatic as well as nervous system growth and development.

We have recently studied the effects of a metal toxicant, cadmium, on somatic and nervous system growth in *Eisenia foetida* raised individually in a chemically-treated mixture of garden soil and horse manure. To prepare the mixture, the chemical ($CdCl_2$) was dissolved in a small volume of water and atomized onto and thoroughly mixed with a known quantity of thawed horse manure (75% water by weight). After mixing, 40 g of moistened soil (25% water by weight) was mixed with 20 g of the treated, moist manure and placed in a petri dish along with a newly hatched worm. Animals were regularly weighed and tested for changes in giant fiber conduction velocity, the latter being a reliable indicator of nervous system growth (O'Gara *et al.*, 1982). Cadmium treatment caused a dose-dependent stunting of somatic growth (Fig. 5); at the highest concentration (500 ppm) weights were less than one-tenth those in the control group. In contrast to the effects of Cd on body growth, there was little or no effect on giant fiber growth, except at the highest concentration (Fig. 6). These

Table 1. Sublethal effects following 48 h contact tests in *Eisenia foetida*.

Chemical Neurophysiological symptoms	Carbamate carbaryl (2.5 ppm)	Organochlorine dieldrin (5 ppm)	Organophosphate diazinon (0.1 μl/ml water)	Pyrethroid fenvalerate (15.5 ppm)
1. MGF conduction velocity	OD	SD	OD	N
LGF conduction velocity	OD	SD	OD	N
SD-significant decrease				
OD-occasional decrease				
N-no change				
2. Touch sensitivity	N	HH	OH	N
HH-highly hypersensitive				
OH-occasional hypersensitivity				
N-normal sensitivity				
3. MGF-mediated muscle potentials	N	D	N/D	N
D-decreased				
N-no effect				
Behavioral/overt symptoms				
4. Peristaltic locomotion	W/A	W/A	W/A	N
A-ataxic, W-weak, N-normal				
5. Other behavioral signs	C	S	C	W
S-shortening W-writhing				
C-coiling				
6. Lesions (see Roberts and Dorough, 1984)	MSS	—	B	—
MSS-mid segmental swelling				
B-bleeding from ends				

results clearly indicate that the nervous system was spared with respect to the growth-stunting effects of Cd.

How can cadmium stunt somatic growth? One possibility is that it somehow causes reduced food consumption, for example by decreasing microbial food content or by causing distastefulness of the food source. One way to test this possibility is to examine somatic and nervous system growth under conditions of decreased food availability (*i.e.*, starvation). Vining *et al.* (1982) studied the effects of starvation on somatic and nervous system growth during postembryonic development in *Eisenia foetida*. Their results showed a dose-dependent stunting of body growth essentially identical to that seen with cadmium treatment. As with cadmium treatment, stunting of giant fiber growth, as indicated by low conduction velocities, occurred only during the most severe food deprivation and was readily reversed by replenishing food. Thus the effects of cadmium treatment were virtually indistinguishable from those of starvation, suggesting that cadmium caused a reduction in food consumption.

Class specificity of neurophysiological and behavioral effects

Pesticides can be classified in various groups depending on their general chemical configuration (*e.g.*, organophosphates, carbamates, organochorines, etc.). Common features of chemical structure within a group often result in common modes of toxic action and symptomology. Conversely, differences in structure between groups may be associated with different modes of action and symptomology. Although we have not yet examined the neurotoxic effects of many pesticides, some preliminary results shown in Table 1 suggest that there may be some class differences with respect to sublethal neurophysiological and behavioral symptoms. These may indicate differences in the fundamental mechanisms of action for each class of pesticide. There is clearly a need for more comprehensive study of this question.

In conclusion, the possibilities of using earthworms as sensitive indicators for ecotoxicity testing appear promising. For the purposes of assessing sublethal neurophysiological effects, the methods we have described herein may be particularly useful. On the other hand, for assessment of other toxicity effects in earthworms, tests involving biochemical assay (Stenersen, 1979), growth (Lofs-Holmin, 1980; Karnak and Hamelink, 1982), segmental regeneration (Drewes *et al.*, 1986) or reproductive capacity (Lofs-Holmin, 1982) may be preferable.

Acknowledgements

We thank Dr. E. Neuhauser for many helpful suggestions, Dr. J. Somers-Reed for providing samples of Diazinon, B. O'Gara for technical assistance, and M. Nims for typing the manuscript. This research was funded, in part, by the Hazardous Materials Assessment Program of the United States Environmental Protection Agency (EPA Cooperative Agreement CR8100060-01). It was administered by the Corvallis Environmental Research Laboratory as part of a research program for evaluating the effectiveness of various bioassays in determination of the biological impact of various hazardous materials.

References

Balter, R.J., Drewes, C.D. and McFall, J.L. 1980. *In vivo* conduction properties of regenerating giant nerve fibers in earthworms. J. Exp. Zool. 211: 395–405.

Davey, S.P. 1963. Effects of chemicals on earthworms: A review of the literature *U.S.D.I. Spec. Sci. Rep.-Wildl. 74: 1–20.*

Drewes, C.D. 1984. Escape reflexes in earthworms and other annelids. *In* Neural Mechanisms in Startle Behavior. pp. 43–91. Edited by R.C. Eaton. Plenum, New York, NY.

Drewes, C.D., Landa, K.B. and McFall, J.L. 1978. Giant nerve fibre activity in intact, freely moving earthworms. J. Exp. Biol. 72: 217–227.

Drewes, C.D. and McFall, J.L. 1980. Longitudinal variations in the efficacy of lateral giant fiber to giant motor neuron transmission in intact earthworms. Comp. Biochem. Physiol. 66A: 315–321.

Drewes, C.D., McFall, J.L., Vining, E.P. and Pallas, S.L. 1980. Longitudinal variations in MGF-mediated giant motor neuron activity and rapid escape shortening in intact earthworms. Comp. Biochem. Physiol. 67A: 659–665.

Drewes, C.D., Callahan, C.A. and Fender, W.M. 1983. Species specificity of giant nerve fiber conduction velocity in oligochaetes. Can. J. Zool. 61: 2688–2694.

Drewes, C.D., Vining, E.P. and Callahan, C.A. 1984a. Non-invasive electrophysiological monitoring: A sensitive method for detecting sublethal neurotoxicity in earthworms. Environ. Toxicol. Chem. 3: 599–607.

Drewes, C.D. and Vining, E.P. 1984b. *In vivo* neurotoxic effects of dieldrin on grant nerve fibers and escape reflex function in the earthworm, *Eisenia foetida*. Pestic. Biochem. Physiol. 22: 93–103.

Drewes, C.D., Zoran, M.J. and Callahan, C.A. 1987. Sublethal neurotoxic effects of the fungicide benomyl on earthworms (*Eisenia foetida*). Pestic. Sci. 19: 197–208.

Edwards, C.A. 1966. Insecticide residues in soils. Resid. Rev. 13: 83–132.

Edwards, C.A. 1973a. Persistent Pesticides in the Environment, 2nd ed. CRC Press, Boca Raton, FL.

Edwards, C.A. 1973b. Pesticide residues in soil and water. *In* Environmental Pollution by Pesticides. pp. 409–458. Edited by C.A. Edwards. Plenum Press, New York, NY.

Edwards, C.A. 1979. Interactions between agricultural practice and earthworms. *In* Soil Biology as Related to Land Use Practices. EPA-560/13-80-038. pp. 3–12. Edited by D.L. Dindal. Office of Pesticide and Toxic Substances, EPA, Washington, D.C.

Edwards, C.A. and Thompson, A.R. 1973. Pesticides and the soil fauna. Resid. Rev. 45: 1–79.

Gish, C.D. 1970. Organochlorine insecticide residues in soils and soil invertebrates from agricultural lands. Pestic. Monit. J. 3: 241–253.

Gish, C.D. and Hughes, D.L. 1982. Residues of DDT, dieldrin and heptachlor in earthworms during two years following application. U.S. Fish Wildl. Serv. Spec. Sci. Rep. 0 241: 1–16.

Goats, G.C. and Edwards, C.A. 1981. Testing the toxicity of environmental chemicals to earthworms. Rotham. Exp. Stat. Rep. 1981 (part 1): 105–106.

Goats, G.C. and Edwards, C.A. 1982. Testing the toxicity of industrial chemicals to earthworms. Rotham. Exp. Stat. Rep. 1982 (part 1): 104.

Karnak, R.E. and Hamelink, J.L. 1982. A standardized method for determining the acute toxicity of chemicals to earthworms. Ecotoxicol Environ. Safety 6: 216–222.

Lofs-Holmin, A. 1980. Measuring growth of earthworms as a method of testing sublethal toxicity of pesticides – experiments with benomyl and trichloroacetic acid (TCA). Swedish J. Agr. Res. 10: 25–33.

Lofs-Holmin, A. 1982. Measuring cocoon production of the earthworm *Allolobophora caliginosa* (Sav.) as a method of testing sublethal toxicity of pesticides. Swedish J. Agr. Res. 12: 117–119.

Lord, K.A., Briggs, G.G., Neale, M.C. and Manlove, R. 1980. Uptake of pesticides from water and soil by earthworms. Pestic. Sci. 11: 401–408.

O'Gara, B., Vining, E.P. and Drewes, C.D. 1982. Electrophysiological correlates of rapid escape reflexes in intact earthworms, *Eisenia foetida* I. Functional development of giant nerve fibers during embryonic and postembryonic periods. J. Neurobiol. 13: 337–353.

Roberts, B.L. and Dorough, H.W. 1984. Relative toxicities of chemicals to the earthworm *Eisenia foetida*. Environ. Toxicol. Chem. 3: 67–68.

Stenersen,J. 1979. Action of pesticides on earthworms. Part III: Inhibition and reactivation of cholinesterases in *Eisenia foetida* (Savigny) after treatment with cholinesterase-inhibiting insecticides. Pestic. Sci. 10: 113–122.

Thompson, A.R. 1973. Pesticide residues in soil invertebrates. *In* Environmental Pollution in Soil Invertebrates. pp. 87–133. Edited by C.A. Edwards. Plenum Press, New York, NY.

Thompson, A.R. and Edwards, C.A. 1974. Effects of pesticides on nontarget invertebrates in freshwater and soil. *In* Pesticides in Soil and Water. pp. 341–386. Edited by W.D. Guenzi. Soil Sci. Soc. Amer., Madison, WI.

Vining, E.P., O'Gara, B. and Drewes, C.D. 1982. Electrophysiological correlates of rapid escape reflexes in intact earthworms. *Eisenia foetida*. II. Effects of food deprivation on the development of giant nerve fibers. J. Neurobiol. 13: 355–367.

Vining, E.P. and Drewes, C.D. 1985. Functional connections are established between giant nerve fibers in grafted earthworms. J. Exp. Zool. 233: 121–125.

Zoran, M.J., Heppner, T.J. and Drewes, C.D. 1986. Teratogenic effects of the fungicide benomyl on posterior segmental regeneration in the earthworm, *Eisenia foetida*. Pestic. Sci. 17: 641–652.

A METHOD TO ASSESS POTENTIAL BIOAVAILABILITY OF CONTAMINANTS

J.M. MARQUENIE[1] and J.W. SIMMERS[2]
[1]*Division of Technology for Society TNO, P.O. Box 217, 2600 AE Delft, The Netherlands;* [2]*Waterways Experiment Station, U.S. Army Engineer Env. Lab., P.O. Box 631, Vicksburg, MS 39180, U.S.A.*

Introduction

In the last decade, a rapid development of biomonitoring and bioassay techniques has occurred. These techniques serve to protect the environment in a biologically relevant way. The main advantages of these techniques over chemical monitoring of the environment are that the measurements include the total response to all chemicals present and, by selection of some of them as guidance, provide direct information about bioavailability. Biomonitoring of marine waters was first advocated by Goldberg *et al.* (1978) through the mussel watch. Active biomonitoring, a technique in which animals are transplanted, was also successfully applied in fresh waters (Marquenie, 1981) and showed a high reproducibility of results. In order to monitor the bioavailability of contaminants in sediments apart from the influence of the overlying waters, the sediments have to be translocated to the laboratory or comparable clean water conditions (Marquenie *et al.*, 1983).

Since bioaccumulation cannot be reliably predicted from 'bulk' analysis of sediment and soil (De Kock *et al.*, 1982), the application of a terrestrial animal bioassay procedure was initiated by the Waterways Experiment Station. The results of terrestrial animal bioassays together with the results of related procedures for marine life (E.P.A., 1978), for animals in estuarine habitats (Rubenstein *et al.*, 1983) and for plants in estuarine and wetland habitats (Folsom *et al.*, 1981) form a biological data base allowing an optimum choice between available disposal strategies.

The present report describes the further development and application of the terrestrial animal bioassay procedure, using the earthworm *Eisenia foetida* as an accumulator for contaminants. This earthworm is a source of food for various animals, and represents the bottom of a food chain. This species has been adopted for OECD and EEC toxicity tests and can be raised in the laboratory easily with sufficiently low background contamination concentrations.

Eisenia foetida was exposed to soil materials that were collected along transects at a confined disposal site for dredged materials, Times Beach, Buffalo, New York State, U.S.A. in order to assess bioavailability of heavy metals and PCBs. The native earthworm species, *Lumbricus rubellus*, occurring at the site was collected, and further experiments were carried out that might explain the differences that were observed in concentrations of contaminants between experimental and native worms.

Times beach confined disposal site

Site description

Between 1972 and 1976, about 7.2×10^5 m^3 dredged materials from Buffalo Harbor

Earthworms in waste and environmental management
edited by C.A. Edwards and E.F. Neuhauser, pp. 367–375

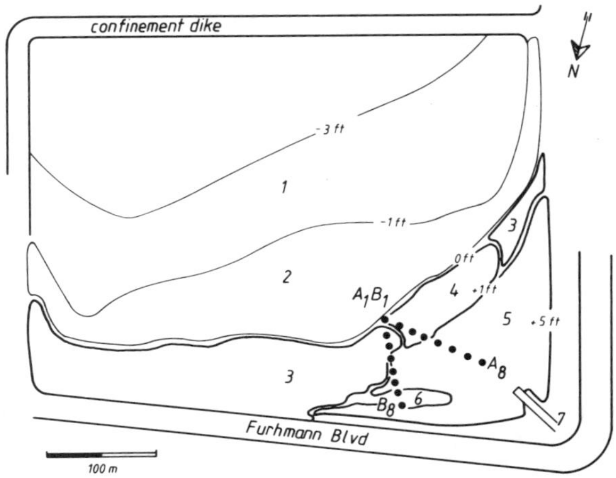

Fig. 1. Times Beach confined disposal site. Transect stations (A8 → B8) and vegetation characteristics. 1 = open water; 2 = *Typha*; 3 = mixed *Phragmites/Typha*; 4 = sedges; 5 = woodland (*Populus*); 6 = *Phragmites*; 7 = disposal pipe.

(NY) were dumped in an artificial lagoon on the shores of Lake Erie. This operation resulted in the creation of an area composed of an aquatic, a wetland and an upland environment. Although the dredged materials were heavily contaminated as a result of the activities of several industries including an oil refinery, steel plants and an aniline dye chemical plant on the waterfront adjacent to the dredging site, a prolific wildlife habitat developed at the disposal site.

This site was selected for investigation because of the recognized ecological value of a continuous sediment/soil interface gradient from a pond with a maximum depth of about 2.00 m to a woodland at about 2.00 m above groundwater levels. This situation also provided an unique opportunity to study the interactions between, on the one hand, the combination of physical conditions and biotic development and, on the other hand, contaminant mobility.

Materials and methods

1. *Bioassay technique*

The earthworm, *Eisenia foetida*, was obtained from laboratory cultures. Prior to all experiments, earthworm samples were analyzed to verify that background contaminant concentrations were sufficiently low. Soil samples were collected at Times Beach and at a nearby reference area. Soil material for the accumulation experiment, was collected at a depth of about 0.60 m (station B6, Fig. 1). Soil samples also were collected at different depths to allow the assessment of the vertical distribution of contaminants and bioavailability. Samples for the determination of the horizontal distribution (upland and wetland gradient) were collected along the transect A8-A1B1-B8: the upper 0.30 m (with the exception of the top humus-rich layer) was collected. In addition, samples were collected at a depth of 0.50–0.80 m at stations A8, A6, B6 and B8.

Randomized subsamples of about 150 worms (20.0 ± 0.5 g) each were exposed for up to 49 days (accumulation experiment) or 28–32 days (all other experiments) to 7 ℓ of soil material in polythene containers at a temperature of 15°C in low light conditions. The lower 3–5 cm were kept saturated with water, whilst the top was kept dry, allowing the worms to find their own optimum moisture conditions. Prior to chemical analysis, worms were purged for 48 hrs and homogenized with an Ultra Turrax modified with titanium blades.

ii. *Analyses*

Wet subsamples of homogenized tissue and thoroughly mixed soil materials were analyzed. Heavy metals in worm tissues were determined in about 0.5 g of tissue by neutron activation analysis, and in soil materials by atomic absorption spectrometry after wet destruction (Cd: ET-AAS; Hg: CV-AAS). Organochlorine contaminants were determined by gas chromatography. Mirex was absent and was used as an internal standard. Soil samples (~ 5 g, spiked with Mirex) were extracted and the extracts were cleaned by chromatography over alumina and dried over sodium sulphate. Tissues (1–4 g) were treated similarly, and proteins were destroyed enzymatically. After clean-up, the extracts were analyzed on a 50 cm fused silica CP Sil 19CB column with electron capture detection. Recovery ranged from 80–90%. Identification was confirmed by mass spectrometry. The moisture content of subsamples was determined by drying for 16 hrs at 105°C, and the total amount of organic materials were estimated by subsequent ashing for 4 hrs at 600°C. Contaminant levels are reported in $\mu g.g^{-1}$ on an ash-free dry weight basis for tissues and on a dry weight basis for soil samples. A listing of all organic contaminants that were analyzed in this program is supplied in Appendix 1.

The horizontal distribution of contaminants and their bioavailability

The horizontal distribution of contaminants and their bioavailability was determined in the top 0.30 m soil layer along the transects. The humus-rich layer and plant materials were excluded in order to minimize variability. Two interconnected transects (with stations 20 m apart) were laid out at Times Beach: from the water front (station A1, B1) into woodland towards the former disposal pipe (A8), and one roughly equidistant from its outlet straight to Furhmann Blvd (B8) (Fig. 1). The results shown in figures present the stations in the order B8 → A8 from top to bottom.

Water content and percent organic matter were obviously interrelated under the standardized conditions of the experiment. Percent organic matter was somewhat lower in the woodland area than in the marshlands, probably due to oxidative processes in the upland environment. There was no relation between percent organic matter and weight increase of worms. Growth of worms was inhibited in the soil from the waterfront (A1, B1).

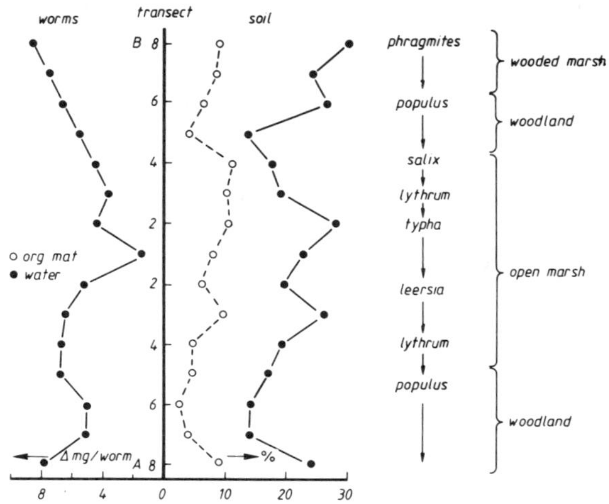

Fig. 2. Transects at Times Beach (B8 → A8, top to bottom). Composition of soil and increase in weight of exposed worms, dominant species in vegetation, and habitat.

Table 1. Metal concentrations ($\mu g \cdot g^{-1}$ dry weight) in Times Beach surface disposed materials and a control (manure)

Station	Cd	Cu	Hg	As
A8	2.10	116	2.10	25.0
A7	1.79	85.9	1.62	15.3
A6	0.76	60	1.52	20.0
A5	1.52	69.1	1.42	19.3
A4	1.87	94.2	1.55	24.1
A3	6.00	223	4.14	43.0
A2	2.73	148	4.22	38.5
A1 B1	7.22	238	4.50	37.7
B2	9.61	334	8.50	72.4
B3	10.8	288	5.42	53.4
B4	10.4	308	5.18	47.8
B5	2.01	88.5	1.18	12.1
B6	5.33	228	4.78	58.8
B7	6.63	224	3.94	36.4
B8	7.74	269	7.45	53.0
Manure 1	0.39	16.5	<0.74	3.40
Manure 2	0.32	21.0	0.069	2.1

i. *Heavy metals*

Accumulation of heavy metals is defined as the concentration of the metals in the worms, increasing over background when placed in the test material. The highest concentrations in the soil samples were found in the open marsh part of the transect (Tables 1 and 2). Apparently the conditions which preserved organic materials in this region also prevented leaching and volatilization of the metals. Exposed worms accumulated only Cd. Remarkably, uptake was relatively reduced in the open marsh

Table 2. Metal concentrations ($\mu g \cdot ^{-1}$ ash-free dry weight) in experimental worms exposed to sediments described in Table 1

Station	Cd	Cu	Hg	As
A8	8.86	27.7	0.482	21.1
A7	11.8	—	1.64	25.2
A6	6.54	17.3	0.981	17.5
A5	9.30	19.1	0.63	14.9
A4	13.3	20.8	0.75	17.2
A3	15.0	33.6	1.13	23.3
A2	11.7	32.1	1.39	24.0
A1 B1	9.01	44.8	0.80	32.2
B2	10.8	57.6	0.805	23.9
B3	11.4	56.2	1.22	24.6
B4	12.3	52.7	1.04	15.9
B5	7.99	28.3	1.28	10.4
B6	17.6	36.2	1.14	35.3
B7	16.0	35.2	1.13	33.0
B8	16.0	46.7	1.77	53.8
Manure	3.04	10.1	0.059	8.72

Table 3. Metal concentrations ($\mu g \cdot g^{-1}$ ash-free dry weight) in native worms at Times Beach and a reference area

Area	Species	Cd	Cu	Hg	As
Times Beach	*Lumbricus rubellus*	113.0	59.7	1.33	30.8
	Lumbricus rubellus	84.4	58.3	1.95	52.9
Reference	*L. rubellus*	17.6	20.2	0.469	8.84
	L. rubellus	22.4	30.2	0.549	11.8
	Allolobophora chlorotica	24.3	11.4	1.76	10.8
	Allolobophora chlorotica	22.5	10.4	2.00	10.4
	Octolasion lacteum	36.5	12.5	1.77	6.47
	Octolasion lacteum	50.7	14.8	2.34	9.59

area where the highest metal concentrations were found. Since the worms do not like anaerobic food sources and hence did not grow and did not accumulate metals, it is speculated that the higher concentrations of organic matter are responsible for this phenomenon and not differences in structure or palatability, because the growth of worms (Fig. 2) varied smoothly along the transect.

Lumbricus rubellus was found only in the woodland (Table 3) and concentrations in this species and two other species of worms are compared with those from a nearby reference site. The concentrations of heavy metals in *L. rubellus* from the confined disposal site were higher than those of the reference site.

A rather large difference in cadmium concentrations was noticed between native worms (Table 3) and experimental worms (Table 2). This difference may be caused by the lifetime exposure of native worms compared with 32 days for experimental worms and the difference in concentrations and bioavailability in materials to which the worms were exposed (layer of humus versus mineral layer).

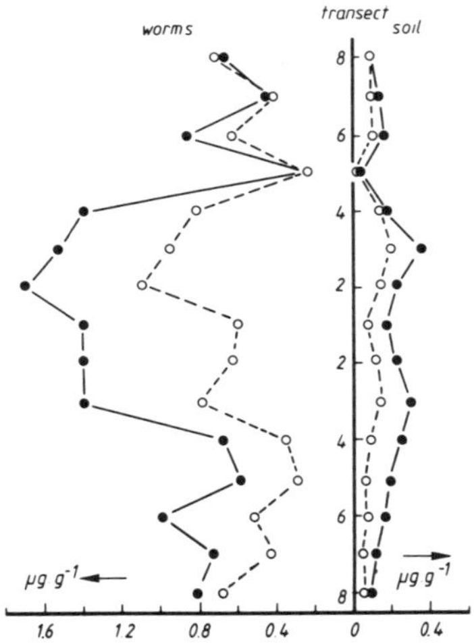

Fig. 3. Concentrations of 2,5,2′,5′ tetrachlorobiphenyl (○) and 2,4,5,2′,5′ pentachlorobiphenyl (●) in soil and in exposed worms along the transect B8 → A8.

ii. *PCB components and HCB*

Higher concentrations of PCBs were found in soil samples from the open marsh than in samples from the woodland (Tables 4 and 5). The prevalent conditions in the open marsh resulted in high concentrations in worms exposed to the soil samples. It seems that the conditions that have reduced bioavailability of the heavy metals, have enhanced the uptake of PCBs. This is shown for two PCB components in Fig. 3. Remarkably, comparable concentrations of PCBs were found in both experimental worms that were exposed to materials from the woodland (Table 5: A8–A5 and B8–B5) and native worms (Table 6) from that area. The concentrations of HCB in experimental worms differed widely, but were generally much higher than those in native worms at Times Beach. The concentrations of HCB were very low in native worms from the reference area, and PCB components were all below the detection limits (Table 6). Further analysis will be needed to clarify whether this similarity is real or coincidental.

Conclusions

The data presented indicate that a contaminated dredged material, once placed in an upland condition, changes with time. Weathering may take place, causing a decrease in contaminant concentrations in the upper layer as the result of leaching and volatilization. That decrease may be enhanced by biological interactions of a successional colonization of plant life and microbial development. On the other hand, it was shown that plant life may increase the concentrations of specific contaminants in the top layers through uptake by the roots and yearly leaf fall. Not only does the concentration change but also the bioavailability for earthworms. This was shown by the application of a terrestrial animal bioassay using the earthworm *Eisenia foetida*.

Table 4. PCB and HCB concentrations ($\mu g \cdot kg^{-1}$ dry weight) in Times Beach surface disposed materials and a control (manure)

Station	PCB Components									
	28	52	49	70	101	87	153	138	180	HCB
A8	43	93	64	78	58	40	34	32	20	57
A7	33	110	76	120	44	27	15	20	7.7	95
A6	110	160	120	160	70	55	15	18	4	50
A5	140	180	140	220	60	39	17	37	7.1	94
A4	94	230	170	220	89	59	26	40	12	160
A3	84	290	190	230	130	89	48	79	21	320
A2	130	220	160	210	110	76	42	41	15	110
A1 B1	54	180	110	150	71	48	23	51	7.7	130
B2	55	220	150	120	140	100	71	71	34	110
B3	64	340	220	290	190	120	79	110	37	330
B4	40	170	97	210	140	72	69	99	33	290
B5	15	36	22	47	26	13	21	28	12	190
B6	50	170	110	120	110	70	46	45	22	130
B7	35	140	87	150	95	49	53	76	28	510
B8	22	93	55	42	83	50	53	52	30	79
Manure	<10	<15	<14	<16	<17	<15	<12	<13	<16	<13
Manure 2	4.8	—	4.9	4.4	3.1	—	9.6	2.3	6.5	6.3

Table 5. PCB and HCB concentrations ($\mu g \cdot kg^{-1}$ ash-free dry weight) in experimental worms to exposed to disposed materials mentioned in Table 4

Station	PCB Components									
	28	52	49	70	101	87	153	138	180	HCB
A8	230	800	460	600	680	310	420	330	120	190
A7	160	740	510	950	420	220	180	230	64	1200
A6	530	1000	760	990	520	310	140	130	<46	350
A5	270	590	440	830	290	140	120	150	42	560
A4	200	680	480	860	360	180	130	210	43	890
A3	190	1400	910	1600	800	430	300	400	81	2000
A2	400	1000	690	830	630	360	280	230	100	510
A1 B1	240	1000	660	1200	590	340	280	360	81	890
B2	260	1700	1100	870	1100	680	470	430	150	620
B3	190	1500	970	1900	960	520	390	540	120	2100
B4	110	1000	570	1500	810	400	410	570	120	1800
B5	37	220	120	500	240	98	200	240	78	920
B6	150	870	540	430	630	360	290	250	100	560
B7	57	460	280	830	430	200	270	340	94	2500
B8	60	680	340	<45	730	350	420	350	150	460
Manure	<32	<48	<45	<51	<54	<48	<39	<42	<51	<19

Differences observed in the contaminant concentrations in the test animals and those in a naturally-occurring native species were thought to be related mainly to the exposure period and the layers in which the worms were exposed rather than to species specific differences.

Table 6. PCB and HCB concentrations ($\mu g \cdot kg^{-1}$ ash-free dry weight) in native worms at Times Beach and a reference area

Area species	PCB Components									
	28	52	49	70	101	87	153	138	180	HCB
Times Beach										
L. rubellus	126	460	220	330	320	160	130	130	50	260
	207	900	460	590	560	280	200	210	160	370
Reference site										
L. rubellus	all values below detection limits = < ~ 40									13
A. chlorotica	all values below detection limits = < ~ 40									12
O. lacteum	all values below detection limits = < ~ 40									8

Acknowledgements

This study was sponsored under the U.S. Army Corps of Engineers Field Verification Program (Contract DAJA 45-84-c-0015; C.C. Calhoun, Program Manager), by MT-TNO and by the RIZA (Dutch Ministry of Transport and Public Works). The authors also thank Glenn Rhett of the WES, Dick Leonard of the Buffalo District, and the Department of Analytical Chemistry of MT-TNO for their enthusiastic cooperation.

References

De Kock, W.Chr. and Marquenie, J.M. 1982. Report EEG 4/81/81. (Comm. Europ. Communities, Brussels, 1982): 178 pp.

EPA Ocean Disposal Bioassay Working Group. 1978. Bioassay procedures for the ocean disposal permit program (EPA-600/9-79-010, 1978): 121 pp.

Folsom, B.L., Jr., Lee, C.R. and Bates, D.J. 1981. Technical Report EL-81-12 (US Army Engineer Waterways Experiment Station, Vicksburg, MS, USA, 1981): 151 pp.

Goldberg, E.D., Bowen, V.T., Farrington, J.W., Harvey, G., Martin, J.H., Parker, P.L., Risebrough, R.W., Robertson, W., Schneider, E. and Gamble, E. 1978. Environ. Conserv. 5: 101–125.

Langston, W.J. 1982. J. Mar. Biol. Ass. U.K. 62: 667–684.

Marquenie, J.M. 1981. Proc. Int. Conf. Heavy Metals in the Environment (CEP Consultants Ltd).

Marquenie, J.M. and Simmers, J.W. 1984. Bioavailability of Heavy Metals, PCB and PCA to the Earthworm *Eisenia foetida*. Proceedings of the International Conference of Environmental Contamination. 820 pp.

Marquenie, J.M., De Kock, W.Chr. and Dineen, P.M. 1983. Proceedings of the International Conference of Heavy Metals in the Environment (CEP Consultants Ltd). 1284 pp.

Rubinstein, N.I., Lores, E. and Gregory, N.R. 1983. Aquat. Toxicol. 3: 249–260.

Appendix 1. Nomenclature of PCB components

PCB components		
28	2,4,4′	– trichlorobiphenyl
52	2,5,2′,5′,	– tetrachlorobiphenyl
49	2,4,2′,5′	– tetrachlorobiphenyl
70	2,5,3′,4′	– tetrachlorobiphenyl
101	2,4,5,2′,5′	– pentachlorobiphenyl
87	2,3,4,2′,5′	– pentachlorobiphenyl
153	2,4,5,2′,4′,5′	– hexachlorobiphenyl
138	2,3,4,2′,4′,5′	– hexachlorobiphenyl
180	2,3,4,5,2′,4′,5′	– heptachlorobiphenyl

RADIOECOLOGICAL STUDIES ON EARTHWORMS AND THEIR VALUE FOR AN ECOTOXICOLOGICAL RISK ASSESSMENT

JUERGEN C. KUEHLE

General aspects of radioecology

The development of radioecology has been influenced strongly from its very beginning by the rapid development of nuclear techniques. The close relation has been caused not only by permanent improvement of radiation detection methods, but also by questions dealing with the effects of nuclear techniques, especially as it was realized that ionizing radiation was becoming an environmental factor. Although the main object has been the evaluation of radiation effects at the ecosystem level, radioecology first developed under two major aspects, namely

1. a 'tracer' aspect under which the migration and 'fate' of radionuclides in natural biogeocenoses has been studied, and
2. an 'effect' aspect under which the effects of ionizing radiation on communities of microorganisms, plants and animals were object of the investigations.

Odum (1965) gave a general review of these two aspects and of the feedback between them and general ecology, and it is the aim of this article to show that there are parallels to recent trends in ecotoxicology, which should stimulate research workers of the two disciplines to compare their viewpoints under methodological and heuristic aspects.

Earthworms and radioecology

Radioecology has developed during the last 25 years in a broad sense, as can be seen from the bibliography of Klement and Schultz (1980) and remained an open system with vigorous exchange of ideas and techniques between related areas of knowledge. Initiatives about the use of isotopes and radiation came from the field of applied sciences for example from agricultural sciences and they influenced radioecology secondarily (L'Annunziata and Legg, 1984).

Soil animals and among them earthworms, have rarely been subject of radioecological investigations, probably because of the fact that the soil medium is a microhabitat of extreme complexity and great variety. Ghilarov and Krivolutsky (1972) summarized radioecological researches in soil zoology. It can be seen that the research activities in the Soviet Union were influenced mainly by the aspects of the effects of irradiation on soil animals (Korganova, 1973; Krivolutsky *et al.*, 1972; Krivolutsky *et al.*, 1983). It is interesting to observe that according to Reichle *et al.* (1971) and Krivolutsky *et al.* (1980) earthworms are important as bioindicators of radioactive contamination of soils.

Earthworms in waste and environmental management
edited by C.A. Edwards and E.F. Neuhauser, pp. 377–388

The other aspect, the 'tracer aspect', has in invertebrate ecology been mainly the subject of entomological research (Kloft, 1984), although some early publications (Odum and Golley, 1963; Ghabbour, 1968) broached studied the participation of pedobionts in the biotic turnover of nutrients, the rates of litter decomposition and interspecific relationships. Concerning earthworm ecology the following radiotracer studies are relevant; they deal:

- with the intake and turnover of radioactive cesium (Crossley *et al.*, 1971);
- with biogeochemical cycles of calcium and strontium (Peredelsky *et al.*, 1957, 1960a, 1960b; Pokarzhevsky and Krivolutsky, 1975; Kühle, 1978);
- with gaseous $^{14}CO_2$ to study the activity of calciferous glands of different earthworm species in different CO_2 atmosphere (Kühle, 1979);
- with ^{59}Fe labelled leaves and other organic materials to study their decomposition (Maldague and Couture, 1972; Couture and Maldague, 1975;
- with autoradiographic techniques to study the intake and migration of ^{32}P labelled substances in soil (Couture and Maldague, 1972) or in organism (Horber and Ghabbour, 1966) and last but not least;
- with different ^{14}C-labelled pesticides as parathion (Führemann and Lichtenstein, 1978), carbofuran (Stenersen *et al.*, 1973) and others to study their metabolism in earthworms.

Ecotoxicological risk asssessment and earthworms

The elements of risk assessment of an environmental hazard are given by three different categories, Kates (1978) which overlap and influence one another:

- by hazard identification, the recognition of a hazard to earthworm biocenoses or to single earthworm species. This questions what the reason is for unnatural changes in earthworm populations?
- by risk estimation, including a quantitative aspect about the threat potential of the hazard which questions how significant are the consequences of a hazard to earthworms for the whole biocenosis of soil organisms or for the homeostasis of a specific ecosystem?
- by social evaluation, which assesses the estimated risk to human society.

The theory of risk assessment is not appropriate here. Nevertheless it is necessary to mention certain aspects of this theory concerning earthworms, because they can help us to understand the following explanations of ecotoxicological strategies in connection with the problem of risk assessment.

The first is the problem of an exact hazard identification. Since soil organisms, in general, live in a non-transparent medium they cannot be used as convenient monitoring systems for an acute toxicological hazard assessment. The methods of studying their population size and population structures are quite different, and in some cases difficult to compare, and the determination of species is time consuming. That is why most of the threats to the environment have not been based in the past on soil organisms but on aquatic organisms, vertebrates or on non-soil inhabiting insects.

Earthworms have been subject of many studies so that we have much information about them (Worden, 1981; Satchell and Martin, 1981, 1985) and owing to their limited number of species, their relatively large size, and wide distribution they should be suitable bioindicators. However, if we study earthworm populations in a certain area, it is difficult to assess the present state of the population studied as 'normal' or 'abnormal'. The reason is that even in comparative studies it is very difficult to formulate the typical characterization of different species, because there is no area in industrialized countries which has not been influenced by anthropogenic stresses.

These stresses may be structural (*e.g.*, by changes in land structure or soil structure by different soil treatments etc.), nutritive ones (by changes in vegetation structure, crop harvest etc.) or toxicological ones or all of these together, each component varying in different degrees. Particularly, in agricultural land, the anthropogenic stress factors can be quite complex, and we have to take into consideration that the present earthworm population of a study area, is a result of the development of these stress factors from natural conditions. Our difficulty is to reconstruct this development up to the present day. This may be helped by comparatively large area studies including natural or natural-like sites to obtain more information about the potential spectrum of species of an area. The survey of the *status quo* is the only base of our knowledge about possible hazards to earthworm populations and it can give us valuable indications about trends. What it normally fails to provide is the exact identification of the causes of certain changes.

We can find in both disciplines the two aspects already mentioned, firstly the 'effect aspect', as when irridation or toxicological effects on organisms are studied and secondly the 'tracer aspect' where the migration and fate of a substance is the central point of interest.

Radioecology and ecotoxicology and their mutual stimulation

In both disciplines, we can distinguish two different points of view for analysis of hazards. The first is characterized by a synthesis and an extrapolation from different laboratory experiments that are part of a compartmental model and which try to describe hazards under defined environmental conditions. The second is characterized by a holistic view, where the distribution of substances and their effects are studied under the conditions of a whole system by systems analysis (Fig. 1).

Concerning the 'effect aspect', there are entirely different methods used between the two disciplines radioecology and ecotoxicology, caused by the different kinds of hazard. They are closely connected in 'single species tests' with the disciplines of radiation biology and toxicology, where the direct effect is studied in hierarchical order from molecule to organism. In contrast to this, studies on a population basis (considering intraspecific relations) or with mixed populations (considering interspecific concurrence or interspecific dependence) are closely connected to population ecology and species enrichment.

Classical synecological methods are used mainly where effects are studied on a system with specific regulation of processes. It could be a future subject to show in which way natural and man-made environmental catastrophes have advanced ecology.

The 'tracer aspect' finds in radioecology and ecotoxicology a common base. It is

Ecotoxicological tests	*System analysis*
1. to study the physico-chemical properties of chemicals 2. to study *toxicity* of environmental chemicals with one earthworm species – *mortality* as parameter (Filter Paper Contact Test, Natural Soil Test, Artificial Soil Test, Artisol Test, Immersion Test et al.) – *population parameters* (natality, development, maturity, life span etc.) – *physiological-biochemical parameters* (neurophysiological response: Corvallis ERC-Test) 3. to study *toxicity* of environmental chemicals with several earthworm species ('*multi-species test*') – *interspecific concurrence* (species of same ecological life type) – interspecific dependence (species of different ecological life type 4. to study *accumulation* and *retention* of chemicals by total contamination of different earthworm species 5. to study *biodegradation* of chemicals by earthworms 6. to study *spacial dispersion* of chemicals by earthworms	1. to describe a present situation (spacial structures, classical synecology) .2 to study self-dynamics and regulations of ecological processes 3. to study defined interference in 'field tests' 4. to study defined interference in simulation models ('microcosm test') 5. to study system specific and species specific contamination

Fig. 1. Analytical and holistic view to study the distribution of environmental chemicals and their effects.

connected with the physico-chemical properties of the substance, with its ecosystem-specific distribution and its species-specific bioavailability, accumulation, retention, biodegradation and transfer (Fig. 1).

Owing to the rapid resolution in radiation detection, in contrast to the limits of chemical analysis of organic compounds, most studies on distribution dynamics of substances, have been made with radiotracers. Today, this is more a question of laboratory experiments than of field studies, because radionuclides currently play only a minor role in environmental pollution. Moreover, field studies with radiotracers are possible only under extreme restrictions. The advantage of studying a defined element occurs also in studies on heavy metals, and so it is understandable that studies about accumulation and retention of substances in earthworms, not only in laboratory tests, but also in the field, are more comprehensive for heavy metals than for organic compounds.

An organic compound can be the subject of different chemical transformations in its system specific routes of distribution. Its bioavailability for a certain earthworm species, if judged from the species-specific point of view, is the same as radioactive contamination. This is the reason why radiotracer studies or radioecological studies can assume an importance for an ecotoxicological hazard assessment.

The NATEC laboratory microcosm experiment

i. *General considerations*

Under the holistic point of view, we have to accept that with the present state of our knowledge about different ecosystems, it is impossible to predict with complete confidence the ecological effects of a chemical. We are able to analyse and characterize systems under natural conditions and assess various anthropogenic pressures. However, if we compare different states of similar ecosystems using methods of ecological system analysis, we can only describe the differences, but not determine the causes of changes.

To attain greater standardization and comparability we study changes of the system in 'field tests'. The procedure of standardization, leads increasingly to a simplification of natural systems to give us more reproducibility and facilitate the description of mathematical models. At the end of the scale, 'holistic tests are extremely simplified artificial simulation models, as given by most of the 'microcosms' of ecology and ecotoxicology. There have been many attempts, especially in aquatic ecology, to study the transfer or the effects of chemicals under simulated conditions, either in an open or in a closed system.

In course of the ecotoxicological program of the BMFT (Federal Ministry of Research and Technology of Germany) the 'Institut für Naturwissenschaftlich-technische Dienste GmbH' (NATEC) in Hamburg developed an environmental test system to study the distribution of certain radiolabelled reference substances, of different physico-chemical properties, under standardized conditions in air, plants, soil and soil water. In connection with this test program, it was possible for the author to study the accumulation of chemicals in earthworms, living in the soil of this microcosm.

ii. *Material and methods*

The NATEC-test system (Fig. 2) consists of a central environmental compartment, built from a cylindrical steel tube (diameter 72 cm) by taking a natural soil core with a depth of 42.5 cm. At its deepest part, the cylinder is provided with an outlet for the percolation of water and the outer wall is coated with a thermal insulation layer. This jacket contains heating and cooling coils by which the steel tube can be kept at a controlled temperatures, and so that a temperature gradient can be set. The upper edge of the steel tube is moulded as a flange, upon which a quartz ring provided with a closable hole and finally the quartz dome are mounted. The upper part of the quartz dome is provided with a circular gap for taking a measurement and control unit for automatic simulation of climatic conditions (light intensity, air temperature, air humidity, wind velocity and rain) as well as an additional outer illumination unit allowing illuminate when the nominal light value is not achieved. The air in the chamber is exchanged continuously (20 minute turnover). From the outside, it is sucked through the chamber and after passing through an absorber unit it is pumped out again. More details about the construction of the system and its climate control are given by Figge and Klahn (1982) and Figge *et al.* (1983).

For each experiment a natural soil core was taken from a defined site. The surface area was divided into four sections, which were cultivated with four different plants (lettuce, radish, beans and the grass *Lolium multiflorum*). During the experiment, a standard climate (20 years average of the climatic parameters of a day in June in Hamburg) was applied for a test period of 30 days. The radiolabelled reference

Fig. 2. The NATEC-test-system.

substances were choosen according to their physico-chemical properties (measured in OECD- and EEC-laboratory intercomparison testing programs).

The following substances have been examined:

I 2,4,5-trichloro [ring-u ^{14}C] phenoxy acetic acid (2,4,5T)
II [u-^{14}C] aniline (ANILINE)
III [u-^{14}C] phenol (PHENOL)
IV 2-nitro [u-^{14}C] phenol (2-N-P)
V 4-nitro [u-^{14}C] phenol (4-N-P)
VI [ring-u-^{14}C] toluene (TOLUENE)
VII [u-^{14}C] trichloroethene (TCE)
VIII bis (2-ethylhexyl)-[7,8-^{14}C] phthalate (DOP)

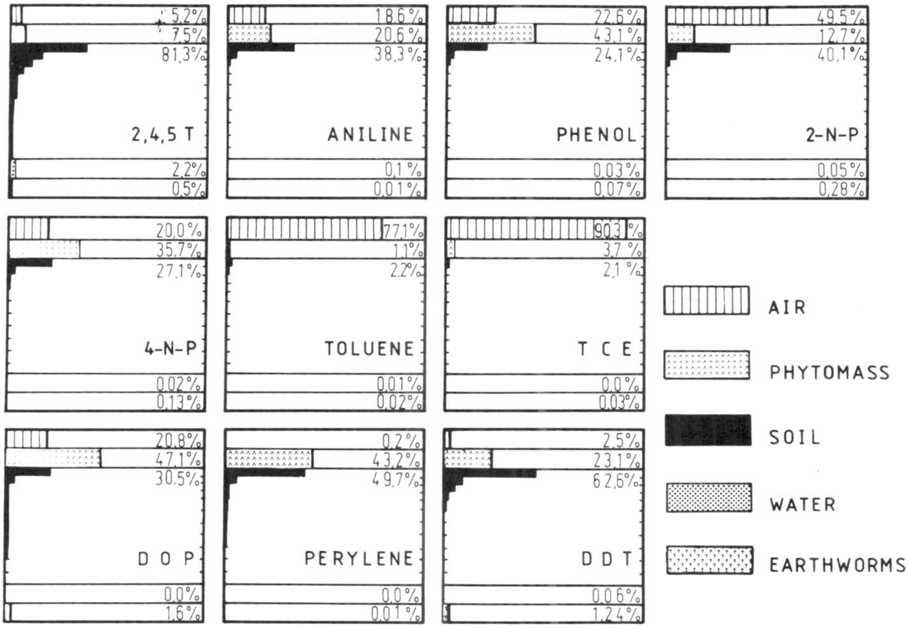

Fig. 3. The distribution of radioactivity (in percentage of the applied radioactivity) in the different system compartments.

IX [6a, 6b, 12a,12b-^{14}C]perylene (PERYLENE)
X 1,1-bis(4-chloro [u-^{14}C]phenyl)-2,2,2-trichloroethane (DDT)

The most important coefficients to characterize the physico-chemical properties of the substances, are:

a. the partition coefficient octanol/water (log $10P_{ow}$)
b. the soil/water adsorption constant: (K_{oc})
c. water solubility in gram/liter: (WS)
d. Oswald's solubility (water/air partition coefficient): (α') (Table 1).

The chemical application was done according to a standard procedure: 0,15 mmol of each substance with about 137 MBq was absorbed on powdery soil material and then sprayed in an aqueous suspension onto the cultivated soil core in the NATEC-test system.

Samples were examined with respect to radioactivity by liquid scintillation counting:

- absorbent materials containing the radioactivity from the air taken daily during the experiment;
- percolating water taken daily during the experiment;
- soil layers (the first from the surface to a depth of 1.5 cm, then soil layers of 3 cm from 1.5 cm to 13.5 cm depth, then a soil layer of 4 cm from 13.5 to 17.5 depth

Table 1. Physico-chemical properties of the test substances, their specific radioactivity in the system compartments and accumulation factors in earthworms (in relation to the first upper soil layer of 1.5 cm)

	I	II	III	IV	V	VI	VII	VIII	IX	X
log10 p^{ow}	0,60	0,96	11,48	1,77	1,90	2,62	3,24	4,88	5,27	6,28
K^{oc}	53	4	27	25	34	210	720	30000	73500	238000
WS (mg/l)	280	34400	82000	1300	14800	520	1100	23×10^9	0,0004	0,0023
α'	4×10^9	$2,3 \times 10^4$	3×10^4	1950	$4,9 \times 10^7$	3,6	2,66	$1,3 \times 10^4$	$5,6 \times 10^3$	910
Spec. activity (dpm/g × 1000)										
plant	836	1021	2184	834	2566	69	166	3074	2782	1186
soil I	313	158	85	82	149	6,7	4	115	172	178
soil mean	50	13	7	7	14	0,8	0,6	28	16	18
Wmax	53994	1161	3365	13620	35089	579	2346	186820	624	205230
Wmean	5580	777	277	1336	8009	71	70	12493	87	2286
AF Wmax/sI	172,5	7,3	39,5	166,1	235,5	86,4	586,5	1624,5	3,6	1152,9
AF W Ø/sI	17,8	4,9	3,3	16,2	53,8	10,6	17,5	108,6	0,5	12,8
Wmax/W Ø	10,0	1,5	12,1	10,2	4,4	8,2	33,5	15,0	7,2	109,5

I = 2,4,5,T; II = aniline; III = phenol; IV = 2-N-P; V = 4-N-P; VI = toluene; VII = TCE; VIII = DOP; IX = perylene; X = DDT.

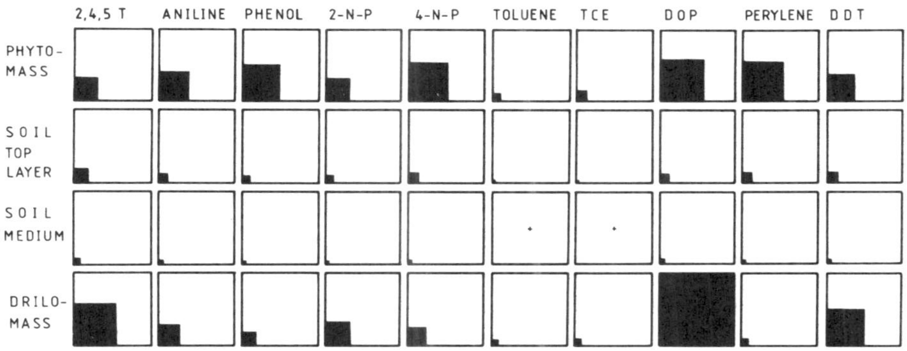

Fig. 4. Specific radioactivity of each system compartment. (The area of one square is about 10 million dpm/g = 0.167MBq/g).

followed by soil layers of 5 cm from 17.5 cm to 42.5 cm depth);
- total phytomass (except some individual parts of the plants for autoradiography);
- earthworms (excluding intestinal contents).

iii. *Results*

The amounts of radioactivity, in percentage of the applied radioactivity, in the different parts of the NATEC test system, are summarized in Fig. 3 and Table 1.

If we compare the mass flow of substances in the different compartments, the part which is distributed in the earthworms is almost negligible (mostly below 1%). As we know from systems ecology, earthworm biomass is quite predominant in relation to other animals, but in relation to phytomass or to biomass it is small and if we consider the abiotic components it becomes negligible. This does not indicate a small importance for earthworms in material distribution (biogenetic transfer) nor of unimportant contamination, but it reflects the mass relations of the compartments and it hints at small accumulation potential. To examine the latter, it is necessary to compare the specific activities (Bq/g) of the different compartments.

The specific activity of each compartments is given in relation to the area of a square with about 10 million dpm/g (= 0.167 MBq/g) in Fig. 4. With the exception of toluene and trichloroethene, the specific activity varies in phytomass between 800,000 and 3 million dpm/g. If we compare these values with those of the first upper soil layer, it can be seen that there is a difference of one order of magnitude, whereas the specific activity in the drilomass is comparible in relation to the order of magnitude with the plants, although this differs from substance to substance.

Although the distribution of the substances can be correlated to their physicochemical properties (Figge *et al.*, 1983, 1985), the specific activities of the earthworms can not be interpreted as a direct correlation. In general they have mean values which are in the same order as the plant material. But if we compare these mean values with individual ones, we can see extreme deviations. The maximum values of specific activity (W_{max}) of individual worms are given in Table 1. It can be seen, that they can reach more than 100 times the mean value (DDT).

In general, substances with a high octanol-water partition coefficient (p_{ow}) and

Table 2. Accumulation factors+ in different organs of *Lumbricus terrestris* after 30 days of applications of 2,4,5- T under conditions of total contamination and the conditions of the NATEC-test system

	Ph	CG	SV	CR	GI	AI	MI	PI	INT
Total	4.18	16.76	1.48	2.59	2.82	28.90	55.37	29.50	2.30
NATEC	1.35	6.48	0.84	0.81	0.99	7.42	10.03	9.40	0.69
% T/N	32.3	38.6	0.57	31.3	35.1	25.7	18.1	31.9	30.0

high water solubility (ws) guarantee high accumulation values (4-nitrophenol, DOP), because of their relatively good penetration into soil, and their water solubility, enables a high contamination of the earthworms and a high degree of accumulation, if the adsorption coefficient is not too low. Substances with a high adsorption coefficient but low water solubility, such as DDT, can accumulate in earthworms, only if there is sufficient contamination on the soil surface or in the upper soil layers.

v. *Conclusions*

The experiments with the NATEC test system and laboratory tests with total contamination give an impression, of how important radiotracer studies or radioecological studies can be, in determining the distribution of substances and the contamination of certain earthworms species. A detailed evaluation of chemical specific bioavailability in different ecosystems as well as species-specific contamination, enables to assess the actual effects of a chemical in the environment, in relation to toxicity tests.

The evaluation of accumulation data, can only be successful, if the different contamination levels in an ecosystem are considered for each indicator organism including earthworms. In addition to field studies, radioecological studies can help to find therefore suitable bioindicators (Kühle, 1983).

Acknowledgements

I am most grateful to Dr. Figge, Dr. Klahn and Dr. Koch from the 'Intitut für naturwissenschaftlich-technische Dienste GmbH' (NATEC) in Hamburg to enable my collaboration in their BMFT-project.

References

Battelle, 1982. Merkblätter über Referenzchemikalien der Projektträgerschaft Umweltchemikalien (KFA Jülich) Battelle-Institut e.V., Frankfurt a.M.

Boehringer, U.R., Kayser, D. and Schmidt-Bleek, F. 1981. Das Umwelt-probenbankprojekt der Bundesrepublik Deutschland Umweltprobenbank 1.2, 13–33 (UBA) Berlin 1981.

Cairns, Jr., J. et al. 1981. Committee to Review Methods for Ecotoxicology – Environmental Studies Board. Testing for effects of chemicals on ecosystems. Washington (Nat. Acad. Press): 103 pp.

Couture, G. and Maldague, M. 1972. Application de la technique autoradiographique à l'étude de l'activité de Lumbricus terrestris. Pedobiologia 12: 128–132.

Couture, G. and Maldague, M. 1975. Application des isotopes radioactifs à l'étude de role agrologique des Lumbricidae. Pedobiologia 15(1): 60–64.

Crossley Jr., D.A., Reichle, D.E. and Edwards, C.A. 1971. Intake and turnover of radioactive cesium by earthworms. Pedobiologia 11: 71–76.

Figge, K. and Klahn, J. 1982. Die Pflanzenstoffwechselbox. GIT Fachz. Lab. 26: 680–685.

Figge, K., Klahn, J. and Koch, J. 1983. Testing of chemicals by evaluation of their distribution and degradation patterns in an environmental standard system. Regul. Toxicol. Pharmacol. 3: 199–215.

Figge, K., Klahn, J. and Koch, J. 1985. Schriftenreihe des Vereins für Wasser-, Boden- und Lufthygiene. Stuttgart (G. Fischer).

Führemann, T.W. and Lichtenstein, E.P. 1978. Release of soil bound methyl carbon-14 parathion residues and their uptake by earthworms and oat plants. J. Agric. Food Chem. 26: 605–610.

Ghabbour, S.I. 1968. Methods for the utilisation of radioisotopes in the study of ecosystems. Rev. Ecol. Biol. Sol. 5: 245–263.

Ghilarov, M.S. and Krivolutsky, D.A. 1971. Radioecological Researches in Soil Zoology. *In* IV. Colloquium Pedobiologiae. pp. 537–545. Edited by J. d'Aguilar *et al.* INRA Dijon 14-19/IX/1970. INRA-Paris.

Horber, E. and Ghabbour, S.I. 1966. Labeling oligochaetes with P-32. Rev. Ecol. Biol. Sol 3: 277–281.

Kates, R.W. 1978. Risk Assessment of Environmental Hazard. SCOPE 8. J. Wiley, Chichester. 112 pp.

Klement, Jr., A.W. and Schultz, V. 1980. Freshwater and Terrestrial Radioecology – A Selected Bibliography. Stroudsburg (Dowden, Hutchinson & Ross Inc.) 587 pp.

Kloft, W.J. 1984. Entomology. *In* Isotopes and Radiation in Agricultural Sciences. II. pp. 51–103. Edited by M.F. L'Annunziata and J.O. Legg.

Korganova, G.A. 1973. Influence of experimental soil pollution with Sr^{90} on the fauna of soil protists. Zool. J. 52: 939–941.

Krivolutsky, D.A., Tichomirova, A.L. and Turchaninova, V.A. 1972. Strukturänderungen des Tierbesatzes (Land- und Bodenwirbelose) unter dem Einfluß der Kontamination des Bodens mit Sr^{90} Pedobiologia 12: 374–380.

Krivolutsky, D.A., Semyashkina, T.M., Mikhal'tsova, Z.A. and Turchaninova, V.A. 1980. Earthworms as bioindicators of radioactive contamination of soil. Sov. J. Ecol. 11: 379–383 (in Russian).

Krivolutsky, D.A., Korganova, G.A. and Tikhomirov, F.A. 1983. Some radioecological problems of terrestrial and soil animals. *In* New Trends in Soil Biology. Proceedings VIIIth International Colloquium of Soil Zoology. Ottignies-Louvain-la-Neuve. pp. 463–469. Edited by P. Lebrun *et al.* Dieu Brichart, Louvain-la-Neuve.

Kühle, J.C. 1978. Radiotraceruntersuchungen zum Calciumstoffwechsel von *Lumbricus terrestris.* Diplomarbeit Universität Bonn Math.-Nat. Fakultät. 107 pp.

Kühle, J.C. 1979. Vergleichende Untersuchungen zur Funktion der Kalkdrüse verschiedener Regenwurmarten bei unterschiedlicher CO_2-Atmosphäre. Verh. Ges. Ökol. VIII (Freising 1978). pp. 411–415.

Kühle, J.C. 1983. Ökotoxikologische Modelluntersuchungen zur Bewertung von Regenwürmern als Bioindikatoren. Verh. Dtsch. Zool. Ges. G. Fischer, Stuttgart. pp. 147–151.

Kühle, J.C. 1988a. Modelluntersuchungen zur strukturellen und ökotoxikologischen Belastung von Regenwürmern in Weinbergen Mitteleuropas. Monogr. Pedobiol. I.

L'Annunziata, M.F. and Legg, J.O. 1984. (ed.) Isotopes and Radiation in Agricultural Sciences, 2 Vol. Academic Press, London.

Lichtenstein, E.P. 1977. Fate of pesticides in a soil-plant microcosm. *In* Proceedings Symposium on Terrestrial Microcosms and Environmental Chemistry. Corvallis Or., June 1977. National Science Foundation, Washington, DC.

Maldague, M. and Couture, G. 1972. Utilisation des litières radioactives par *Lumbricus terrestris.* Proceedings 4th International Congress Soil Zoology. pp. 193–200.

Metcalf, R.L. 1977. Model ecosystem approach to insecticide degradation: a critique. Ann. Rev. Entomol. 22: 241–261.

Metcalf, R.L., Sangha, G.K. and Kapoor, I.P. 1971. Model ecosystem for the evaluation of pesticide degradability and ecological magnification. Environm. Sci. Technol. 5: 709–713.

Metcalf. R.L., Cole, L.K., Wood, S.G., Mandel, D.J. and Milbrath, M.L. 1979. Design and Evaluation of a Terrestrial Model Ecosystem for Evaluation of Substitute Pesticide Chemicals. EPA-600/3-79-004, Washington, DC. 298 pp.

Moriarty, F. 1983. Ecotoxicology. The Study of Pollutants in Ecosystems. Academic Press, London. 233 pp.

Neely, W.B. 1980. Chemicals in the Environment. M. Dekker, New York. 245 pp.

Odum, E.P. 1965. Feedback between radiation ecology and general ecology. Health Phys. 11: 1257–1262.

Odum, E.P. and Golley, F.B. 1963. Radioactive tracers as an aid to the measurement of energy flow at the population level in nature. *In* Radioecology. pp. 403–410. Edited by V. Schultz and A.W. Klement Jr. Reinhold Publ., New York.

Odum, H.T. 1970. A Tropical Rainforest. 3 Vol. Office of Information Sciences – US Atomic Energy Comm.

Peredelsky, A.A., Poryadkova, N.A. and Rodionova, L.Z. 1957. The role of earthworms in purification of soil contaminated with radioactive isotopes. Dokl. Akad. Nauk. 115: 809–812.

Peredelsky, A.A. 1960a. Effect of earthworms and wireworms on absorption by plants of the radioactive isotopes Ca-45 and Sr-90 from soil. Dokl. Akad. Nauk. 134: 1450–1452.

Peredelsky, A.A. 1960b. Dispersion of radioactive isotopes in the soil by earthworms. Dokl. Akad. Nauk. 135: 185–188.

Petersen, H., Luxton, M. 1982. A comparative analysis of soil fauna populations and their role in decomposition processes. Oikos 39: 288–388.

Pokarzhevsky, A.D. and Krivolutsky, D.A. 1975. The role of pedobionts in the biogeochemical cycles of calcium and strontium–90 in the ecosystem. *In* Progress in Soil Zoology. pp. 249–254. Edited by J. Vanek. Academia, Prag.

Reichle, D.E. et al. 1971. Effects of beta-gamma radiation of earthworms under simulated fallout conditions. *In* Survival of Food Crops and Livestock in the Event of Nuclear War. pp. 527–534. Oak Ridge Tenn. Ut-AEC.

Satchell, J.E. and Martin, K. 1981. A Bibliography of Earthworm Research. Inst. Terr. Ecol. Grange-over-Sands, Cu. 178 pp.

Satchell, J.E. and Martin, K. 1985. A Bibliography of Earthworm Research. Supplement 1. Inst. Terr. Ecol. Grange-over-Sands, Cu. 258 pp.

Stenersen, J., Gilman, A. and Vardanis, L. 1973. Carbofuran: its toxicity to and metabolism by the earthworm (*Lumbricus terrestris*). Agric. Food Chem. 21: 166–171.

Worden, D.D. 1981. Workshop on the role of earthworms in the stabilization of organic residues. Vol. II Bibliography. Kalamazoo, Mich. (Beech Leaf Press). 120 pp.

INDEX OF AUTHORS

GENERAL SUBJECT INDEX

SPECIES INDEX